Lecture Notes in Computer Science 15599

The series Lecture Notes in Computer Science (LNCS), including its subseries Lecture Notes in Artificial Intelligence (LNAI) and Lecture Notes in Bioinformatics (LNBI), has established itself as a medium for the publication of new developments in computer science and information technology research, teaching, and education.

LNCS enjoys close cooperation with the computer science R & D community, the series counts many renowned academics among its volume editors and paper authors, and collaborates with prestigious societies. Its mission is to serve this international community by providing an invaluable service, mainly focused on the publication of conference and workshop proceedings and postproceedings. LNCS commenced publication in 1973.

Mohammed Alser · Mukul S. Bansal ·
Yury Khudyakov · Serghei Mangul ·
Ion I. Mandoiu · Marmar R. Moussa ·
Murray Patterson · Sanguthevar Rajasekaran ·
Pavel Skums · Shibu Yooseph ·
Alexander Zelikovsky
Editors

Computational Advances in Bio and Medical Sciences

13th International Conference, ICCABS 2025
Atlanta, GA, USA, January 12–14, 2025
Revised Selected Papers

Springer

Editors
Mohammed Alser
Georgia State University
Atlanta, GA, USA

Mukul S. Bansal
University of Connecticut
Storrs, CT, USA

Yury Khudyakov
Centers for Disease Control
Atlanta, GA, USA

Serghei Mangul
Ștefan cel Mare University of Suceava
Suceava, Romania

Ion I. Mandoiu
University of Connecticut
Storrs, CT, USA

Marmar R. Moussa
University of Oklahoma
Norman, OK, USA

Murray Patterson
Georgia State University
Atlanta, GA, USA

Sanguthevar Rajasekaran
University of Connecticut
Storrs, CT, USA

Pavel Skums
University of Connecticut
Storrs, CT, USA

Shibu Yooseph
Claremont McKenna College
Claremont, CA, USA

Alexander Zelikovsky
Georgia State University
Atlanta, GA, USA

ISSN 0302-9743 ISSN 1611-3349 (electronic)
Lecture Notes in Computer Science
ISBN 978-3-032-02488-6 ISBN 978-3-032-02489-3 (eBook)
https://doi.org/10.1007/978-3-032-02489-3

This Springer imprint is published by the registered company Springer Nature Switzerland AG
The registered company address is: Gewerbestrasse 11, 6330 Cham, Switzerland

Preface

This volume contains the papers presented at ICCABS 2025, the 13th International Conference on Computational Advances in Bio and Medical Sciences, held on January 12–14, 2025 in Atlanta, Georgia, USA. ICCABS 2025 was hosted by the Department of Computer Science at Georgia State University with Alexander Zelikovsky from Georgia State University and Sanguthevar Rajasekaran from University of Connecticut as General Chairs, and Murray Patterson from Georgia State University and Shibu Yooseph from Claremont McKenna College as Program Committee Chairs. ICCABS has the goal of bringing together researchers, scientists, and students from academia, laboratories, and industry to discuss recent advances on computational techniques and applications in the areas of biology, medicine, and drug discovery.

There were a total of 75 conference submissions. Following a rigorous single-blind review process in which each submission was reviewed by at least two reviewers, the committee decided to accept 15 extended abstracts for oral presentation and publication in the post-proceedings volume. The conference program also included invited talks presented as part of four co-located workshops: 8 talks presented at the 13th Workshop on Computational Advances for Next Generation Sequencing (CANGS 2025), 22 talks presented at the 12th Workshop on Computational Advances in Molecular Epidemiology (CAME 2025), 3 talks presented at the 6th Workshop on Computational Advances for Single-Cell Omics Data Analysis (CASCODA 2025), and 15 talks presented at the 1st Workshop on Metagenomics Research and Applications (CAMeRA 2025). Workshop speakers were invited to submit extended abstracts to the post-proceedings volume, and, after following the same review process used for the main conference, 11 additional extended abstracts (3 from the CANGS workshop, 6 from the CAME workshop, and 2 from the CAMeRA workshop) were selected for publication.

The technical program of ICCABS 2025 also featured keynote talks by four distinguished speakers: Yury Khudyakov from the Centers for Disease Control and Prevention gave a talk on "Conceptual challenges of viral molecular epidemiology," Srinivas Aluru from Georgia Institute of Technology gave a talk on "Genome graphs: algorithms and applications," Yana Bromberg from Emory University gave a talk on "Learning the language of functional signatures of Life," and Gerardo Chowell from Georgia State University gave a talk on "A Methodological workflow for epidemic forecasting."

We would like to thank all keynote speakers and authors for presenting their work at the conference. We would also like to thank the Program Committee members and external reviewers for volunteering their time to review and discuss the submissions. Additionally, we would like to extend special thanks to the Steering Committee members for their continued leadership. Last but not least, we would like to thank our sponsors, especially Georgia State University and the US National Science Foundation (NSF) for their support in making ICCABS 2025 a successful event.

Mohammed Alser
Mukul S. Bansal
Yury Khudyakov
Serghei Mangul
Ion I. Măndoiu
Marmar R. Moussa
Murray Patterson
Sanguthevar Rajasekaran
Pavel Skums
Shibu Yooseph
Alexander Zelikovsky

Organization

Steering Committee

Srinivas Aluru	Georgia Institute of Technology, USA
Reda A. Ammar	University of Connecticut, USA
Tao Jiang	University of California, Riverside, USA
Vipin Kumar	University of Minnesota, USA
Ming Li	University of Waterloo, Canada
Sanguthevar Rajasekaran (Chair)	University of Connecticut, USA
John Reif	Duke University, USA
Sartaj Sahni	University of Florida, USA

General Chairs

Sanguthevar Rajasekaran	University of Connecticut, USA
Alex Zelikovsky	Georgia State University, USA

Program Chairs

Murray Patterson	Georgia State University, USA
Shibu Yooseph	Claremont McKenna College, USA

Workshop Chairs

Mohammed Alser	Georgia State University, USA
Mukul S. Bansal	University of Connecticut, USA
Yuri Khudyakov	Centers for Disease Control & Prevention, USA
Ion Mandoiu	University of Connecticut, USA
Serghei Mangul	University of Suceava, Romania
Marmar Moussa	University of Oklahoma, USA
Murray Patterson	Georgia State University, USA
Pavel Skums	University of Connecticut, USA
Alex Zelikovsky	Georgia State University, USA

Web Chair

Mohammed Alser	Georgia State University, USA

Program Committee

Derek Aguiar	University of Connecticut, USA
Sahar Al Seesi	Southern Connecticut State University, USA
Max Alekseyev	George Washington University, USA
Mohammed Alser	Georgia State University, USA
Mukul S. Bansal	University of Connecticut, USA
Simone Ciccolella	Università degli studi di Milano-Bicocca, Italy
Jaime Davila	St. Olaf College, USA
Jorge Duitama	Universidad de los Andes, Colombia
Richard Edwards	University of New South Wales, Australia
Oliver Eulenstein	Iowa State University, USA
Daniel Gibney	University of Texas at Dallas, USA
Danny Krizanc	Wesleyan University, USA
M. Oguzhan Kulekci	Indiana University Bloomington, USA
Manuel Lafond	Université de Sherbrooke, Canada
Ion Mandoiu	University of Connecticut, USA
Serghei Mangul	University of Suceava, Romania
Marmar Moussa	University of Oklahoma, USA
Murray Patterson (Chair)	Georgia State University, USA
Maria Poptsova	HSE University, Russia
Sanguthevar Rajasekaran	University of Connecticut, USA
Pavel Skums	Georgia State University
Yanni Sun	City University of Hong Kong, China
Sing-Hoi Sze	Texas A&M University, USA
Sharma V. Thankachan	University of Central Florida, USA
Ugo Vaccaro	University of Salerno, Italy
Balaji Venkatachalam	Google, USA
Jianxin Wang	Central South University, China
Fang Xiang Wu	University of Saskatchewan, Canada
Shibu Yooseph (Chair)	Claremont McKenna College, USA
Alex Zelikovsky	Georgia State University, USA
Shaojie Zhang	University of Central Florida, USA
Wei Zhang	University of Central Florida, USA
Cuncong Zhong	University of Kansas, USA

Organizing Committee

Mohammed Alser	Georgia State University, USA
Mukul S. Bansal	University of Connecticut, USA
Ion Mandoiu	University of Connecticut, USA
Serghei Mangul	University of Suceava, Romania
Marmar Moussa	University of Oklahoma, USA
Murray Patterson	Georgia State University, USA
Pavel Skums	University of Connecticut, USA
Shibu Yooseph	Claremont McKenna College, USA
Alex Zelikovsky	Georgia State University, USA

Additional Reviewers

Abdelnaby, Mohamed
Adeniyi, Ezekiel A.
Aguiar, Derek
Alomair, Ryan
Ansari, Md Istiaq
Banerjee, Aranya
Baul, Sudipto
Boldirev, Grigore
Farooq, Hafsa
Hossen, Md Helal
Jiang, Qibing
Juyal, Silvia
Kafikang, Maryam
Markin, Alexey
Nemira, Alina
Nihalani, Rahul
Oladapo, Olajumoke
Shakya, Pramesh
Weiner, Samson

Contents

A Benchmarking Study of Random Projections and Principal Components for Dimensionality Reduction Strategies in Single Cell Analysis

Mohamed Abdelnaby and Marmar R. Moussa(✉)

School of Computer Science, University of Oklahoma, Norman, OK, USA
marmar.moussa@ou.edu

Abstract. Principal Component Analysis (PCA) has long been a cornerstone in dimensionality reduction for high-dimensional data, including single-cell RNA sequencing (scRNA-seq). However, PCA's performance typically degrades with increasing data size, can be sensitive to outliers, and assumes linearity. Recently, Random Projection (RP) methods have emerged as promising alternatives, addressing some of these limitations. This study systematically and comprehensively evaluates PCA and RP approaches, including Singular Value Decomposition (SVD) and randomized SVD, alongside Sparse and Gaussian Random Projection algorithms, with a focus on computational efficiency and downstream analysis effectiveness. We benchmark performance using multiple scRNA-seq datasets including labeled and unlabeled publicly available datasets. We apply Hierarchical Clustering and Spherical K-Means clustering algorithms to assess downstream clustering quality. For labeled datasets, clustering accuracy is measured using the Hungarian algorithm and Mutual Information. For unlabeled datasets, the Dunn Index and Gap Statistic capture cluster separation. Across both dataset types, the Within-Cluster Sum of Squares (WCSS) metric is used to assess variability. Additionally, locality preservation is examined, with RP outperforming PCA in several of the evaluated metrics.

Our results demonstrate that RP not only surpasses PCA in computational speed but also rivals and, in some cases, exceeds PCA in preserving data variability and clustering quality. By providing a thorough benchmarking of PCA and RP methods, this work offers valuable insights into selecting optimal dimensionality reduction techniques, balancing computational performance, scalability, and the quality of downstream analyses.

1 Background and Motivation

Single-cell RNA sequencing (scRNA-seq) has revolutionized the field of genomics, enabling cellular and molecular profiling of gene expression at the individual cell level. This technology's particular strength lies in uncovering cellular variability within samples and tissues, providing insights into dynamic biological

M. Alser et al. (Eds.): ICCABS 2025, LNCS 15599, pp. 1–15, 2026.
https://doi.org/10.1007/978-3-032-02489-3_1

processes, disease progression, and immune responses [1,2]. However, scRNA-seq data, specifically the 'count' matrices are inherently high-dimensional and sparse, posing significant challenges for data analysis and interpretation [3].

Dimensionality reduction is a crucial step in scRNA-seq analysis that aims to transform high-dimensional data into a lower-dimensional latent space while preserving essential biological aspects. While distance-based methods, e.g. t-distributed stochastic neighbor embedding (tSNE) [4] or Uniform Manifold Approximation and Projection (UMAP) [5] focus on preserving locality (i.e. similarity between cells), other techniques have a strength in preserving or capturing variability; Principal Component Analysis (PCA) is one of the most widely used techniques for this purpose [6]. PCA reduces dimensionality by identifying linear combinations of variables (i.e. principal components) that capture the maximum variance of the data. It has been effectively applied in various scRNA-seq studies for visualization, clustering, and trajectory inference [7–9].

Despite its widespread use, PCA has limitations when applied to large complex scRNA-seq data. PCA assumes strictly linear relationships among variables and struggles to capture non-linear structures inherent in biological data [10]. Additionally, PCA is sensitive to outliers and noise, which are common in scRNA-seq data due to technical variability and sparsity [11,12], moreover, the PCA algorithm is typically computationally intensive for large sets [13]. Random Projection (RP) methods have emerged as promising alternatives. Based on the Johnson-Lindenstrauss lemma [14], RP techniques reduce dimensionality by projecting data onto a lower dimensional subspace using a random matrix while aiming at approximately preserving pairwise distances [15]. While RP methods have shown success in fields like machine learning and signal processing [16], their application in scRNA-seq data analysis is only starting to gain attention.

Previous studies in scRNA-seq have primarily focused on comparing PCA with non-linear dimensionality reduction techniques like t-SNE and UMAP [17, 18]. Although PCA has been extensively benchmarked [19], these studies did not include RP methods, highlighting an opportunity for further investigation into the potential advantages of RP in scRNA-seq data analysis. Only a couple of recent studies have explored the use of RP in scRNA-seq, demonstrating its efficiency in handling large, complex datasets [20,21], yet, to our knowledge there are no systematic comprehensive benchmarking studies that contrast PCA and RP techniques. Such evaluation can shed the light on not only time or computing complexity of implementations, but also suitability for downstream analyses, hence the motivation for this work.

2 Approach

In this study, we systematically benchmark multiple PCA algorithms, including standard (or full) PCA and randomized SVD-based PCA, against two common RP methods: Sparse Random Projection (SRP) and Gaussian Random Projection (GRP). We assess their computational efficiency and effectiveness in downstream analysis tasks using both labeled (i.e. with known ground truth of cell populations' annotation) and unlabeled scRNA-seq datasets. By providing a comprehensive evaluation of PCA and RP methods, our work aims to:

- Benchmark the practical time complexity of PCA methods (full SVD and randomized SVD) and RP methods (Sparse and Gaussian) to determine their scalability on increasingly large scRNA-seq datasets.
- Investigate the effectiveness of each method in downstream analysis, specifically clustering.
- Evaluate each method's ability to preserve data structure or locality as well as variability.

Clustering performance is evaluated using Hierarchical Clustering and Spherical K-Means algorithms. For labeled datasets, we measure clustering accuracy using the Hungarian algorithm and Mutual Information. For unlabeled datasets, we use the Dunn Index and Gap Statistic to assess cluster separation. We also examine the preservation of data variability using the Within-Cluster Sum of Squares (WCSS) metric.

Our findings (see Sects. 4 and 5) demonstrate that RP methods not only offer significant computational speed-ups over PCA but also rival and, in some cases, surpass PCA in preserving latent structure, enhancing clustering performance. This study expands the toolbox of dimensionality reduction techniques available for scRNA-seq data analysis and underscores the importance of method selection and evaluation in the face of growing data complexity.

3 Methods

As previously described, we evaluated two types of PCA: full SVD and randomized SVD and two common RP techniques were explored: Sparse Random Projection (SRP) and Gaussian Random Projection (GRP). SRP uses sparse random matrices, leading to faster computations and reduced memory usage [22]. GRP employs dense random matrices with entries drawn from a Gaussian distribution, providing theoretical guarantees on distance preservation. All methods were applied across varying target component sizes. For the commonly targeted range (a few, e.g. 5 to 25 components), we varied our tests in steps of 1, for the less explored range (25 to 1000), we varied our tests in steps of 25. We evaluated the practical implementation computational efficiency and clustering performance of these methods using labeled and unlabeled scRNA-seq datasets as described in Sect. 3.3. Below, we give a summary of PCA and RP definitions when applied to scRNA-Seq data:

3.1 PCA

Standard PCA. Standard PCA computes the principal components using the full SVD of the count matrix. Let $\mathbf{X} \in \mathbb{R}^{m \times n}$ be the count matrix, where m is the number of observations or cells and n is the number of features or genes. The SVD of $\mathbf{X}$ is expressed as: $\mathbf{X} = \mathbf{U}\mathbf{\Sigma}\mathbf{V}^\top$ where: $\mathbf{U} \in \mathbb{R}^{m\times m}$ is an orthogonal matrix containing the left singular vectors, $\mathbf{\Sigma} \in \mathbb{R}^{m\times n}$ is a diagonal matrix with singular values on the diagonal, and $\mathbf{V} \in \mathbb{R}^{n\times n}$ is an orthogonal matrix containing the right singular vectors. The top k principal components are obtained by selecting the first k columns of $\mathbf{V}$, denoted as $\mathbf{V}_k$. The data projected onto these principal components is given by: $\mathbf{T} = \mathbf{X}\mathbf{V}_k$ [23].

Randomized PCA. Randomized SVD-based PCA approximates the principal components efficiently by using randomization to reduce dimensionality followed by computing the SVD. This method leverages a random Gaussian matrix $\mathbf{\Omega} \in \mathbb{R}^{n\times k}$ to approximate the range of $\mathbf{X}$: $\mathbf{Y} = \mathbf{X}\mathbf{\Omega}$, where $\mathbf{Y} \in \mathbb{R}^{m\times k}$. An SVD is then computed on the smaller matrix $\mathbf{Y}$: $\mathbf{Y} = \tilde{\mathbf{U}}\tilde{\mathbf{\Sigma}}\tilde{\mathbf{V}}^\top$. The approximate principal components are obtained from $\tilde{\mathbf{V}}$, and the data projected onto these components is: $\mathbf{T} = \mathbf{Y}\tilde{\mathbf{V}}$. This approach significantly reduces computational complexity while claiming to maintain accuracy. [24]

3.2 RP

RP methods reduce the dimensionality of data by projecting it onto a random lower-dimensional subspace and is thought to be computationally efficient while claiming to preserve the structure of high-dimensional data. The two RP methods evaluated are:

Sparse RP. Sparse RP uses a sparse random matrix $\mathbf{R} \in \mathbb{R}^{n\times k}$, where k is the target dimension. The entries of $\mathbf{R}$ are defined as:

$$r_{ij} = \begin{cases} \sqrt{\frac{s}{k}} & \text{with probability } \frac{1}{2s} \\ 0 & \text{with probability } 1 - \frac{1}{s} \\ -\sqrt{\frac{s}{k}} & \text{with probability } \frac{1}{2s} \end{cases}$$

where s controls the sparsity of the matrix $\mathbf{R}$ (not to be confused with the count matrix sparsity). This scaling ensures that the expected value of r_{ij}^2 is $\frac{1}{k}$, which is crucial for preserving distances during projection. The transformed data $\mathbf{Z}$ is then computed as: $\mathbf{Z} = \mathbf{X}\mathbf{R}$ Sparse RP methods are thought to be advantageous over GRP in computational efficiency and memory savings, especially when s is large. [22]

Gaussian RP. Gaussian RP uses a dense random matrix $\mathbf{G} \in \mathbb{R}^{n\times k}$ with entries drawn independently from a Gaussian distribution: $g_{ij} \sim \mathcal{N}\left(0, \frac{1}{k}\right)$ The projected data $\mathbf{Z}$ is obtained as: $\mathbf{Z} = \mathbf{X}\mathbf{G}$ [15] GRP is thought to be more effective in preserving the pairwise distances of the original data due to the properties of Gaussian distributions, which could translate into better performance in clustering tasks.

Our study is set to explore these properties and claims of strength of each of the methods in practical scRNA-Seq data setting.

3.3 Datasets

Four publicly available scRNA-seq datasets were used to evaluate the effectiveness of the dimensionality reduction methods. These include:

1. **Sorted PBMC Dataset** (from [25]): This dataset includes 2,882 cells and 7,174 genes and serves as a **labeled** set with 7 annotated distinct cell populations, providing a baseline for clustering methods.
2. **50/50 Mixture Dataset (Jurkat:293T Cell Mixture)** (from [26]): This dataset contains approximately 3,400 cells, with an approximately 50% distribution of Jurkat and 50% of 293T fairly homogeneous cell lines. This is a **labeled** dataset with two ground truth labels representing both cell lines.
3. **Targeted PBMC Dataset** (from [27]): This dataset utilizes a panel of putative immune-related genes (approximately 1000 genes after QC) and contains unannotated 10,497 cells (**unlabeled**). Beside for unbiased clustering, this dataset was additionally used with varying sizes ranging from 1000 to 10,000 cells to evaluate scalability.
4. **COVID-19 T Cell Dataset** (from [28]): This data focuses on human TCells in the context of bronchoalveolar immune cell from COVID-19 patients and healthy subjects and is an unlabeled dataset.

The **Sorted PBMC Dataset**, **50/50 Mixture Dataset (Jurkat:293T Cell Mixture)**, and **COVID-19T Cell Dataset (Liao 2020)** were all downloaded from the SC1 Tool. The **Targeted PBMC Dataset** was downloaded from the 10x Genomics website.

3.4 Validation Metrics

As described in Sect. 2. Clustering performance is evaluated using two clustering methods to evaluate PCA and RP robustness when controlling for the effect of the clustering algorithm:

- **Agglomerative Hierarchical Clustering** was used with Ward linkage algorithm and Cosine distance, constructing dendrograms or trees that were cut to the known number of clusters in case of labeled sets.
- **Spherical K-Means Clustering** was used also with Cosine distance, grouping cells based on directional similarities.

To test accuracy and effectiveness or quality of the clustering analysis as one of the main downstream analyses for scRNA-seq data, we measured the following metrics:

- **Accuracy (Hungarian algorithm)**- used for known Ground Truths, maximizing the sum of label matches between predicted and true labels.
- **Mutual Information**- used for known Ground Truths, given as $I(X;Y) = \sum_{x,y} P(x,y) \log \frac{P(x,y)}{P(x)P(y)}$, where $P(x,y)$ is joint probability of a true label x and predicted y. Higher values indicate better match.
- **Dunn Index**- used for unknown Ground Truths and measures the ratio between inter-cluster distance and intra-cluster compactness. It is given as $D = \frac{\min_{i \neq j} d(C_i, C_j)}{\max_k \delta(C_k)}$, where $d(C_i, C_j)$ is the distance between cluster centers C_i and C_j, and $\delta(C_k)$ is the diameter of cluster C_k. Higher values mean better separation.
- **Gap Statistics**- also used for unknown Ground Truths and is given by $\text{Gap}(k) = \frac{1}{B} \sum_{b=1}^{B} \log(W_b) - \log(W_k)$, where W_k is the within-cluster dispersion for k clusters, and W_b represents the expected dispersion under a null reference distribution. Here too, higher values indicate better separation.

We also examine the preservation of data variability using the Within-Cluster Sum of Squares (WCSS) defined as $\text{WCSS} = \sum_{k=1}^{K} \sum_{x \in C_k} \|x - \mu_k\|^2$, where K is the number of clusters, C_k is the set of cells in cluster k, x is a data point in C_k, and μ_k is the centroid of cluster k.

4 Results

4.1 Dimensionality Reduction for Downstream Analysis

We examined two aspects of the dimensionality reduction properties of the examined methods; first how they are used for visualizing the data and second and more importantly, how well cells cluster in the reduced low dimensional space:

Visualization. Figure 1 visualizes the Sorted PBMC dataset using PCA and GRP using only the first two components each. Although both methods show overlap in the projections, GRP provides less clear visualization compared to PCA, this is expected since with the PCA the first few components capture more of the data latent properties like variability while this distinction between the earlier and later components is meaningless for RP methods.

Evaluating Clustering Accuracy and Quality. We used clustering effectiveness as means of evaluating how well the reduced, low dimensional latent space produced from different methods is suited for use in downstream analyses. Figure 2 shows the results for the labeled sets displaying the Dunn Index and Mutual Information for all labeled sets and all methods over varying number of components.

Furthermore, for unlabeled data, we examined the 'goodness' of clustering by evaluating the Dunn Index and Gap Statistics values as described in Methods. Figure 3 illustrates these results, again for all evaluated methods and varying number of components used for clustering.

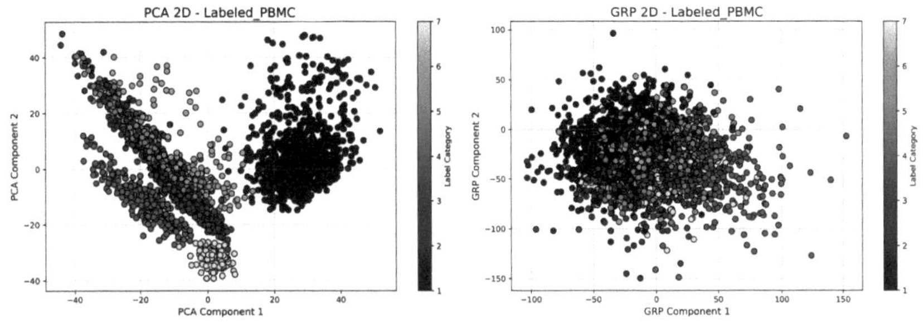

Fig. 1. Visualization for the Sorted (color codes used for labels) PBMC dataset using PCA (left) with Full SVD and GRP (right), both using the first two components only.

4.2 Variability Preservation

Since capturing variability or heterogeneity is one of the main insights PCA and RP methods provide, we set to evaluate whether preserving variability would directly be correlated with perhaps lower accuracy performance, however when evaluating the WCSS metric to measure the heterogeneity 'preserved' by each dimensionality reduction method for each dataset, see Fig. 4, we see that RP methods indicate higher variability preservation while still achieving higher accuracies, especially when using > 25 components for clustering.

4.3 Locality Preservation

We compared the ability of PCA and RP to preserve locality or pairwise similarity of cells by projecting the Sorted PBMC dataset and the Jurkat - 293T 50-50 Mixture dataset into a 2D space using UMAP. Figure 5 shows that SRP results in visually similar UMAP to PCA, preserving essential data relationships crucial for downstream analyses, such as clustering or other. To quantify the locality preservation further, we calculated cluster accuracy metrics for PCA and RP when using 500 embeddings or components each followed by applying UMAP to project further to a 3 dimensions UMAP space. SKMeans Clustering accuracy and Mutual Information metrics using the resulting UMAP components are given in Table 1.

4.4 Computational Efficiency

Figure 6 (a to d) demonstrate the execution times of the dimensionality reduction methods across all the datasets. Additionally, we conducted an experiment in which we varied the size of the targeted PBMCs by Gibbs sub-sampling cells to create multiple datasets of varying sizes ranging from 1000 to 10,000 cells with steps of 1000; using these datasets we assessed the execution time in relevance to dataset size. Figure 6 (e) highlights the scalability of RP methods across increasing dataset sizes, with SRP consistently being the most efficient method, even for largest sample sizes.

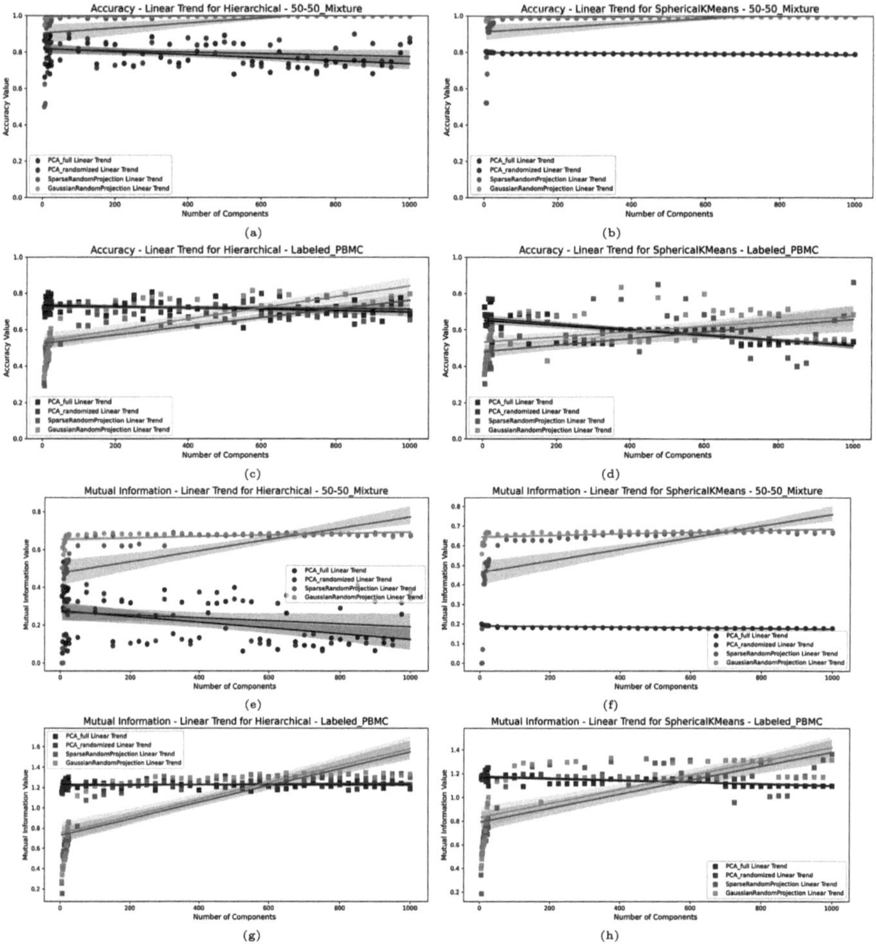

Fig. 2. a) to d): Accuracy for a) Jurkat - 293T 50-50 Mixture dataset with Hierarchical Clustering b) Jurkat - 293T 50-50 Mixture with SKMeans c) Labeled PBMC with Hierarchical Clustering d) Labeled PBMC with SKMeans. e) to h): Mutual Information for e) Jurkat - 293T 50-50 Mixture with Hierarchical Clustering f) Jurkat - 293T 50-50 Mixture with SKMeans g) Labeled PBMC with Hierarchical Clustering h) Labeled PBMC with SKMeans.

5 Discussion

Our findings indicate that RP outperforms PCA in clustering accuracy across several datasets. As shown in Fig. 2, both SRP and GRP variants of RP achieved higher accuracy compared to PCA (utilizing Full and Randomized SVD) on the Jurkat-293T 50-50 Mixture dataset, using both Hierarchical and SKMeans clustering methods. In the labeled PBMC dataset, PCA performed slightly better in

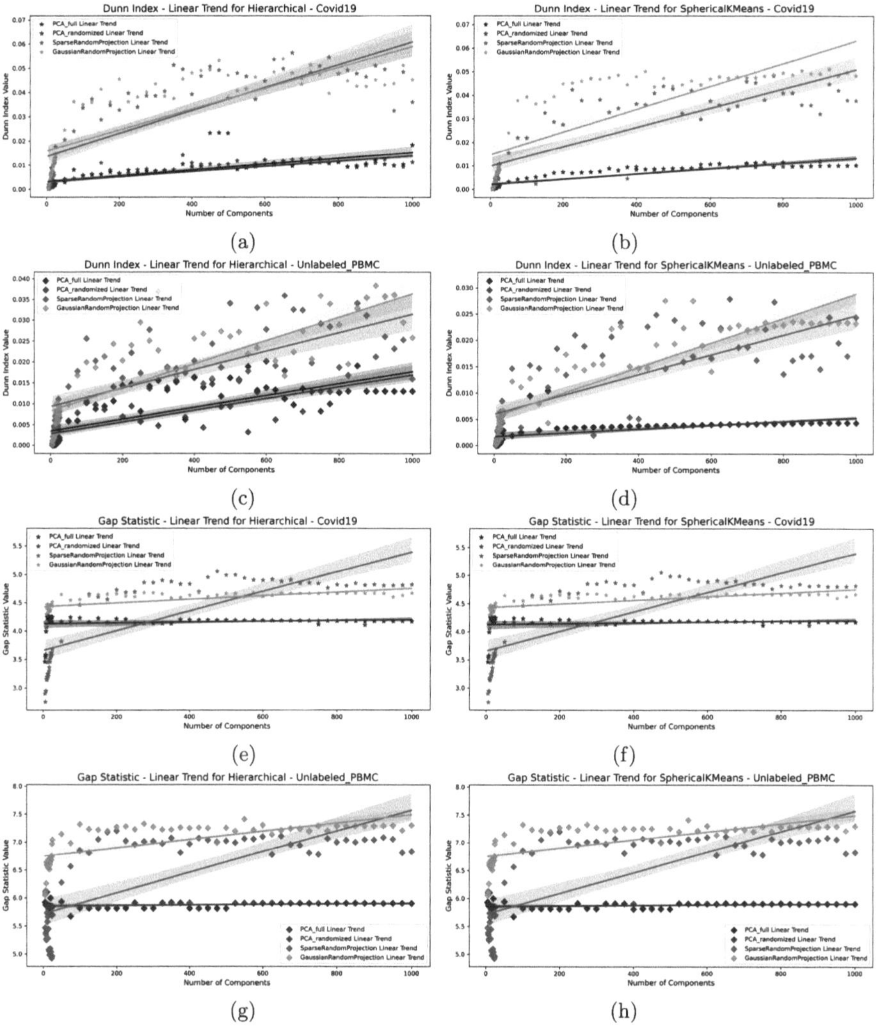

Fig. 3. a) to d): Dunn Index for a) Covid19 dataset with Hierarchical Clustering b) Covid19 dataset with SKMeans c) Unlabeled PBMC dataset with Hierarchical Clustering d) Unlabeled PBMC dataset with SKMeans. e) to h): Gap Statistic for e) Covid19 dataset with Hierarchical Clustering f) Covid19 dataset with SKMeans g) Unlabeled PBMC dataset with Hierarchical Clustering h) Unlabeled PBMC dataset with SKMeans.

lower-dimensional spaces, but as dimensionality increased, RP either matched (for Hierarchical clustering) or outperformed PCA (for SKMeans).

This trend is similarly reflected in when considering he Mutual Information Index, suggesting RP's superiority in the 50-50 Mixture dataset. While RP initially under-performs in lower dimensions in the labeled PBMC dataset, it begins

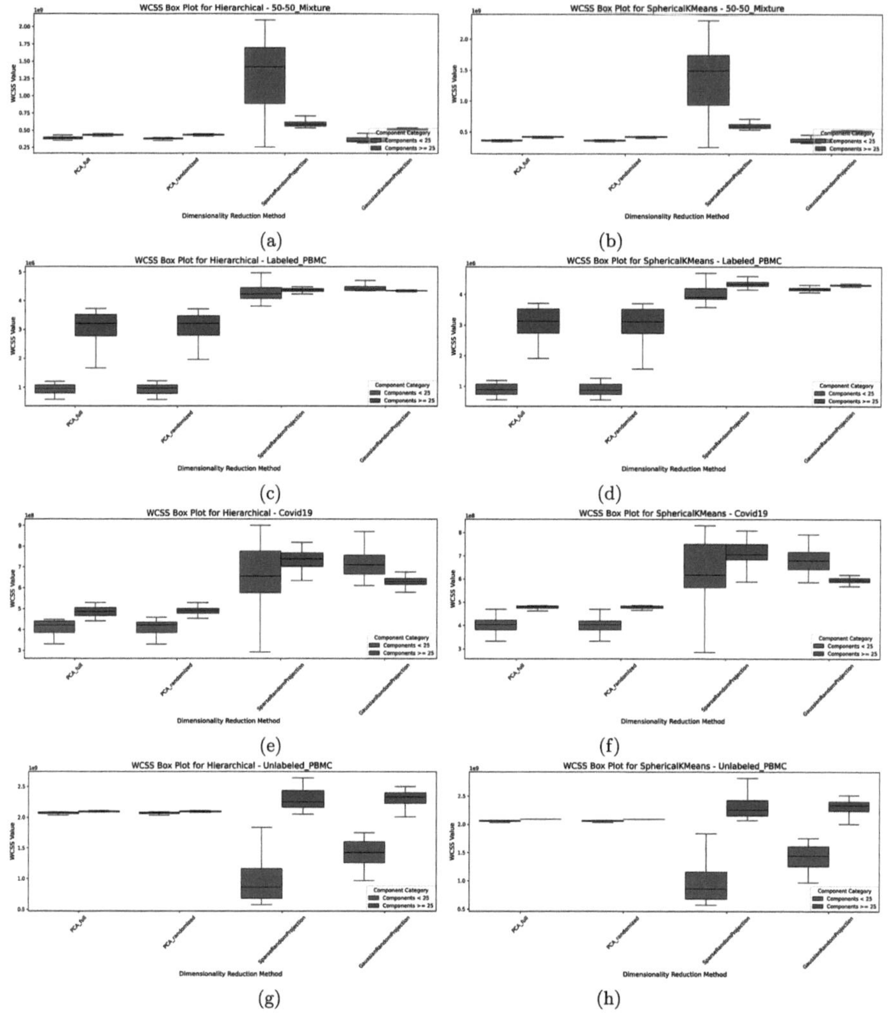

Fig. 4. a) to d): WCSS (Variability Measure) for a) Jurkat - 293T 50-50 Mixture with Hierarchical Clustering b) Jurkat - 293T 50-50 Mixture with SKMeans c) Labeled PBMC with Hierarchical Clustering d) Labeled PBMC with SKMeans. e) to h): WCSS (Variability Measure) for e) Covid19 with Hierarchical Clustering f) Covid19 with SKMeans g) Unlabeled PBMC with Hierarchical Clustering h) Unlabeled PBMC with SKMeans.

to surpass PCA as more components are added, suggesting RP's increasing reliability with higher dimensions.

For the unlabeled datasets, RP again demonstrates improved performance over PCA. Figure 3 shows that RP consistently outperformed PCA in Dunn index values on the Covid19 dataset across both clustering algorithms. In the unlabeled PBMC dataset, RP and PCA had nearly identical performance in

Table 1. Clustering Performance Metrics at 500 Components and after UMAP to 3D

Method	Metrics at 500 Components	
	Accuracy	Mutual Information
PCA Full	0.7893	0.1784
PCA Randomized	0.7893	0.1784
GRP	0.9970	0.6713
SRP	0.9921	0.6499
Method	Metrics After UMAP to 3D	
	Accuracy	Mutual Information
PCA Full + UMAP	0.9967	0.6707
PCA Randomized + UMAP	0.9949	0.6593
GRP + UMAP	0.9878	0.6329
SRP + UMAP	0.9973	0.6739

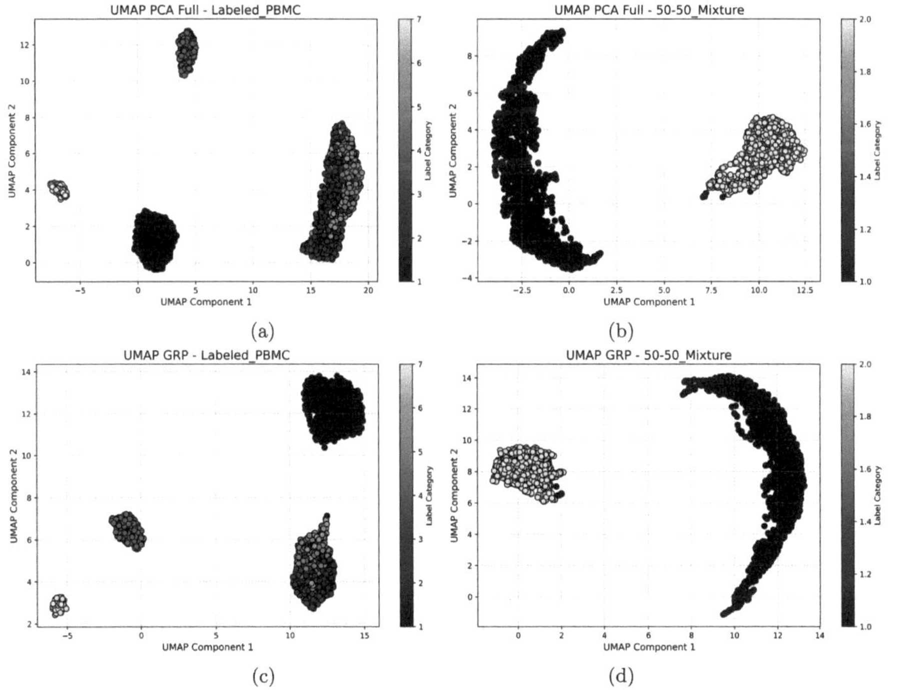

Fig. 5. UMAP 2D projection of the Sorted PBMC dataset with dimensionality reduction of PCA and GRP of 500 components. a) PCA with full SVD for the Labeled PBMC dataset b) PCA with full SVD for the Jurkat - 293T 50-50 Mixture dataset c) GRP for the Labeled PBMC dataset d) GRP for the Jurkat - 293T 50-50 Mixture dataset.

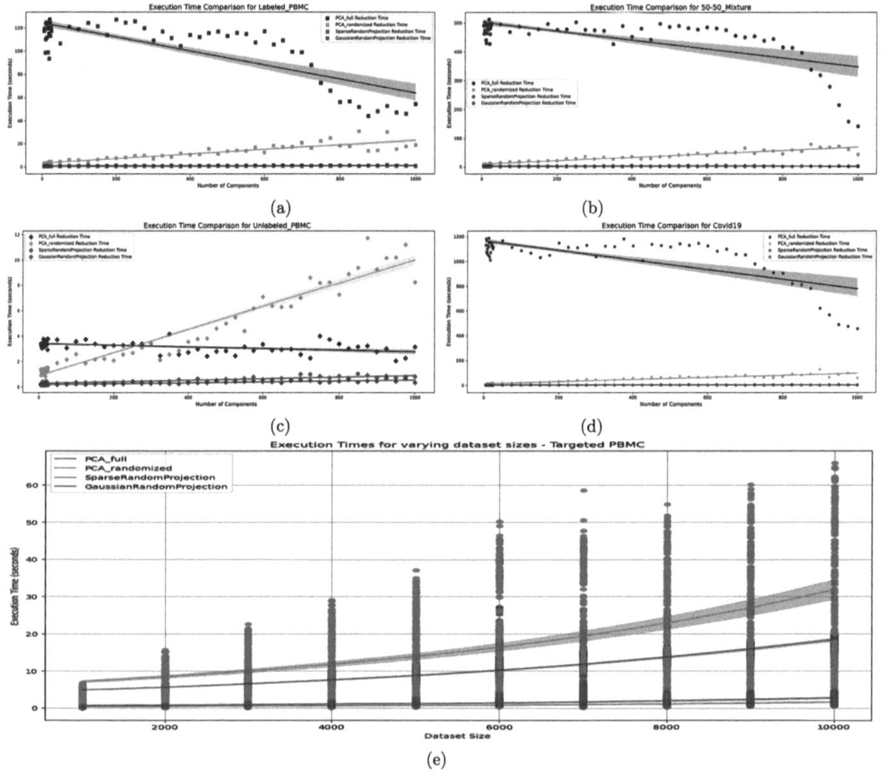

Fig. 6. Execution time vs. number of components for each dimensionality reduction technique on the a) Sorted PBMC dataset, b) Jurkat - 293T 50-50 Mixture, c) Targeted PBMC dataset, and d) Covid19 data. In e) we show Execution time vs. dataset sizes (1000 to 10,000 samples) for each of the dimensionality reduction techniques.

lower dimensions, but RP gained an advantage as dimensionality increased for both clustering algorithms.

In Fig. 3, we evaluated clustering separation with the Gap Statistic on the unlabeled datasets, where RP again consistently outperforms PCA. These results support the effectiveness of RP, particularly in higher dimensions and across diverse datasets, underscoring its potential as a superior dimensionality reduction method for clustering tasks compared to PCA.

In Fig. 4, the box plots reveal that RP methods generally display higher WCSS variability compared to PCA, especially in the range using higher number of components (>25) (Note the median line in all box plots). This increased WCSS median value reflects a broader spread within the clusters formed by RP, which, while resulting in slightly looser clusters, does not detract from RP's overall clustering accuracy. Interestingly, as dimensionality increases, the variability of the WCSS measure itself (inter-quartile range of the box plots) decreases for RP, suggesting that RP's clustering performance becomes more stable and consistent at higher dimensions, still preserving more variability than PCA (higher median line).

Indeed, for all labeled data as well as for the Covid19 dataset, the box plots further show RP's tendency toward higher WCSS across dimensions. Despite this variability, RP consistently performs well in clustering accuracy, as supported by other evaluation metrics. For the Unlabeled PBMC dataset, a targeted gene panel data set, both PCA and RP show relatively similar WCSS values in higher dimensions, but RP's variability is decreased increases when only a few components are used, reflecting the importance of all genes in the panel for capturing the latent variability of this dataset.

Finally, Fig. 6 illustrates RP's clear and significant advantage in terms of execution time. We measured the execution time across all datasets for different numbers of components; RP, especially GRP, consistently and significantly outperformed both PCA methods, highlighting the strength of this method from the computational complexity perspective. We noticed an interesting phenomenon, where PCA with Full SVD shows some decrease in execution time at higher dimensions. This can be explained by the effect of sparsity in the produced embeddings with higher components (i.e. when higher dimensions are calculated, the per component mean value is lower as shown in Fig. 7. Another interesting observation, in (c), the lower dimension targeted panel dataset, the randomized PCA performs better than Full SVD PCA up to a certain point before becoming less efficient, highlighting the overhead of performing randomization when not needed for lower dimensional data.

All in all, and across all datasets, RP demonstrates superior performance in execution time and downstream analysis, highlighting its practical advantages and value in single cell RNA-Seq analysis.

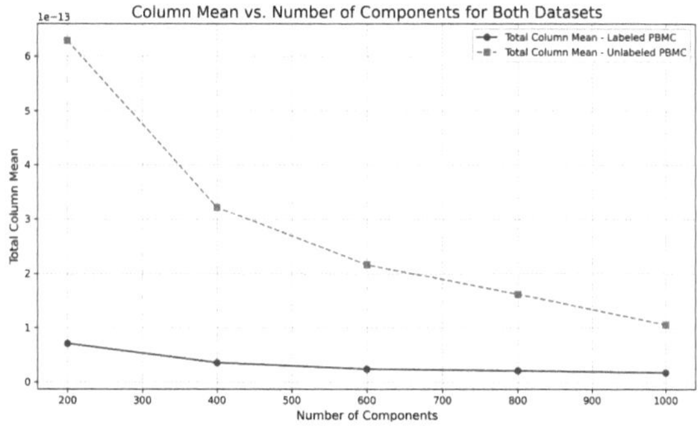

Fig. 7. Mean value per component vs. Number of Components calculated for Labeled and Unlabeled PBMC Datasets, highlighting the impact of PCA Full SVD on execution time.

6 Conclusion

This study demonstrates that RP can outperform PCA in scRNA-Seq analysis, especially, clustering tasks across various datasets, particularly as more dimensions are considered. RP, especially the SRP and GRP variants, achieved higher clustering accuracy and faster execution times compared to PCA. Although PCA performed slightly better in lower dimensions on some datasets, RP consistently excelled in higher-dimensional spaces, showing strong results across accuracy metrics like the Mutual Information Index and Dunn Index. While RP sometimes resulted in broader cluster spreads, this variability did not compromise clustering performance, particularly in high-dimensional settings.

Acknowledgments. This work is supported by the following awards: NSF-2341725, NIH-NCI K25CA270079, and OU-BIC2.0.

Data and Code Availability. The data and code used for this analysis are available on GitHub: https://github.com/moussa-lab/BenchmarkingPCA-RP or upon reasonable request to the authors.

References

1. Tang, F., et al.: mRNA-Seq whole-transcriptome analysis of a single cell. Nat. Methods **6**(5), 377–382 (2009)
2. Saliba, A.E., Westermann, A.J., Gorski, S.A., Vogel, J.: Single-cell RNA-seq: advances and future challenges. Nucleic Acids Res. **42**(14), 8845–8860 (2014)
3. Andrews, T.S., Hemberg, M.: Identifying cell populations with scRNASeq. Mol. Aspects Med. **59**, 114–122 (2018)
4. Van der Maaten, L., Hinton, G.: Visualizing data using t-SNE. J. Mach. Learn. Res. **9**(11) (2008)
5. McInnes, L., Healy, J., Melville, J.: Umap: uniform manifold approximation and projection for dimension reduction. arXiv preprint arXiv:1802.03426 (2018)
6. Jolliffe, I.T., Cadima, J.: Principal component analysis: a review and recent developments. Philos. Trans. A Math. Phys. Eng. Sci. **374**(2065), 20150202 (2016)
7. Van den Berge, K., et al.: Trajectory-based differential expression analysis for single-cell sequencing data. Nat. Commun. **11**(1), 1201 (2020)
8. Horning, A.M., et al.: Single-cell RNA-seq reveals a subpopulation of prostate cancer cells with enhanced cell-cycle-related transcription and attenuated androgen response. Cancer Res. **78**(4), 853–864 (2018)
9. Moussa, M., Măndoiu, I.I.: Computational cell cycle analysis of single cell RNA-Seq data. In: Jha, S.K., Măndoiu, I., Rajasekaran, S., Skums, P., Zelikovsky, A. (eds.) ICCABS 2020. LNCS, vol. 12686, pp. 71–87. Springer, Cham (2021). https://doi.org/10.1007/978-3-030-79290-9_7
10. Hinton, G.E., Salakhutdinov, R.R.: Reducing the dimensionality of data with neural networks. Science **313**(5786), 504–507 (2006)
11. Hubert, M., Rousseeuw, P.J., Vanden Branden, K.: ROBPCA: a new approach to robust principal component analysis. Technometrics **47**(1), 64–79 (2005)
12. Zappia, L., Phipson, B., Oshlack, A.: Splatter: simulation of single-cell RNA sequencing data. Genome Biol. **18**(1) (2017)
13. Tian, L., et al.: Benchmarking single cell RNA-sequencing analysis pipelines using mixture control experiments. Nat. Methods **16**(6), 479–487 (2019)
14. Freksen, C.B.: An introduction to Johnson-Lindenstrauss transforms. arXiv preprint arXiv:2103.00564 (2021)
15. Bingham, E., Mannila, H.: Random projection in dimensionality reduction. In: Proceedings of the Seventh ACM SIGKDD International Conference on Knowledge Discovery and Data Mining. ACM, New York, NY, USA (2001)
16. Fern, X.Z., Brodley, C.E.: Random projection for high dimensional data clustering: a cluster ensemble approach. In: Proceedings of the Twentieth International Conference on Machine Learning. ICML'03, pp. 186–193. AAAI Press (2003)

17. Kobak, D., Berens, P.: The art of using t-SNE for single-cell transcriptomics. Nat. Commun. **10**(1), 5416 (2019)
18. Becht, E., et al.: Dimensionality reduction for visualizing single-cell data using UMAP. Nat. Biotechnol. **37**(1), 38–44 (2018)
19. Tsuyuzaki, K., Sato, H., Sato, K., Nikaido, I.: Benchmarking principal component analysis for large-scale single-cell RNA-sequencing. Genome Biol. **21**(1), 9 (2020)
20. Vrahatis, A.G., Tasoulis, S.K., Georgakopoulos, S.V., Plagianakos, V.P.: Ensemble classification through random projections for single-cell RNA-seq data. Information (Basel) **11**(11), 502 (2020)
21. Wan, S., Kim, J., Won, K.J.: SHARP: hyperfast and accurate processing of single-cell RNA-seq data via ensemble random projection. Genome Res. **30**(2), 205–213 (2020)
22. Li, P., Hastie, T.J., Church, K.W.: Very sparse random projections. In: Proceedings of the 12th ACM SIGKDD International Conference on Knowledge Discovery and Data Mining. ACM, New York, NY, USA (2006)
23. Tharwat, A.: Principal component analysis - a tutorial. Int. J. Appl. Pattern Recogn. **3**(3), 197–240 (2016)
24. Erichson, N.B., Voronin, S., Brunton, S.L., Kutz, J.N.: Randomized matrix decompositions using r. arXiv preprint arXiv:1608.02148 (2016)
25. Moussa, M., Măndoiu, I.I.: Sc1: a tool for interactive web-based single-cell RNA-seq data analysis. J. Comput. Biol. **28**(8), 820–841 (2021). https://doi.org/10.1089/cmb.2021.0051, pMID: 34115950
26. 10x Genomics: 10x genomics datasets (2024). https://www.10xgenomics.com/datasets?query=&page=1&configure%5BhitsPerPage%5D=50&configure%5BmaxValuesPerFacet%5D=1000. Accessed 27 Oct 2024
27. 10x Genomics: Pbmcs from a healthy donor: targeted, immunology panel, single cell dataset by cell ranger v4.0.0 (2020). https://www.10xgenomics.com/datasets/pbm-cs-from-a-healthy-donor-targeted-immunology-panel-3-1-standard-4-0-0
28. Liao, M., et al.: Single-cell landscape of bronchoalveolar immune cells in patients with COVID-19. Nat. Med. **26**(6), 842–844 (2020)

Resistance Genes are Distinct in Protein-Protein Interaction Networks According to Drug Class and Gene Mobility

Nazifa Ahmed Moumi[1], Connor L. Brown[2], Shafayat Ahmed[1], Peter J. Vikesland[3], Amy Pruden[3], and Liqing Zhang[1](✉)

[1] Department of Computer Science, Virginia Polytechnic Institute and State University, Blacksburg, VA, USA
lqzhang@vt.edu
[2] Genetics, Bioinformatics, and Computational Biology, Virginia Polytechnic Institute and State University, Blacksburg, VA, USA
[3] Department of Civil and Environmental Engineering, Virginia Polytechnic Institute and State University, Blacksburg, VA, USA

Abstract. With growing calls for increased surveillance of antibiotic resistance as an escalating global health threat, improved bioinformatic tools are needed to track antibiotic resistance genes (ARGs) across One Health domains. Most studies to date profile ARGs using sequence homology, but such approaches provide limited information about the broader context or function of the ARG in bacterial genomes. Here we introduce a new pipeline, PPI-ARG-finder, for identifying ARGs in genomic data that employs machine learning analysis of Protein-Protein Interaction Networks (PPINs) as a means to improve predictions of ARGs while also providing vital information about the genetic context, such as gene mobility. A random forest model was trained to effectively differentiate between ARGs and nonARGs and was validated using the PPINs of ESKAPE pathogens (*Enterococcus faecium*, *Staphylococcus aureus*, *Klebsiella pneumoniae*, *Acinetobacter baumannii*, *Pseudomonas aeruginosa*, and *Enterobacter cloacae*), which represent urgent threats to human health because they tend to be multi-antibiotic resistant. The pipeline exhibited robustness in discriminating ARGs from nonARGs, achieving an average area under the precision-recall curve of 88%. We further identified that the neighbors of ARGs, i.e., genes connected to ARGs by only one edge, were disproportionately associated with mobile genetic elements, which is consistent with the understanding that ARGs tend to be more mobile compared to randomly sampled genes in the PPINs. PPI-ARG-finder showcases the utility of PPINs in discerning distinctive characteristics of ARGs within a broader genomic context and in differentiating ARGs from nonARGs through network-based attributes and interaction patterns.

M. Alser et al. (Eds.): ICCABS 2025, LNCS 15599, pp. 16–27, 2026.
https://doi.org/10.1007/978-3-032-02489-3_2

1 Introduction

The increasing prevalence of antibiotic-resistant infections poses a significant health threat [1]. According to the Centers for Disease Control and Prevention (CDC), more than 2.8 million people are infected with antibiotic-resistant pathogens annually, resulting in approximately 35,000 deaths [2,3]. Of particular concern is the evolution and spread of novel resistance phenotypes, which results from the horizontal transfer of antibiotic resistance genes (ARGs) to new species or strains or through previously unknown ARGs emerging in the genome [4].

The advent of next-generation DNA sequencing over the past decade represents a promising approach to support One Health surveillance of antibiotic resistance, i.e., across humans, animals, plants/crops, and the environment. DNA sequences can be compared against publicly-available databases to profile ARGs and thereby compare genotypic resistance patterns. However, the incompleteness of public databases is an inherent limitation of this approach, particularly if there is interest in detecting and monitoring previously unidentified ARGs [5]. False positives are also possible, due to local sequence similarity [6]. Finally, simple read-matching homology-based profiling of this nature ignores the context of putative ARGs and other genes of importance in potentiating and mobilizing antibiotic resistance.

Machine learning approaches, especially deep learning, can help to improve the prediction of ARGs, including novel ARGs, relative to simple sequence homology-based comparisons [7,8]. However, such approaches fail to tap into broader contextual information available in the genome to precisely predict ARG function [9]. These approaches struggle to effectively encompass complex interactions between various ARGs as well as with other genomic contexts, e.g., coexpression, co-localization, and genetic background [10]. They also intrinsically ignore information about other genes that may be involved in the mobility of the ARG or in the expression of its phenotype [11]. Focusing analysis on proteins, instead of nucleotide sequences, could present advantages in this regard because it is the proteins encoded by ARGs that perform the ultimate function of conferring resistance to antibiotics, e.g., by target modification, enzymatic degradation of the antibiotic, or pumping the antibiotic out of the cell. In particular, analysis of protein-protein interaction networks (PPINs) could help to address this need, by enhancing our understanding of the proteins involved in resistance, and their interactions within an organism [12].

The overall objective of this study was to develop and validate a PPIN-based pipeline for characterizing ARGs in whole genome data. We hypothesized that ARGs would exhibit distinct patterns in network topology relative to nonARGs in the PPIN, which can then be recognized by machine-learning algorithms to predict their resistance mechanisms. Of particular interest was to assess whether ARGs are likely to be mobile, based on the strength of their networks with mobile genetic elements (MGEs). To validate the pipeline, we analyzed whole genome sequences of representative "ESKAPE" pathogens (*Enterococcus faecium*, *Staphylococcus aureus*, *Klebsiella pneumoniae*, *Acinetobacter baumannii*, *Pseudomonas aeruginosa*, and *Enterobacter cloacae*), which represent an urgent clinical threat because of their tendency to be multi-antibiotic resistant due to carriage of multiple ARGs on MGEs [13–15]. The findings of this study highlight the potential of PPIN-based analysis as a new and accurate means of classifying ARGs, providing more comprehensive characterization than typical read-matching approaches. The approach here also overcomes the limitations of publicly-available databases, enabling the discovery of previously unknown ARGs.

2 Materials and Methods

We propose a framework for identifying ARGs in PPINs based on network topology (Fig. 1). This approach compares network features of potential ARGs to randomly sampled genes associated with basic, housekeeping functions of cellular life (i.e., non-resistance related). We further validate by examining the potential mobility of predicted ARGs across a range of ARG drug classes and pathogens.

2.1 Data Collection

The protein-protein interaction networks (PPINs) for ESKAPE pathogens were collected from STRING-DB, a comprehensive biological database of known and predicted protein-protein interactions [16]. STRING provides two types of interaction networks: functional associations and physical interactions. Functional associations indicate that two proteins are involved in a shared biological process or function. In contrast, physical interactions refer to proteins that directly bind to each other in vivo, forming stable or transient complexes to carry out specific functions. Each interaction in STRING is associated with a confidence score, reflecting the reliability of the data. We chose the physical interaction network over the functional network due to the generally higher confidence scores associated with direct physical interactions.

Table 1. Network information: total number of proteins/nodes, interactions/edges, ARGs, and neighboring nodes to the ARGs in the PPINs for ESKAPE pathogens from STRING-DB.

	EF	*SA*	*AB*	*KP*	*PA*	*EC*
#Proteins (Nodes)	1,867	1,834	2,229	3,688	3,897	3,384
#Interactions (Edges)	26,874	21,669	26,171	35,520	63,007	35,778
#ARGs	182	181	216	468	447	388
#Neighbours to the ARG nodes	857	797	825	1,177	1,504	1,264

Here, *EF: E. faecium*, *SA: S. aureus*, *KP: K. pneumoniae*, *AB: A. baumannii*, *PA: P. aeruginosa*, *EC: E. cloacae*.

Proteins that are likely to encode antibiotic resistance were selected from the set of proteins in the PPINs by using DIAMOND alignment with 14,872 total

Table 2. Summary of features based on the locations of the ARG nodes within the PPIN.

Feature	Function	Description
1 N index	$\frac{k_i^p}{k_i}$	Proportion of the number of links to other ARGs
2 N index	$\frac{\sum_{j \in N_i} k_j^p}{\sum_{j \in N_i} k_j}$	Proportion of links to other ARGs in the second neighbors of a node
Average distance to ARGs	$\frac{\sum_{j \in M} d_{ij}}{\lvert M \rvert}$	The communication efficiency of a node to ARGs
Nearest ARG distance	$\min_{j \in M} d_{ij}$	The distance from a node to its closest ARG in the network
Positive topology coefficient	$\frac{\sum_{j \in M_i} \frac{C_{ij}}{min(k_i,k_j)}}{\lvert M_i \rvert}$	A variant of classical topological coefficient

k_i^p denotes the number of edges lined to an ARG; k_i denotes the degree of the node; N_i denotes a node-set consisting of all the neighbors of node i; M denotes a node set consisting of all the ARGs; d_{ij} denotes the shortest path between two nodes; C_{ij} denotes the number of nodes that are connected to both nodes i and j; M_i denotes a node set consisting of nodes that share neighbors with node i.

ARGs from DeepARG (Table 1) [7,18]. To create a negative control set of non-ARGs, we selected housekeeping genes, which are unlikely to be involved in drug resistance since they encode enzymes essential for basic metabolic functions [19]. We specifically chose genes from 184 Gene Ontology (Biological Process) terms that capture key characteristics of housekeeping genes, such as involvement in metabolic pathways, nucleotide excision repair, and the aerobic respiration chain [20,21]. By listing the proteins from the ESKAPE pathogens from UniProt that contain at least one of these GO terms in its functionality, the designated housekeeping genes were found for these microorganisms.

2.2 Feature Selection

The PPIN can be represented as an undirected graph denoted by $G(V, E)$, where V represents the set of protein vertices and E represents the set of edges. Proteins are connected by an edge if they co-occur within a gold-standard protein complex [16]. To discern between ARGs and nonARGs in the PPIN, we selected ten network topology-based features calculated using NetworkAnalyzer (AverageShortestPathLength, BetweennessCentrality, ClosenessCentrality, ClusteringCoefficient, Degree, Eccentricity, NeighborhoodConnectivity, Radiality, Stress, TopologicalCoefficient) in Cytoscape and five ARG node-based features (Table 2) [22].

2.3 Random Forest Classifier

The constructed networks for ESKAPE pathogens, consisting of a relatively small number of proteins/nodes, prompted the adoption of ensemble learning models, such as random forest to achieve robust performance. Random forest includes a built-in feature selection mechanism calculating the decrease in node impurity weighted by the probability of reaching that node, enabling identification of the most relevant features crucial for accurately distinguishing ARGs from

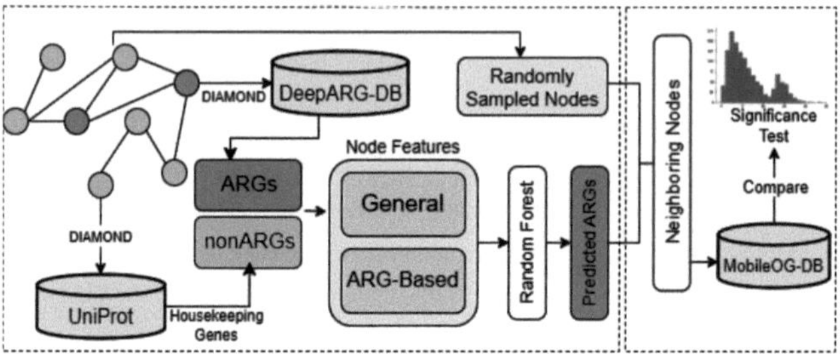

Fig. 1. Overview of the PPI-ARG-finder pipeline. A random forest classifier is used to predict the ARGs in the Protein-Protein Interaction Networks (PPIN). NonARGs are randomly sampled and annotated using MobileOG-DB for comparing their mobility relative to ARGs.

nonARGs. Moreover, random forest is well-suited for analyzing high-dimensional biological data due to its robustness to noise and outliers [23].

To enhance the prediction performance of the models, we implemented the following sampling technique. In this approach, we considered M as the number of instances in the minority class (ARG nodes) and N as the number of instances in the majority class (nonARG nodes) within the training dataset, with M significantly smaller than N. During each iteration, we randomly sampled M instances from the majority class. Subsequently, we combined these M instances with all instances from the minority class to train one random forest model. This sampling process was repeated k times to train k separate models and, eventually, all of the predictions were ensembled together. As such, we achieved a balanced 1:1 ratio between ARGs and nonARGs by creating several undersampled subsets within the nonARGs class. Each of these subsets served as the training data for separate random forest models and the predictions were aggregated through a majority voting approach.

We split our dataset into an 80–20 ratio for training and testing and ensured that there were no ARGs shared between the testing and training sets. We assessed the performance of our random forest classifier by training six models on six different pathogens with five-fold cross-validation. We varied the number of trees in each model from 10 to 400 and selected the optimal number of trees to be used for each model.

A null hypothesis framework was established to evaluate statistical differences in the accuracy achieved in ARG identification. This involved analyzing the distribution of evaluation metric scores derived from ten randomly selected positive sets, which represented ARGs within the network. For each organism under study, we systematically carried out a random sampling procedure to generate the same number of proteins as the actual ARGs identified specifically for that organism. This random sampling process was repeated ten times, yielding ten distinct sets of randomly selected positive proteins for each organism.

Subsequently, we trained separate random forest models by treating these randomly selected sets as the ARGs or positive sets, while the remaining proteins were considered nonARGs or negative sets. The sampling strategy was replicated for this analysis to ensure a 1:1 ratio between the positive and the negative sets. The prediction results obtained from these models were then aggregated using the average method. This approach enabled us to construct a robust statistical framework for assessing the performance of the models by comparing their results to the average evaluation scores derived from these randomized positive sets.

2.4 Mobility Analysis of Nodes in PPINs

Antibiotic resistance in bacteria has been driven in large part due to the rapid spread of mobile ARGs, especially among ESKAPE pathogens. To both validate the predictions and explore connections promoting the spread of ARGs, we inspected the mobility of the ARGs by identifying MGE hallmark genes in the PPINs. We used mobileOG-DB as a MGE protein reference database. Connected proteins in the PPIN were classified into two subcategories derived from mobileOG-db, transfer and integration/excision [24]. These represent proteins mediating intracellular (between cell) and intercellular (within cell) transfer of MGEs, respectively.

As validation, we examined whether PPI-predicted ARGs are more mobile compared to the other nodes. This was performed by comparing the number of putatively mobile ARGs (i.e., ARG nodes with first-order neighbors assigned to the integration/excision or transfer category proteins) to the randomly sampled housekeeping genes. We hypothesized that the first-order neighbors of ARGs would be disproportionately associated with MGE-hallmark genes relative to the nonARGs.

We combined the protein sets from the six pathogens and identified genes encoding resistance to a specific drug class of interest. We then extracted the neighbors of these genes. We randomly selected an equal number of proteins to the number of ARGs from the combined protein set. Next, from the neighboring nodes of these randomly selected proteins, we again randomly sampled an equal number of proteins to match the number of neighbors to the ARGs for that specific drug class. This particular sampling technique was employed to ensure that the network structure remained consistent between the actual neighboring proteins of ARGs and the proteins chosen randomly. This sampling process was repeated 100 times. For each sample, we identified nodes aligning to MGE-hallmarks using DIAMOND. Significance tests were performed to compare the relative mobility with the randomly picked protein sets, as well as across different groups of genes resistant to different drug classes.

3 Results

3.1 Overall Network Topology and Model Accuracy

The results of our analysis revealed an average area under the precision-recall curve (AUPRC) of 0.8816 across all of the ESKAPE pathogens, indicating relatively good overall performance (Fig. 2(a)). However, the performance was better

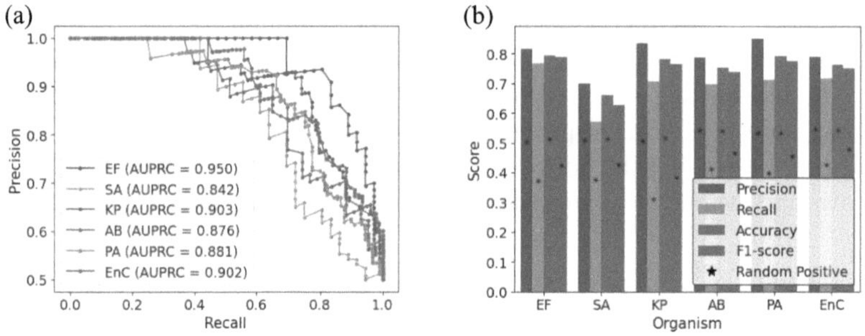

Fig. 2. (a) Area Under the Precision-Recall Curve (AUPRC) and (b) precision, recall, accuracy, and F1 score across the six pathogens (EF: *Enterococcus faecium*, SA: *Staphylococcus aureus*, KP: *Klebsiella pneumoniae*, AB: *Acinetobacter baumannii*, PA: *Pseudomonas aeruginosa*, EnC: *Enterobacter cloacae*.

for some species than others. The model trained on the network from *S. aureus* has the lowest AUPRC of 0.813, likely due to its PPIN having the least number of proteins and also the least number of ARGs. By contrast, the model trained on the network from *K. pneumoniae* demonstrated relatively better performance across all evaluation metrics (AUPRC: 0.914). Comparatively, *K. pneumoniae*'s PPIN had one of the largest numbers of nodes as well as the number of ARGs.

Across various organisms, we consistently observed lower accuracy, recall, precision, and F1-score—averaging around 45% (Fig. 2(b))—when identifying randomly sampled positive nodes, compared to models trained specifically on ARGs. However, when we evaluated the model's performance using the actual set of ARGs unique to each organism, the evaluation metrics improved significantly, nearly doubling the values observed with models trained on randomly sampled nodes.

In our multi-drug class classification, certain drug classes—namely macrolides, fluoroquinolones, multidrug resistance, and tetracyclines—were notably prevalent among ESKAPE pathogens, accounting for 40% of the total ARGs. Notably, our model was able to classify the ARGs belonging to these commonly found drug classes, as evidenced by the darker shades of color along the diagonal in the heatmap of the confusion matrix (Fig. 3). In particular, glycopeptide, peptide, and macrolide drug classes were consistently found among the top ten correctly classified drug classes in all six pathogen-specific models. The majority of genes belonging to these drug classes were accurately identified by our models.

We also observed that some genes from less common drug classes (kasugamycin, trimethoprim, oxazolidinone, elfamycin, etc.) were misclassified as nonARGs. This can be attributed to the limited number of representative genes from these classes in our training set. As a result, our model may not have learned sufficient interaction patterns to accurately classify these less prevalent drug classes.

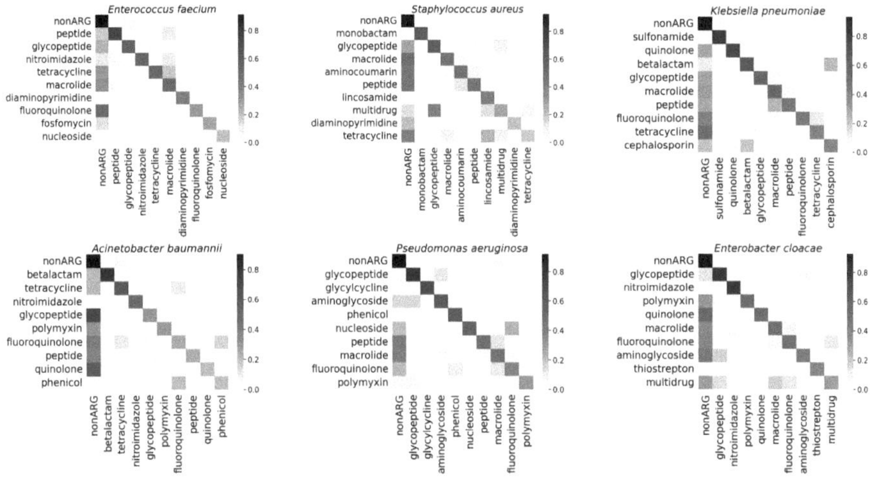

Fig. 3. Normalized confusion matrix showing classification performance for genes encoding resistance across multiple drug classes, with true drug classes on the y-axis and predicted drug classes on the x-axis. The figure highlights the top 10 correctly predicted drug classes for each ESKAPE pathogen.

Feature importance analysis revealed that, in predicting ARGs, the five most informative features contributing to the accuracy of the model included stress centrality, clustering coefficient, betweenness centrality, neighborhood connectivity, and average distance to ARGs. Of these, stress centrality and betweenness centrality, are centrality-based features, highlighting the importance of information connectivity across the PPIN [25,26]. They determine the essentiality of a particular protein in carrying out an associated function by connecting bridges within the network. By incorporating feature values from all ESKAPE pathogens, we observed that ARGs exhibit higher values for stress, closeness centrality, clustering coefficient, and neighborhood connectivity compared to nonARGs. In the PPIN, the stress of a node indicates its role as a key mediator of interactions and communication within the network. Stress measures the number of shortest paths that pass through a given node, highlighting its importance in coordinating diverse biological functions [25].

3.2 Integration/Excision and Conjugation Proteins in the PPINs Show a Significant Association with ARGs

We performed mobility analysis of the 17 prevalent drug classes, including multidrug, tetracycline, fluoroquinolone, macrolide, glycopeptide, peptide, macrolide, aminoglycoside, monobactam, aminocoumarin, cephalosporin, fosfomycin, nucleoside, beta-lactam, polymixin, quinolone, phenicol, sulfonamide which are known to be associated with MGEs in ESKAPE pathogens [13]. The plots indicate a significant and consistent association between ARGs from these

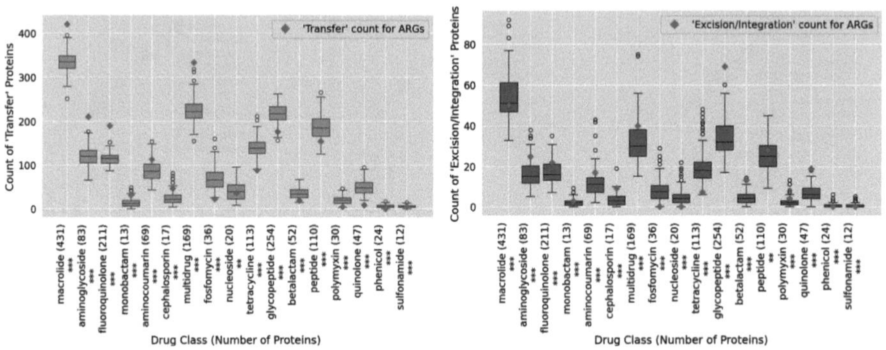

Fig. 4. The distribution of 'Transfer' and 'Excision/Integration' tags among randomly sampled proteins across various drug classes for the ESKAPE pathogens. The red diamond markers highlight the actual counts of the two tags, with asterisks denoting statistical significance (p-value $<$ 0.05 based on t-test). (Color figure online)

prevalent drug classes and transfer/excision elements in the PPINs (Fig. 4). While drug classes including aminoglycoside, aminocoumarin, cephalosporin, monobactam, and multidrug showed a significant association with MGEs, drug classes such as sulfonamide, fosfomycin, nucleoside, phenicol, tetracycline, and glycylcycline exhibited a significantly lower association.

Drug classes like bicyclomycin, bacitracin, bleomycin, chloramphenicol, elfamycin, fosmidomycin, kasugamycin, lincosamide, mupirocin, and oxazolidinone were excluded from the analysis due to their limited representation, with each class comprising less than 1% of the total ARGs individually.

Genes among the beta-lactam, glycopeptide, peptide, polymyxin, and quinolone drug classes exhibited a significantly higher presence of excision/integration elements and a lower presence of transfer elements compared to the randomly sampled nodes (Fig. 4). This suggests that these drug classes may be more prone to gene excision and integration processes, potentially leading to increased resistance. In contrast, none of the drug classes displayed the reverse pattern of low excision/integration elements and high transfer elements.

4 Discussion and Conclusion

PPINs have been used to identify key proteins involved in resistance mechanisms among ESKAPE pathogens and pinpoint essential clusters that highlight resistance pathways [27,28]. Additionally, PPINs were employed to shed light on the intricacies of *S. aureus* pathogenesis within the context of antibiotic resistance [29]. However, to the best of our knowledge, PPIN and its topological properties have not previously been employed to develop a machine-learning model for distinguishing ARGs from nonARGs in specific bacterial strains.

Our research expands on our prior work, which leveraged PPIN to identify ARGs in *Escherichia coli* and *A. baumannii* and analyze their mobility patterns [30]. We have now developed a pipeline, PPI-ARG-finder, that extends this approach to differentiate ARGs from nonARGs, providing insights into ARG mechanisms and mobility patterns across all ESKAPE pathogens. Additionally, a multiclass classification model was developed as a part of the pipeline to determine the specific drug class to which a resistance gene confers resistance. Importantly, we demonstrated that the ARGs can be classified into different drug classes based solely on their network features, without relying on sequence similarities. Overall, our study revealed that genes conferring resistance to different drug classes exhibit distinct behaviors within the PPIN with respect to network topology and mobility.

The study also delves into the role of MGEs in spreading antibiotic resistance. The mobility analysis aligns with the initial assumption that ARGs tend to exhibit higher mobility compared to nonARGs [31,32]. Moreover, we found that ARG mobility patterns can vary with drug classes; for example, genes encoding resistance to aminoglycosides, aminocoumarins, cephalosporins, monobactams, and multidrugs are more likely to be mobile, correlating with higher association with MGEs. In contrast, genes encoding resistance to drug classes like sulfonamides, fosfomycin, and phenicol show less mobility, consistent with previous research that reports their chromosomal localization [33,34].

We conducted our analysis on six highly relevant pathogens (ESKAPE) known for their mobile forms of antibiotic resistance. In the future, PPI-ARG-finder could be adapted to other virulent and antibiotic-resistant bacterial pathogens, leveraging their respective PPINs to predict ARGs based solely on interaction patterns and to further analyze their mobility characteristics. Additionally, our pipeline can be applied to metagenomic data with a preprocessing step to predict interactions among proteins in different organisms within a metagenomic sample. An intriguing aspect of our analysis is the identification of a significant number of proteins falsely predicted as ARGs by our models. This raises questions about their potential roles in resistance mechanisms and whether they could represent novel ARGs, necessitating further investigation. The PPI-ARG-finder pipeline developed in this study can also be applied to identify and study other resistance genes, such as metal resistance genes or biocide resistance genes.

Our pipeline establishes that the network features from the PPIN have some discriminatory power in distinguishing between ARGs and nonARGs. This approach allows for the potential identification of novel ARGs by observing and following the established patterns of well-characterized ARGs within the PPIN. By leveraging PPINs, we gain insights into the mechanisms underlying ARG resistance, and their mobility between bacterial species, and potentially uncover new strategies for combating antibiotic resistance.

Acknowledgments. This work is supported in part by funds from the National Science Foundation (NSF: #2319521, #2125798, and #2004751).

References

1. Talebi Bezmin Abadi, A., Rizvanov, A.A., Haertlé, T., Blatt, N.L.: World health organization report: current crisis of antibiotic resistance. BioNanoScience **9**(4), 778–788 (2019). https://doi.org/10.1007/s12668-019-00658-4
2. Hernando-Amado, S., Coque, T.M., Baquero, F., Martínez, J.L.: Defining and combating antibiotic resistance from one health and global health perspectives. Nat. Microbiol. **4**(9), 1432–1442 (2019)
3. Centers for Disease Control and Prevention and others: Antibiotic resistance threats in the United States. US Department of Health and Human Services, Centres for Disease Control (2019)
4. Pal, C., Bengtsson-Palme, J., Kristiansson, E., et al.: The structure and diversity of human, animal, and environmental resistomes. Microbiome **4**, 54 (2016)
5. Boolchandani, M., D'Souza, A.W., Dantas, G.: Sequencing-based methods and resources to study antimicrobial resistance. Nat. Rev. Genet. **20**(6), 356–370 (2019)
6. Berglund, F., Österlund, T., Boulund, F., Marathe, N.P., Larsson, D.G.J., Kristiansson, E.: Identification and reconstruction of novel antibiotic resistance genes from metagenomes. Microbiome **7**(1), 52 (2019)
7. Arango-Argoty, G., Garner, E., Pruden, A., Heath, L.S., Vikesland, P., Zhang, L.: DeepARG: a deep learning approach for predicting antibiotic resistance genes from metagenomic data. Microbiome **6**(1), 23 (2018)
8. Ahmed, S., Emon, M.I., Moumi, N.A., Zhang, L.: LM-ARG: identification & classification of antibiotic resistance genes leveraging pre-trained protein language models. In: 2022 IEEE International Conference on Bioinformatics and Biomedicine (BIBM), pp. 3782–3784 (2022)
9. Inda-Díaz, J.S., Lund, D., Parras-Moltó, M., Johnning, A., Bengtsson-Palme, J., Kristiansson, E.: Latent antibiotic resistance genes are abundant, diverse, and mobile in human, animal, and environmental microbiomes. Microbiome **11**(1), 44 (2023)
10. Clausen, P.T.L.C., Zankari, E., Aarestrup, F.M., Lund, O.: Benchmarking of methods for identification of antimicrobial resistance genes in bacterial whole genome data. J. Antimicrob. Chemother. **71**(9), 2484–2488 (2016)
11. Nielsen, T.K., Browne, P.D., Hansen, L.H.: Antibiotic resistance genes are differentially mobilized according to resistance mechanism. GigaScience **11**, giac072 (2022)
12. Hanes, R., Zhang, F., Huang, Z.: Protein interaction network analysis to investigate stress response, virulence, and antibiotic resistance mechanisms in listeria monocytogenes. Microorganisms **11**(4), 930 (2023)
13. De Oliveira, D.M.P., et al.: Antimicrobial resistance in ESKAPE pathogens. Clin. Microbiol. Rev. **33**(3), e00181-19 (2020)
14. Denissen, J., et al.: Prevalence of ESKAPE pathogens in the environment: antibiotic resistance status, community-acquired infection and risk to human health. Int. J. Hyg. Environ. Health **244**, 114006 (2022)
15. Pendleton, J.N., Gorman, S.P., Gilmore, B.F.: Clinical relevance of the ESKAPE pathogens. Expert Rev. Anti Infect. Therapy **11**(3), 297–308 (2013)
16. von Mering, C., et al.: STRING: known and predicted protein-protein associations, integrated and transferred across organisms. Nucleic Acids Res. **33**(Database issue), D433–D437 (2005)
17. Chen, J., et al.: Increasing confidence of protein-protein interactomes. In: Genome Informatics International Conference on Genome Informatics, vol. 17, no. 2, pp. 284–297 (2006)

18. Buchfink, B., Xie, C., Huson, D.H.: Fast and sensitive protein alignment using DIAMOND. Nat. Methods **12**(1), 59–60 (2015)
19. Kumar, R.R., Prasad, S.: Metabolic engineering of bacteria. Indian J. Microbiol. **51**(3), 403–409 (2011)
20. Petit, C., Sancar, A.: Nucleotide excision repair: from E. Coli Man. Biochimie **81**(1–2), 15–25 (1999)
21. Anraku, Y., Gennis, R.B.: The aerobic respiratory chain of Escherichia Coli. Trends Biochem. Sci. **12**, 262–266 (1987)
22. Doncheva, N.T., Assenov, Y., Domingues, F.S., Albrecht, M.: Topological analysis and interactive visualization of biological networks and protein structures. Nat. Protoc. **7**(4), 670–685 (2012)
23. Qi, Y.: Random forest for bioinformatics. In: Zhang, C., Ma, Y. (eds.) Ensemble Machine Learning: Methods and Applications, pp. 307–323. Springer, New York (2012). https://doi.org/10.1007/978-1-4419-9326-7_11
24. Brown, C.L., et al.: mobileOG-Db: a manually curated database of protein families mediating the life cycle of bacterial mobile genetic elements. Appl. Environ. Microbiol. **88**(18), e00991-22 (2022)
25. Gilbert, M., et al.: Comparison of path-based centrality measures in protein-protein interaction networks revealed proteins with phenotypic relevance during adaptation to changing nitrogen environments. J. Proteomics **235**, 104114 (2021)
26. Li, M., Zhang, H., Wang, J., Pan, Y.: A new essential protein discovery method based on the integration of protein-protein interaction and gene expression data. BMC Syst. Biol. **6**(1), 15 (2012)
27. Priyamvada, P., Debroy, R., Anbarasu, A., Ramaiah, S.: A comprehensive review on genomics, systems biology and structural biology approaches for combating antimicrobial resistance in ESKAPE pathogens: computational tools and recent advancements. World J. Microbiol. Biotechnol. **38**(9), 153 (2022)
28. Miryala, S.K., Ramaiah, S.: Exploring the multi-drug resistance in Escherichia Coli O157:H7 by gene interaction network: a systems biology approach. Genomics **111**(4), 958–965 (2019)
29. Otarigho, B., Falade, M.O.: Analysis of antibiotics resistant genes in different strains of staphylococcus aureus. Bioinformation **14**(3), 113–122 (2018)
30. Moumi, N.A., Brown, C.L., Vikesland, P.J., Pruden, A., Zhang, L.: Protein-protein interaction network analysis reveals distinct patterns of antibiotic resistance genes. In: 2022 IEEE International Conference on Bioinformatics and Biomedicine (BIBM), pp. 73–76 (2022)
31. Stokes, H.W., Gillings, M.R.: Gene flow, mobile genetic elements and the recruitment of antibiotic resistance genes into gram-negative pathogens. FEMS Microbiol. Rev. **35**(5), 790–819 (2011)
32. Partridge, S.R., Kwong, S.M., Firth, N., Jensen, S.O.: Mobile genetic elements associated with antimicrobial resistance. Clin. Microbiol. Rev. **31**(4), e00088-17 (2018)
33. Rizzo, L., et al.: Urban wastewater treatment plants as hotspots for antibiotic resistant bacteria and genes spread into the environment: a review. Sci. Total Environ. **447**, 345–360 (2013)
34. Li, W., et al.: Population-based variations of a core resistome revealed by urban sewage metagenome surveillance. Environ. Int. **163**, 107185 (2022)

DuoHash: Fast Hashing of Spaced Seeds with Application to Spaced K-mers Counting

Leonardo Gemin, Cinzia Pizzi(✉), and Matteo Comin(✉)

Department of Information Engineering, University of Padua, Padua, Italy
{cinzia.pizzi,matteo.comin}@unipd.it

Abstract. Alignment-free genomic sequence analysis has facilitated high-throughput processing within numerous bioinformatics workflows. A central task in alignment-free applications is hashing k-mers, commonly used for indexing, querying, and fast similarity searches. Recently, spaced seeds—a specialized pattern designed to accommodate errors or mutations—have increasingly replaced k-mers, enhancing sensitivity in various applications. However, spaced seed hashing is computationally intensive, introducing significant delays. This paper addresses the challenge of efficient spaced seed hashing and presents DuoHash, a framework that enables the efficient computation of several hash functions. Our experimental results demonstrate that the proposed method substantially outperforms existing algorithms, achieving speedups of up to 11x. To illustrate practical utility, we further applied DuoHash to the problem of spaced k-mers counting. The code of DuoHash is available at https://github.com/CominLab/DuoHash/.

Keywords: spaced seeds · hashing · counting k-mers · alignment-free

1 Introduction

Alignment-free methods are at the basis of numerous state-of-the-art tools for sequence analysis [2]. The sheer volume of data generated by modern sequencing technologies make alignment-based approaches increasingly inadequate for the high-throughput processing required by today's sequence analysis applications, which deal with massive datasets.

Compared to alignment-based techniques, k-mer-based approaches are significantly faster by several orders of magnitude. However, this comes at the cost of a reduced sensitivity, as these methods rely on exact matches across all k positions. To address this limitation, several variants of the exact k-mer matching paradigm have been introduced. Noteworthy examples include approaches that accounts for longest matches [1,26], weighted patterns [23], or methods that permit non-consecutive matches within k-mers [16].

M. Alser et al. (Eds.): ICCABS 2025, LNCS 15599, pp. 28–39, 2026.
https://doi.org/10.1007/978-3-032-02489-3_3

The concept of *spaced seeds*, fixed-length patterns that incorporate wildcards at specific positions, was first introduced in [16], significantly increasing the likelihood of detecting relevant similarities. Since then, spaced seeds have helped to enable many successful algorithms in various bioinformatics fields. Examples include homology search [13,19], protein classification [20]; read mapping [24]; phylogenetic tree reconstruction [14,25]; clustering and classification of metagenomic reads [3,6,21,27]; oligonucleotide design [10]; and protein-protein interaction prediction [15].

While alignment-free techniques using spaced seeds are faster than alignment-based methods, they experience a significant slowdown in comparison to equivalent k-mer-based solutions. This is because k-mer indexing takes advantage of the fact that consecutive k-mers share a substantial portion of their sequence, allowing for more efficient hashing [18]. In contrast, the projection of two consecutive segments of a sequence based on spaced seeds may have minimal overlap due to the placement of wildcards [8], which limits the potential for similar optimization.

Several methods have been proposed for the hashing for spaced seed (see next section), but they all focus on optimizing the computation of a specific hash function. In this paper we address the problem of efficient hashing of spaced seeds. In particular we propose a framework that enable the efficient computation of several hash functions. Moreover, to show its applicability in practice, we applied the proposed method to the problem of spaced k-mers counting.

1.1 Related Works

Unlike traditional k-mer, spaced seeds allow for the introduction of gaps (wildcard positions) within the nucleotide or amino acid sequence, thus allowing for a more flexible comparison of sequences. In formal terms, a spaced seed Q is a string on the alphabet $\{0, 1\}$, where the `1`s correspond to the matching positions:

$$Q = \{x \mid x \in \{0,1\}^*,\ \#_1(x) = w\},$$

where k is the *length*, or *span*, and w is the *weight* of the spaced seed.

One can also represent the spaced seeds by their *shape* Q, which is the set of positions of the `1` in the spaced seed [12]. In this case the weight of Q is defined as $k = |Q|$, while the length, or span, is equal to $s(Q) = \max(Q) + 1$.

Spaced k-mers, also called Q-grams, are fragments of a nucleotide sequence x that respect the pattern dictated by a spaced seed Q. Given a string x, the spaced k-mer $x[i + Q]$ is a string defined as follows:

$$x[i + Q] = \{x_{i+k} \mid k \in Q\},$$

where $i \in \{0, 1, \ldots, |x| - s(Q)\}$.

Example 1. Let us consider the sequence `CTTGTCGTTGACT` and the spaced seed `111010101`. Then the Q-gram at position 0 of x is defined as

x	C T T G T C G T T G [...]
Q	1 1 1 0 1 0 1 0 1
$x[0+Q]$	C T T T G T

The other Q-grams are: x[1 + Q] = TTGCTG, x[2 + Q] = TGTGTA, x[3 + Q] = GTCTGC, and x[4 + Q] = TCGTAT.

The problem of spaced seed hashing focuses on the design of efficient methods to exploit similarities between different Q-grams, aiming to reduce the number of encoding and shift operations needed for computing spaced seed-based hashes of a DNA sequence.

Fast Spaced Seed Hashing (FSH) [8], leverages the similarity between adjacent hash values of the same DNA sequence to compute each hash more efficiently. For this purpose, it reuses parts of previously computed hash values, extracting them through a mask, and then combining the results with the encoding of the remaining positions. Fast Indexing for Spaced Seed Hashing (FISH) [7], instead exploits block-indexing. It decomposes the spaced seed into unit blocks of consecutive '1's, which the algorithm treats as k-mers of varying lengths that can be quickly hashed. Iterative Spaced Seed Hashing (ISSH) [22] builds on FSH by recovering information from more than one previous hash. Using a greedy strategy, ISSH iteratively searches for hashes that maximize the number of positions recovered, considering only the characters that remain to be encoded.

All these methods compute an encoding of the spaced Q-grams, in which each DNA character is encoded with two bits. In general, a hash function applied to a sequence will map that sequence to an integer. Regardless of the size of the input, a hash function should typically produce a fixed-size output. However, the encodings produced by ISSH, FISH and FSH depend on the length of the spaced seed, and it does not map to a fixed-size output. Thus, all these methods are not producing in output the hashing values of spaced seeds, but just the 2-bit encoding of the spaced Q-mers. Another problem with the above methods is that they only deal with the forward strand, ignoring the reverse complement.

A method that addresses these issues is ntHash [18] that computes a proper hash function, for both forward and reverse complement. The method of ntHash is based on a recursive function, known as a rolling hash function, which calculates the hash value of the current k-mer h_i from the hash value of the previous k-mer h_{i-1} via a recursive formula. To initialise the hash calculation, the first k-mer is calculated as follows:

$$h_0 = \bigoplus_{j=0}^{k-1} \mathrm{rol}^{k-1-j} h(x[j]).$$

In this formula, $\mathrm{rol}^j(\cdot)$ is a left-cyclic rotation, $\oplus$ is the exclusive OR (XOR) operator, and $h(\cdot)$ is a seed table where the nucleotide characters, $\Sigma = \{\mathtt{A}, \mathtt{C}, \mathtt{G}, \mathtt{T}\}$, are assigned to different 64-bit random integers.

The hash value of each successive k-mer is calculated recursively:

$$\begin{aligned} h_i &= f(h_{i-1}, x[i+k-1], x[i-1]) \\ &= \mathrm{rol}^1\, h_{i-1} \oplus \mathrm{rol}^0\, h(x[i+k-1]) \oplus \mathrm{rol}^k\, h(x[i-1]) \end{aligned}$$

With the introduction of ntHash2 [11], the algorithm was further improved to handle spaced seeds.

2 Method: DuoHash

In this section we present DuoHash, a novel approach that overcomes the limitations of FHS, FISH and ISSH by efficiently computing a general hash function.

The major contribution of DuoHash is the introduction of a new strategy to calculate the hashing of a nucleotide sequence, both forward and reverse, significantly reducing the calculation time thus increasing the overall efficiency. While our approach can be applied to any hash function, in our experiments we will use the same hash function as ntHash, for comparison purposes.

The basic idea behind the DuoHash algorithm is the use of tables with pre-computed hashes, including all possible combinations of four DNA bases, in order to speed up the overall hash computation. This approach drastically reduces the number of operations. Let us consider an example of a look-up table.

Example 2. Example of a look-up table.

sequence	encoding	hashing
AAAA	00000000	0x53EC3F8647623EED
CAAA	00000001	0x3B2DEE31FC53472D
GAAA	00000010	0xB622149EFBEB046D
TAAA	00000011	0xFD19ADB0B6403FFD
ACAA	00000100	0x678CD75D9AFA820D
...	...	...
TTTT	11111111	0x9400B260ACBDFF13

Given the nucleotide sequence x = AGGCCCACTGGAAGTTGTAGCCACCG and the spaced seed 11110111011100111011101111 , the Q-gram $x[0+Q]$ and its binary encoding are:

x	AGGCCCACTGGAAGTTGTAGCCACCG
Q	11110111011100111011101111
$x[0+Q]$	AGGC CAC GGA TTG AGC ACCG

$$encoding(x[0+Q]) = 1001010001100010111100101001000101101000$$

There are $k = 5$ groups of 8 bits (corresponding to the encoding of 4 bases). Each group is used as an index i to access the pre-calculated hash tables:

i	encoding	hashing
0	01101000	0x15609AFAC162C235
1	10010001	0x3DA45F3F050E3E0D
2	11110010	0x8841C2559987C40B
3	01100010	0x8249A46E23AF65F5
4	10010100	0x6105665363A7FB2D

The hashing value is then calculated using the following formula:

$$\text{hashing} = \bigoplus_{i=0}^{k-1} rol^{4\cdot(k-i-1)}\text{look-up}[i]$$

Figure 1 shows the computation of the hash value for the Q-gram x[0+Q].

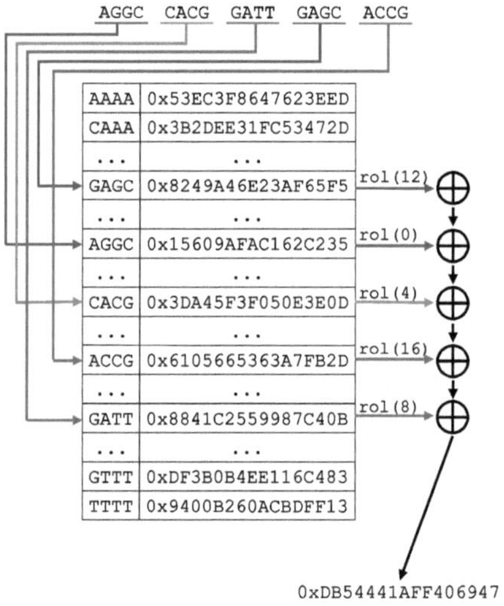

Fig. 1. Schematic and visual procedure for using look-up tables to calculate the hashing value of the Q-gram **AGGCCACGGATTGAGCACCG**.

To handle both forward and reverse hashing, and cases where the last group of 4 bases is not completely filled, we need 8 look-up tables in total. We call n-f-Hash, with $n = 1, 2, 3, 4$, the four tables for the forward strand. The value of n indicates that the table manages groups of exactly n bases. Each table contains 4^n values. For each of these values, the corresponding rotations are also pre-calculated, avoiding rotation operations at runtime. Similarly, we have four tables n-r-Hash for the reverse strand.

Taking into account that spaced seeds with a maximum weight of 32 are allowed, and that the values in the look-up tables are in groups of 4 bases, the possible shifts are 8. Considering that each table contains 4^n values, that for each value the corresponding 8 shifts are also pre-calculated, and that each value is a 64-bit integer (8 Bytes), each table requires a memory space equal to 4^{n+3} Bytes. In total, the look-up tables occupy 21.25 kBytes. Algorithm 1 in the Appendix describes how the look-up tables are pre-calculated.

DuoHash starts by computing the encoding of the nucleotide sequence. For this purpose the already optimized tool ISSH can be used. The encoding is then stored within a structure called **Hash**, which also contains two variables set for forward and reverse hashing. Their value is computed by the `getHashes` function, as shown in Algorithm 1. The function gets two parameters: the structure `Hash` and the value $s(Q)$. The variable `encoding` is temporarily broken down into bytes, each of which represents the encoding of 4 bases. Using the byte value as an index, the function accesses a series of tables of pre-calculated hash values.

```
Function getHashes(Hash, s(Q))
    bytes ← s(Q)/4;                    // Hash = ⟨encoding, forward, reverse⟩
    for i ← 0 to bytes do
        curr_byte ← i-th byte of encoding;
        forward ← forward ⊕ 4_f_Hash[curr_byte][bytes − i − 1];
        reverse ← reverse ⊕ 4_r_Hash[curr_byte][i];
    end
    if s(Q) mod 4 ≠ 0 then
        curr_byte ←last byte of encoding;
        forward ← rol^(s(Q) mod 4) forward;
        if s(Q) mod 4 = 3 then
            forward ← forward ⊕ 3_f_Hash[curr_byte][0];
            reverse ← reverse ⊕ 3_r_Hash[curr_byte][bytes];
        end
        else if s(Q) mod 4 = 2 then
            forward ← forward ⊕ 2_f_Hash[curr_byte][0];
            reverse ← reverse ⊕ 2_r_Hash[curr_byte][bytes];
        end
        else if s(Q) mod 4 = 1 then
            forward ← forward ⊕ 1_f_Hash[curr_byte][0];
            reverse ← reverse ⊕ 1_r_Hash[curr_byte][bytes];
        end
    end
end
```

Algorithm 1: DuoHash: getHashes function

The new strategy is based on the idea of avoiding the repetitive calculation of the same values by exploiting pre-computed look-up tables, which drastically reduces the number of operations required to obtain the hashes. The approach also takes into account the need to handle variable-length sequences efficiently. Indeed, handling $s(Q) \bmod 4$ makes it possible to deal with cases where the length of the sequence is not an exact multiple of 4.

A further significant advantage of this implementation is the ease with which the hashing function can be modified. Thanks to the modular structure, it is possible to update the `getHashes` function to switch from a rolling hash function to any other hashing function, without having to modify other parts of the code. For example, one could replace the rolling hash function with a hash function based on the algorithm of FNV [5], known for its simplicity and efficiency. Similarly, one can easily modify the pre-computed look-up tables to implement new features, like to convert the encodings into nucleotide sequences, in order to save the spaced-seed in a FASTA file, creating a dataset that can be used by third-party tools such as JellyFish [17]. This makes DuoHash extremely flexible and easily adaptable to new requirements or hashing algorithms.

3 Results

Two distinct groups of artificial datasets, called "L" and "R", were used in the experiments. They were designed to vary either the length or the number of reads. Their details are summarized in the Appendix. The number of reads in the "L" datasets varies from a minimum of 500,000 to a maximum of 5,000,000, while their length is always 80bp. The datasets in "R" all have 500,000 reads, while the length of the reads varies from 250 to 5,000bp.

To enable an accurate comparison with the ntHash2 tool, it was necessary to select the spaced seeds so that they were symmetrical. For ntHash2, in fact, the symmetry of the spaced seeds is fundamental for the calculation of the reverse hashing. It is important to emphasise that the DuoHash does not require spaced seed symmetry, which is a considerable advantage in terms of flexibility. Each spaced seed set is composed of three sets of three spaced seeds, designed to meet specific criteria: maximization of the probability of success [21], minimisation of overlap complexity [9], maximisation of sensitivity [9]. A total of six seedsets of different weights and lengths were used, as shown in Table 1. The complete set of spaced seeds is reported in the Appendix. The heterogeneity of the seedsets makes it possible to assess the efficiency of the tools in various situations.

Table 1. Spaced Seed sets used in the experiments.

Spaced Seed	Description
W10L15	spaced seeds of weight 10 and length 15
W14L31	spaced seeds of weight 14 and length 31
W18L31	spaced seeds of weight 18 and length 31
W22L31	spaced seed of weight 22 and length 31
W26L31	spaced seeds of weight 26 and length 31
W32L45	spaced seeds of weight 32 and length 45

3.1 Analysis of the Time Performances

In this section we discuss computational time performances in terms of speedup obtained by DuoHash w.r.t. ntHash2. To ensure the accuracy of the results, each configuration was tested 10 times, and the average of these values was taken.

Figure 2 shows the results for the spaced seed with weight 26 and length 31 (W26L31). The speed-up achieved by DuoHash tends to remain constant,

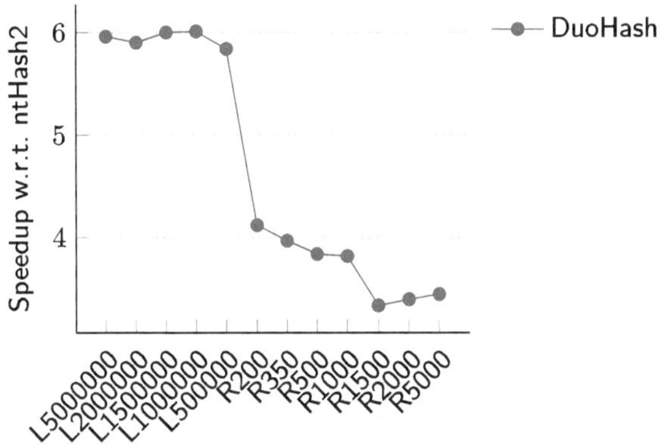

Fig. 2. DuoHash speed-up for seedset W26L31 on each datasets.

or decrease slightly. This trend is particularly evident for datasets in the "L" group, where the speed-up is highest. For the "L" group datasets - characterised by reads of varying number, but constant length - the method demonstrates a constant and significant speed-up in all scenarios tested. This indicates that the software scales efficiently with the data volume when the length of the reads remains unchanged. For datasets in the "R" group that maintain a constant number of reads, but vary in length - the speed-up shows a greater variability. In this group, the speed-up generally tends to decrease slightly as the length of the reads increases. Despite this, the overall performance of the DuoHash remains robust, demonstrating its ability to effectively handle different lengths of reads.

From the results, it clearly emerges that the `DuoHash` method prevails over ntHash2, with a speed-up between 3.34× and 11.23×.

3.2 Performance Evaluation with Varying Seed Weight

The weight of a spaced seed is a critical factor that can influence the sensitivity and specificity of the software. In Fig. 3 results for four seed sets with a fixed length of 31bp and different weights are reported: W14L31, W18L31, W22L31 and W26L31.

For the "L" group datasets, the speed-up shows some fluctuations. In general, there is a slight decrease in speed-up as the seed weight increases. However, there are significant upward deviations for seedset W18L31, indicating that this particular seedset performs exceptionally well with a 11.23× speed-up.

For the datasets in the "R" group, the speed-up follows a similar trend to that observed in the "L" group. The general trend is a decrease in speed-up with increasing seed weight. However, the upward deviations observed for seedset W18L31 in the "L" group datasets are much less pronounced in the "R" group datasets. This suggests that, although the seedset W18L31 continues to perform well, its relative advantage is reduced when it comes to variable read lengths.

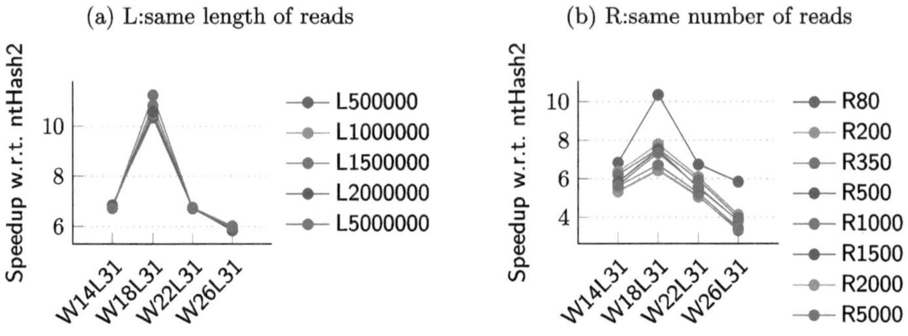

Fig. 3. `DuoHash` speed-up on seedset with varying weight and dataset of L and R.

The maximum speed-up achieved for the "R" group datasets was 10.35× with seedset W18L31.

3.3 Performance Evaluation with Varying Seed Length

The length of the spaced seed is another crucial parameter that can influence the time performances. To assess the impact of spaced seed length, seedsets with different lengths and fixed density were used: W10L15, W22L31 and W32L45. As there are only three seedsets, a variation in one of them can significantly influence the overall trend. Although the pattern of behavior is clear and consistent in all the cases we analyzed (see Fig. 4), we cannot speak of a well-defined trend.

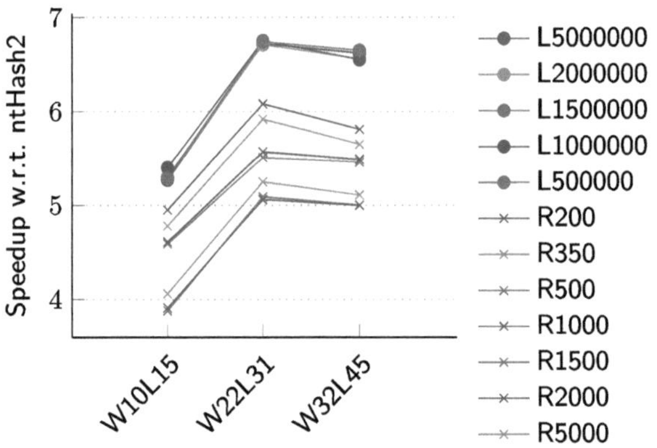

Fig. 4. Speed-up graph for method `DuoHash` among seedset with varying length and dataset of both "L" and "R" groups.

For the "L" group datasets, we observe that for W10L15 the speed-up is approximately 20 per cent than those of W22L31 and W32L45. This trend is observed on all datasets. This behaviour might suggest that, beyond a certain seed length, the efficiency of the tool remains stable. The maximum speed-up achieved for the L-group datasets was 6.75× with seedset W22L31.

For the R-group datasets, the speed-up value for W10L15 is about 20 per cent smaller than the speed-up of the other two seedsets (W22L31 and W32L45), which maintain constant values. For this group of datasets, a maximum speed-up of 6.08× was achieved with seedset W22L31.

3.4 Analysis of the time performances in spaced k-mer Counting

In this section, a detailed comparison will be made between two tools for spaced k-mer counting: DuoHash and MaskJelly [4]. Both tools are set to output a FASTA file containing the spaced k-mer extracted from the nucleotide sequence provided as input. The generated FASTA file will serve as input for JellyFish, a software tool known for counting k-mer. This comparison will allow us to evaluate the efficiency and performance of the two tools in the context of preprocessing sequences for k-mer counting.

The entire process was evaluated in terms of execution time, in order to determine which tool offers better performance in preprocessing sequences. The speed-up of DuoHash compared to MaskJelly is depicted in Fig. 5(a). DuoHash offers a significant improvement in terms of execution time compared to MaskJelly on all four datasets we considered, with an average speed-up ranging between 5.19× and 6.46×.

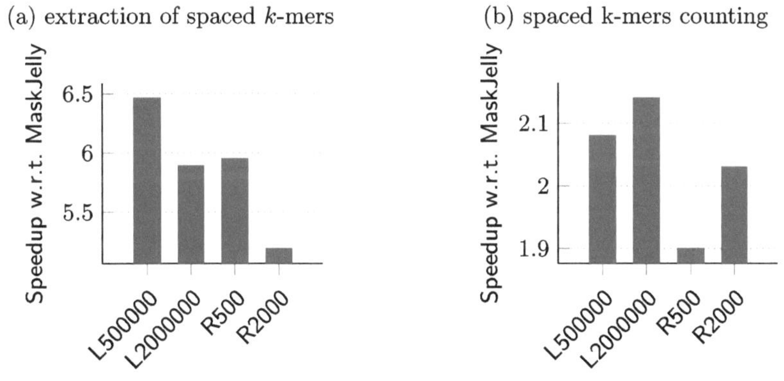

Fig. 5. Speed-up graph for DuoHash with respect to MaskJelly for the extraction of spaced k-mers (left) and overall counting (right).

o gain a comprehensive understanding of the overall performance improvement provided by DuoHash, it is also crucial to compare the speed-up of the entire process, which includes both the pre-processing of the nucleotide sequences and the subsequent k-mer counting by JellyFish. Figure 5(b) illustrate this comparison, showing the speedup for each dataset. The integration of DuoHash significantly accelerates the complete process, offering a speed-up ranging between

1.90× and 2.14×. This enhancement underlines DuoHash's efficiency not only in the preprocessing stage but also in the context of the entire k-mer counting workflow, making it a highly effective tool for genomic sequence analysis.

4 Conclusions

We presented DuoHash, a highly efficient tool for hashing spaced k-mer for nucleotide sequences, significantly improving performance over existing tools such as ntHash2. Performance tests have shown that DuoHash offers a significant speed-up, with an observed maximum speed-up of about 11× compared to ntHash2. High speed-ups are particularly evident in datasets characterised by reads of constant length, where DuoHash showed a marked superiority. These results were achieved thanks to significant innovations in the data architecture, such as the use of optimised hash structures and pre-computed look-up tables, which drastically improved the calculation performance. The flexibility of the system, with the possibility of easily modifying the hash function and handling both symmetric and asymmetric spaced seeds, makes DuoHash a versatile tool for different types of genomic analysis. For instance, the compatibility with JellyFish makes DuoHash a powerful tool for pre-processing nucleotide sequences for spaced k-mer counting.

Acknowledgments. M.C. and C.P. are supported by the Project funded under the National Recovery and Resilience Plan (NRRP), Mission 4 Component 2 Investment 1.4 - Call for tender No. 3138 of 16 December 2021, rectified by Decree n.3175 of 18 December 2021 of Italian Ministry of University and Research funded by the European Union NextGenerationEU.

References

1. Apostolico, A., Guerra, C., Pizzi, C.: Alignment free sequence similarity with bounded hamming distance. In: 2014 Data Compression Conference (DCC), Snowbird, UT, USA, pp. 183–192 (2014)
2. A.Zielezinski, Vinga, S., Almeida, J., et al.: Alignment-free sequence comparison: benefits, applications, and tools. Genome Biol. **18**, 186 (2017)
3. Břinda, K., Sykulski, M., Kucherov, G.: Spaced seeds improve k-mer-based metagenomic classification. Bioinformatics **31**(22), 3584 (2015)
4. Ebler, J., Ebert, P., Clarke, W.E., et al.: Pangenome-based genome inference allows efficient and accurate genotyping across a wide spectrum of variant classes. Nat. Genet. **54**(4), 518–525 (2022)
5. Fowler, G., Noll, L.C., Vo, K.P.: Fowler-noll-vo (fnv) hash function (2005). http://www.isthe.com/chongo/tech/comp/fnv/index.html
6. Girotto, S., Comin, M., Pizzi, C.: Metagenomic reads binning with spaced seeds. Theoret. Comput. Sci. **698**, 88–99 (2017)
7. Girotto, S., Comin, M., Pizzi, C.: Efficient computation of spaced seed hashing with block indexing. BMC Bioinform. **19**(15), 441 (2018)

8. Girotto, S., Comin, M., Pizzi, C.: FSH: fast spaced seed hashing exploiting adjacent hashes. Algorithms Mol. Biol. **13**(1), 8 (2018)
9. Hahn, L., Leimeister, C.A., Ounit, R., Lonardi, S., Morgenstern, B.: rasbhari: Optimizing spaced seeds for database searching, read mapping and alignment-free sequence comparison. PLOS Comput. Biol. **12**(10), 1–18 (2016)
10. Ilie, L., Ilie, S., Khoshraftar, S., Bigvand, A.M.: Seeds for effective oligonucleotide design. BMC Genomics **12**(280) (2011)
11. Kazemi, P., Wong, J., Nikolić, V., Mohamadi, H., Warren, R.L., Birol, I.: nthash2: Recursive spaced seed hashing for nucleotide sequences. Bioinformatics **38**(20), 4812–4813 (2022)
12. Keich, U., Li, M., Ma, B., Tromp, J.: On spaced seeds for similarity search. Discret. Appl. Math. **138**(3), 253–263 (2004)
13. Kucherov, G., Noé, L., Roytberg, M.A.: A unifying framework for seed sensitivity and its application to subset seeds. J. Bioinform. Comput. Biol. **4**(2), 553–569 (2006)
14. Leimeister, C.A., Boden, M., Horwege, S., Lindner, S., Morgenstern, B.: Fast alignment-free sequence comparison using spaced-word frequencies. Bioinformatics **30**(14), 1991 (2014)
15. Li, Y., Ilie, L.: Sprint: ultrafast protein–protein interaction prediction of the entire human interactome. BMC Bioinform. **18**(485) (2017)
16. Ma, B., Tromp, J., Li, M.: Patternhunter: faster and more sensitive homology search. Bioinformatics **18**(3), 440 (2002)
17. Marçais, G., Kingsford, C.: A fast, lock-free approach for efficient parallel counting of occurrences of k-mers. Bioinformatics **27**(6), 764–770 (2011)
18. Mohamadi, H., Chu, J., Vandervalk, B.P., Birol, I.: ntHash: recursive nucleotide hashing. Bioinformatics, btw397 (2016)
19. Noé, L., Martin, D.E.K.: A coverage criterion for spaced seeds and its applications to support vector machine string kernels and k-mer distances. J. Comput. Biol. **21**(12), 947–963 (2014)
20. Onodera, T., Shibuya, T.: The gapped spectrum kernel for support vector machines. In: Perner, P. (ed.) MLDM 2013. LNCS (LNAI), vol. 7988, pp. 1–15. Springer, Heidelberg (2013). https://doi.org/10.1007/978-3-642-39712-7_1
21. Ounit, R., Lonardi, S.: Higher classification sensitivity of short metagenomic reads with clark-s. Bioinformatics **32**(24), 3823 (2016)
22. Petrucci, E., Noé, L., Pizzi, C., Comin, M.: Iterative spaced seed hashing: closing the gap between spaced seed hashing and k-mer hashing. J. Comput. Biol. **27**(2), 223–233 (2020). https://doi.org/10.1089/cmb.2019.0298
23. Pizzi, C., Ukkonen, E.: Fast profile matching algorithms – a survey. Theoret. Comput. Sci. **395**(2), 137–157 (2008)
24. Rumble, S.M., Lacroute, P., Dalca, A.V., Fiume, M., Sidow, A., Brudno, M.: Shrimp: accurate mapping of short color-space reads. PLOS Comput. Biol. **5**(5), 1–11 (2009)
25. Röhling, S., Linne, A., Schellhorn, J., Hosseini, M., Dencker, T., Morgenstern, B.: The number of k-mer matches between two dna sequences as a function of k and applications to estimate phylogenetic distances. PLoS One **15** (2020)
26. Ulitsky, I., Burstein, D., Tuller, T., Chor, B.: The average common substring approach to phylogenomic reconstruction. J. Comput. Biol. **13**(2), 336–50 (2006)
27. Wood, D., Lu, J., Langmead, B.: Improved metagenomic analysis with kraken 2. Genome Biol. **20**(257) (2019)

Unsupervised Learning for Tertiary Structure Prediction of Protein Molecules: Systematic Review

Kazi Lutful Kabir(✉)

George Mason University, Fairfax, VA 22030, USA
kkabir@gmu.edu

Abstract. Tertiary structures of molecules represent high-dimensional data containing spatial information of hundreds (even thousands) of atoms. Unsupervised learning techniques can be applied to such spatial data to uncover hidden organizations that can be subjected to further evaluation. Such techniques have already been employed in a number of relevant applications e.g., tracking the conformational changes in a set of biomolecular structures, detecting biologically active tertiary structures from computed structures of proteins, analyzing molecular dynamics simulation of peptides, and so on. This paper presents a comprehensive review of clustering techniques for tertiary (3D) molecular structure data focusing on protein molecules. In fact, the article systematically organizes and analyzes the existing approaches in terms of data representation, methodology, proximity measure, and evaluation metric. Besides, it highlights key open challenges and proposes future research directions to advance this domain.

Keywords: Clustering · Protein Tertiary Structure · Proximity Measure

1 Introduction

The tertiary structures of molecules represent the three-dimensional arrangement of atoms, which are highly complex and dynamic, particularly in molecules like proteins they take on different configurations under physiological conditions. Understanding their dynamic behavior requires organizing these structures into structural states, which can be addressed using unsupervised learning, specifically clustering. Proteins exhibit fast transitions within the same state and slower transitions between different states, making clustering a suitable tool to summarize their behavior and identify states relevant to cellular interactions. Clustering, as an optimization problem, lacks universal evaluation metrics, with methods differing based on data representation, proximity measures, and optimization processes. The fundamental steps include selecting appropriate data representation, proximity measures, techniques, and evaluation metrics. Applications in computational biology range from capturing conformational changes

M. Alser et al. (Eds.): ICCABS 2025, LNCS 15599, pp. 40–53, 2026.
https://doi.org/10.1007/978-3-032-02489-3_4

in protein structures [11] to detecting macrostates in molecular dynamics simulations [26]. This article reviews and categorizes the existing research, highlighting findings and limitations in clustering methods for protein tertiary structures. The rest of this paper is organized as follows. Firstly, the key concepts are briefly summarized in Sect. 2. Section 3 presents an area taxonomy that has been identified and then summarizes existing methods along the identified taxonomy. The article concludes with a summary of future directions (in Sect. 4) and concluding remarks in Sect. 5.

2 Preliminaries

2.1 Representation of Protein Molecular Structures

Cartesian Coordinates: Tertiary structures of protein molecules are generally represented as ordered sequences of 3D coordinates for their constituent amino acids. For a molecule with N atoms, a naive representation places it as a point S in a $3N$-dimensional space with $S = (x_1, y_1, z_1, \ldots, x_N, y_N, z_N)$. Structural data, such as that stored in the Protein Data Bank (PDB) [9], includes lists of atoms along with their 3D spatial coordinates. To reduce dimensionality, representations often focus on specific subsets of atoms, such as using only the alpha-carbon (C_α) atoms or the backbone atoms (C_α, C, N, and O).

Dihedral Angles: Instead of Cartesian coordinates, one can use the backbone dihedral/torsion angles (ϕ and ψ angles per amino acid) as features. A dihedral angle is an angle between two planes; the plane formed by the atoms $i-2, i-1, i$ and the plane formed by the atoms $i-1, I, i+1$ where $i-2, i-1, i, i+1$ are four sequentially bonded atoms (Fig. 1). The backbone of a protein (which links the backbone atoms) has 3 different torsion angles- phi (ϕ): rotation around N–C_α bond in an amino acid, psi (ψ): rotation around C_α–C bond in an amino acid, omega (ω): rotation around C–N bond linking two consecutive amino acids.

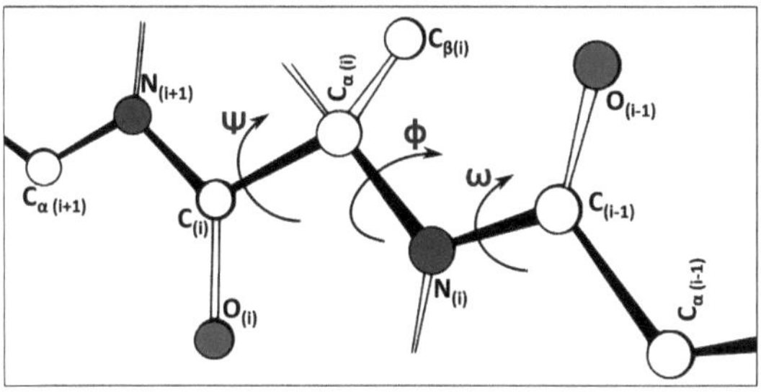

Fig. 1. Backbone Dihedral Angles.

Shape-Based Features: Ultrafast Shape Recognition (USR) metrics are used to characterize the 3D shapes of ligands and can be applied to featurize tertiary molecular structures. These metrics rely on moments of the distance distribution of atoms to compare molecular shapes. USR identifies four reference points from a structure: the molecular centroid (ctd), the closest atom to the centroid (cst), the farthest atom from the centroid (fct), and the farthest atom from fct (ft). The geometry and shape are captured through the mean, variance, and skewness of the distance distributions for these points, resulting in a set of 12 features [42].

Contact Map: A contact map is a binary two-dimensional $L \times L$ square matrix, M that represents the distance between all possible residue pairs of a tertiary structure of a protein (L denotes the residue length). The element $M(i, j)$ is 1 if the distance between the residues i and j is less than a predefined threshold and 0 otherwise. Besides, dimensionality reduction techniques (such as PCA [6], isometric feature mapping [38]) are often employed to simplify the representation of molecular structures.

2.2 Proximity/Distance Measures

To compare the tertiary structures of proteins, several similarity/dissimilarity measures are available. Among them, the most prominent ones include:

Root Mean Square Deviation (RMSD): The RMSD between pairs of equivalent atoms is widely used to capture the degree of similarity between two optimally superimposed tertiary structures of a protein. $RMSD = \sqrt{\frac{1}{N}\sum_{1}^{N} \mid S_A^i - S_B^i \mid^2}$ where N is the number of atoms, S_A^i and S_B^i represent the coordinate vectors for i-th atom of the structure A and structure B respectively (after optimal superimposition). One can consider several settings to compute RMSD such as: only over C_α atoms [31] or backbone atoms [11,32] or overall atomic coordinates ($cRMSD$ [22]) of the structures.

Global Distance Test-Total Score (GDT-TS): It determines the similarity between two structures with their corresponding superimposed residues. $GDT-TS(S_A, S_B) = \frac{P_1+P_2+P_4+P_8}{4}$ where P_t denotes the percentage of residues from structure S_A to be superimposed with the corresponding residues from structure S_B having chosen distance threshold, t ($t \in \{1, 2, 4, 8\}$). GDT-TS value ranges from 0 to 1. The larger score indicates better similarity.

Template-Modeling (TM) Score: TM-Score determines the global structural similarity of a structure concerning the reference structure in terms of the distances of each pair of residues. $TM - Score = max\left[\frac{1}{N}\sum_{i=1}^{N}\frac{1}{1+(\frac{d_i}{d_0})^2}\right]$ where N denotes the number of residues, d_i represents the distance of the i-th pair of residues after alignment. The range of TM-Score values is $(0, 1]$, with a higher value indicating better similarity. Besides, metrics like contact map overlap (CMO) [22], Tanimoto coefficient (T_c) [19], Local-Global Alignment (LGA) Score [43], and C-score [44] have also been used as proximity measures in different occasions.

3 Systematic Review

Taxonomy-Based Survey

Taxonomy provides a systematic way to organize the methods and tools developed in a particular area and most importantly it helps to identify the research gaps in the area. However, no such effort has been found in the literature for unsupervised learning of molecular structures of proteins. Hence an attempt is made here based on the comprehension of the research landscape in this domain. Table 1 represents an overview of the methods categorized under four major heads: representation, proximity measure/ distance function used, the types of clustering techniques employed, and the evaluation metrics applied.

3.1 Representation

Tertiary structures of proteins are three-dimensional objects, they have shape and occupy volume in space as they are composed of atoms occupying positions. The atoms are not free-floating and connect to each other with links/bonds. Hence the obvious key question one would have to answer first is that if we want to cluster three-dimensional objects, how do we represent them? What do we encode that will allow us to recognize any inherent organization? The most popular and intuitive way is to consider the structures as ordered sequences of 3D coordinates and represent them using Cartesian coordinates. One has to make a decision regarding whether to keep all of the atoms (all-atoms [22]) or a specific type of atom (C_α) [28,30,31] or a group of atoms (backbone atoms [11]). While suitable for classic distance metrics like Euclidean distance, this representation is extremely high-dimensional, leading to issues such as the curse of dimensionality and reduced performance for clustering algorithms. On the other hand, the changes in molecular structures can be considered as the outcome of rotations around bonds that connect atoms. In fact, the comparison of structures (at room temperature) reveals that changes in angles are due to some specific ones (dihedral angles) [37]. Dihedral angle-based representation ensures dimensionality reduction by a factor of 7 over the Cartesian coordinates [33]. The most intuitive distance function for this representation would be L1-norm.

Table 1. Overview of the Unsupervised Learning Methods for Protein Tertiary Structures

Based on	Categories	Instances from Literature
Representation	Cartesian Coordinates	C_α **atoms**: *Calibur* [31], *ONION* [30], *SCUD* [28]; **Backbone atoms**: Boutonnet et al. [11]; **All-atoms**: *ClusCo* [22]
	Dihedral Angles	*ICON* [37]
	Shape-based	Zaman et al. [42], Kabir et al. [26]
	Contact Map	Han et al. [20]
Proximity	Similarity	**TM-Score**: *ONION* [30], *Calibur* [31]
	Distance	**RMSD**: *SCAR* [10], *Durandal* [7], SCUD [28], *SPICKER* [45]
Clustering Type	Partitional	*Pleiades* [21], *SCAR* [10], *ONION* [30], *ClusCo* [22]
	Hierarchical	*HCPM* [18], Estrada et al. [15], *bcl − Cluster* [5]
	Graph-based	Akhter et al. [4], Kabir et al. [25], Zhou et al. [46]
	Factorization-based	SNMF-DS [24], NMF-MAD and NMF-Rank [2], NTF-REL [23]
Evaluation	Application-specific	*SCAR* [10], *SCUD* [28], *ONION* [30], *Calibur* [31]
	Application-agnostic	Li et al. [29], Zhou et al. [46], ClusPro [13]

However, it is necessary to go beyond the L1-norm to design more meaningful distance functions as all angles are not equally important. If we interpret angles as rotations, changes in angles at the beginning of the chain of atoms bring a larger impact (in terms of the swept volume in 3D) than changes in angles at the end of the chain. Besides, changes in atomic positions or angles ultimately result in changes to the shape of the structure. And it is possible to come up with a coarse representation of shape. USR metrics summarize the distance distribution of atoms from four reference points via mean, variance, and skewness [42]. This mechanism ensures dimensionality reduction by a good margin but fails to capture subtle structural changes. Contact maps capture the internal geometry of the structure and bring forth a more reduced representation of protein tertiary structures (in comparison to their entire three-dimensional atomic coordinates) and are also invariant to rotations and translations. Moreover, a contact map can also be represented by an ordered graph [20]. However, this representation throws out subtle information regarding structures.

3.2 Proximity/Distance Measure

Proximity measures or distance functions (for comparing protein tertiary structures) are primarily focused on Cartesian coordinate-based representation. The most popular and intuitive distance function is RMSD which is a variant of Euclidean distance. The RMSD value is dependent on the size of the molecule (number of atoms). The computation is time-demanding as it requires that the structures be aligned first to remove differences due to rigid-body motions (whole-body translation and whole-body rotation in 3D). Similarity metric GDT-TS aims to address the dependency of RMSD on the number of atoms. It first finds the subsets of atoms within certain thresholds of RMSD (1Å, 2Å, 4Å, 8Å) and then reports an average of these percentages over the thresholds. GDT-TS is more accurate than RMSD at capturing structural differences but it is also computationally demanding. On the other hand, TM-score weights shorter distances between corresponding atoms more strongly than longer distances. It ensures more sensitivity to global topology rather than local structure deviations. The magnitude of the TM-score is length-independent for random structure pairs.

3.3 Clustering Techniques

Clustering algorithms identify groups of observations that are more similar to each other than to the observations of other groups. For the clustering of protein tertiary structures, the strategy followed by most systems (e.g., *ROSETTA* [36], *I-TASSER* [41], *SPICKER* [45], *SCUD* [28], *Calibur* [31]) can be summarized by the following steps:

1. For a given set of structures, choose a threshold, t_h for the proximity measure
2. The structure with the most neighbors within distance t_h from it is extracted and is reported as the structure with the highest rank

3. This structure and all of its neighbors form the first cluster and are removed
4. Repeat steps (3) and (4) until no further clusters are found

Techniques for clustering protein tertiary structure fall into four major branches: partitional, hierarchical, graph-based methods, and factorization-based methods.

Partition-Based Methods: `k-means` is the most popular partitional clustering method. While dealing with the structure data, `k-means` tries to solve the following problem: Given n structures $S_1, S_2, S_3, ..., S_n$, `k-means` attempts to group the structures into k clusters $(A_1, A_2, A_3, ..., A_k)$ to optimize the objective function of k-means. In fact, `k-means` is considered as the baseline in [3,42]. *Pleiades* is a k-means-based method for clustering protein structures that uses a 31-dimensional tuned Gauss integral (GIT) vector representation of the structures to approximate the RMSD. The Euclidean distance (having a fair correlation with the RMSD) measures the proximity between two structures [21]. While faster than RMSD-based k-means, GIT representation may map different structures to the same vector. RMSD-based k-means, though slower, produce more accurate clusters, showing that k-means can offset GIT's limitations. Pleiades' key advantage is its computational speed [21]. *SCAR* is a k-means-inspired clustering method that uses Relative RMSD (RRMSD) as a universal proximity measure (enabling robust clustering, with centroids representing global consensus topology) and structure packing number for adaptive local cluster cutoff that considers the variability in the cluster's internal dispersion. It includes a refinement step with centroid realignment via singular value decomposition (SVD). Overlapping clusters are resolved by removing the one with the lower packing density (a measure of cluster cohesion). The method minimizes inter-cluster correlation by ensuring centroid distances exceed individual cluster dispersion [10]. *ONION* operates similarly to k-means, with a slightly modified objective function minimizing RMSD between structures and optimal centroids obtained via superimposition [30]. It determines the optimal k (number of clusters) using a Gaussian mixture model with Schwartz's Bayesian information criterion (BIC). ONION outperforms Pleiades with a polynomial-time approximation scheme. On the other hand, *ClusCo* was mainly developed as a high-throughput tool for all-versus-all comparison of protein structures with different proximity measures using parallel `k-means` clustering. *Calibur* uses heuristics-based preprocessing to speed up clustering in three ways: grouping of structures into proximity groups to avoid pairwise RMSD computation (triangular inequality), using efficiently computable upper and lower bounds to skip RMSD calculation whenever possible, and discarding structures with low similarity to other structures (before clustering). *Calibur* is faster than SPICKER, but SPICKER produces better structures by focusing on low-energy regions and self-adjusting RMSD cutoffs. SPICKER outperforms SCAR in terms of the RMSD for top clusters. In [39], *SPICKER* is used along with additional steps (filtering and cluster reduction based on multidimensional scaling). *SCUD* avoids pairwise RMSD by using a random reference and sets protein-dependent cutoffs to balance cluster sizes [28]. *Durandal* [7] accelerates clustering using triangular inequality and initializes distances with a random reference, incorporating quaternion-based characteristic polynomial (QCP)-oriented RMSD for efficiency via information gain-based approach [8]. MUFOLD-CL [44], using the D-Score metric, is the fastest and produces superior structures compared to SPICKER, Pleiades, and Calibur. *SK-means* [40] combines *SPICKER* and k-means to improve the selection of initial cluster centers, addressing a key limitation of basic k-means. This method outperforms SPICKER in terms of average TM-score for the reported best structure. SK-means has quadratic polynomial time complexity. Li et al. [29] propose an ensemble clustering method based on k-medoids for protein structures. The method generates multiple clustering outcomes through repeated k-medoid runs with random initializations, then combines them using a voting-based approach. It uses the TM score to build a distance matrix and identify cluster centers. It outperforms SPICKER in final structure selection. The method is also utilized in [20]. On the other hand, Ferone and Marateas [16] propose graded Possibilistic c-Medoids, a medoid-based variation of graded possibilistic clustering combining fuzzy c-means (FCM) and possibilistic c-means (PCM). Unlike standard fuzzy clustering, GPCMdd allows membership sums greater than 1, enabling a soft transition between probabilistic and possibilistic distance functions. This makes it more robust to outliers and noise. Its high computational cost limits its feasibility for large structure datasets.

Hierarchical Clustering Methods: Strategies for hierarchical clustering generally fall into two major types: agglomerative and divisive. *HCPM* [18] uses an agglomerative clustering strategy with an average-link measure to calculate distances between clusters, using cRMSD as the average inter-atomic distances. The cluster representative is the structure nearest to the average distance map. The merging distance cut-off is determined using the plateau-center finding approach in a sigmoid plot. HCPM can also explore local energy minima in a protein energy landscape [17]. *UQlust* [1] combines structural profiles with consensus ranking and profile hashing for efficient hierarchical agglomerative clustering of protein structures. It projects 3D coordinates into single-dimensional structural profiles by assigning each residue to a specific state, facilitating the comparison of structures to detect common substructures. UQlust employs two heuristics: profile hashing-based clustering and reference-based partitioning, offering better time and space efficiency compared to ClusCo. Additionally, Boutonnet et al. [11] came up with a multiple linkage hierarchical clustering algorithm to analyze protein structural changes. *bcl-Cluster* [5] employs an agglomerative hierarchical clustering algorithm using pre-calculated pairwise distances between structures. It supports various proximity measures, including GDT, longest continuous segment, MaxSub, RMSD, largest common substructure, and Tanimoto coefficient. Integrated with PyMOL, bcl-Cluster can provide detailed outputs containing dendrograms, molecular structures, cluster sizes, and color-coded results based on numerical descriptors. *Probabilistic Hierarchical Clustering* [15] combines fuzzy c-means clustering (FCM) and a divisive hierarchical algorithm with dynamic identification of the number of clusters. The clustering cutoff is probabilistically determined based on variability. FCM starts with randomly chosen centroids and iteratively computes the degree of belonging for each structure using a normalized inverse distance from the centroid. New centroids are calculated as the weighted mean of structures, and the process repeats until convergence. Then divisive hierarchical clustering splits the structures into two subsets, temporarily removing redundant structures. It selects partitions based on a probability proportional to size and inversely proportional to internal variance, continuing subdivision until partition means are statistically significant. *Parallel Ward Clustering* [14] employs a massively parallel CUDA implementation of the nearest neighbor chain algorithm for hierarchical Ward clustering of protein structures, using atom-based RMSD and rigid-body RMSD. For atom-based RMSD clustering, three strategies are analyzed: Threads per Atomic Coordinate (TAC), Thread Blocks per Centroid (TBC), and Threads per Cluster Centroid (TCC). Among these, the TCC approach demonstrates significant speedup over multi-threaded CPU implementations and surpasses ClusCo in performance, while also enabling the computation of the full hierarchical tree. Additionally, ClusPro [13] applies a pairwise RMSD-based hierarchical algorithm to cluster protein structures after an initial filtering step based on desolvation and electrostatic properties.

Graph-Based Clustering Methods: For graph-based techniques, the first step is to represent the set of structures in terms of a graph. An intuitive way is to employ the nearest-neighbor graph. The proximity of the molecular structures under consideration can be encoded in the structure space via a nearest-neighbor graph (nngraph) where the vertex set is populated with the structures, and the edge set is populated by inferring a local neighborhood over each observation. The distance between two structures can be measured in terms of a suitable proximity measure (and an appropriate threshold for the same) after optimal superimposition with respect to a reference structure. In fact, such graph embedding of molecular structures of proteins has been considered in [4, 25] under the context of template-free protein structure prediction. Besides, Zhou et al. [46] employ techniques for constructing amino acid network-based graphs, also known as residue interaction graphs (RIG), where nodes represent amino acids through all atoms, side-chain atoms, or only C_α or C_β atoms. Edges in the graph are characterized by similarity measures such as physical distance or residue interaction energy between atomic units or side chains. Another work [25] utilizes nearest-neighbor graphs (nngraphs) and community detection algorithms, originally developed for social networks, to group protein structures. On the other hand, the `Basins-Select` [4] detects basins in the energy landscape by considering the potential energies of protein structures. It constructs a nearest-neighbor graph (nngraph) and extracts basins by first locating points of attraction or focal minima, which are vertices representing local energy minima. Each local minimum represents a basin. The other vertices are assigned to basins by following a negative gradient descent. This process continues until a local minimum is reached, with all vertices converging to the same minimum assigned to the corresponding basin using the Structural Bioinformatics Library (SBL) [12] to decompose the structure nngraph into basins

Table 2. Comparison of Different Methods

Name	Key Characteristics	Comparison found in literature	Evaluation Criteria
SPICKER [45]	– shrink the dataset (least energy structures from subsets) – pairwise RMSD cutoff for cluster membership – structure with the most neighbors as cluster center	better than SCAR by the quality of the top cluster	RMSD from an experimentally known structure
SCAR [10]	– relative RMSD (RRMSD) – global cluster cutoff to determine cluster membership – cluster-wise cutoff for refinement	demonstrates the effectiveness of RRMSD over RMSD	Avg. RMSD (cluster)
ONION [30]	– centroid approximation by MULTIPLE STRUCTURE SUPERIMPOSITION – similar to kmeans (by objective function)	– slightly better than SPICKER in single model selection – faster than Calibur and SPICKER	TM-Score
Calibur [31]	– auxiliary grouping of structures (threshold selection with C_α RMSD) – preliminary screening via lower and upper bounds – filtering of highly dissimilar structures from the dataset	– faster than SPICKER – slightly better than SPICKER	– Avg. TM-Score w.r.t. the experimentally known structure (cluster) – C_α RMSD w.r.t. the experimentally known structure (single structure)
ClusCo [22]	– high-throughput comparison of protein structures – different similarity measures (RMSD, GDT-TS, TM-Score, MaxSub)	– clustering results comparable to Calibur – faster than SPICKER and Calibur but slower than Durandal	comparison with ref. structure by different similarity measures (RMSD, GDT-TS, TM-Score, MaxSub)
Durandal [7]	– randomly chosen reference for distance matrix computation – lower and upper bound strategy to favor distance ranges over exact measures – uses the triangular inequality to accelerate exact clustering	faster than Calibur and SCUD	RMSD
Pleiades [21]	– Gauss integral representation for the tertiary structures of proteins – approximation of RMSD via Euclidean distance	– (slightly better) comparable to Calibur by all-atom RMSD – faster than Calibur	RMSD
SCUD [28]	– RMSD from randomly chosen reference as proximity measure – most neighboring structure: cluster representative – Representative structures are ranked by cluster size	slower than Durandal	RMSD
SK-means [40]	– integrates SPICKER with k-means – centroid of the largest cluster: best structure	better than SPICKER	RMSD
Ensemble method based on k-medoids [29]	– k-medoids algorithm is run several times (with different initializations) – voting to combine the clustering outcomes – largest cluster as the best cluster	– better than SPICKER – time-consuming	RMSD
Graph Clustering Methods [4,25]	– encoding of proximity via nearest neighbor graph – community detection/ energy landscape analysis	landscape analysis-based methods perform better than community-based methods	Purity
MUFOLD-CL [44]	– D-Scores based measures having high correlation with RMSD and TM-Score – projection-based clustering – largest cluster as the best cluster	– faster than most of the approaches – better than SPICKER, Pleiades, Calibur and a bit worse than ONION (avg. RMSD of prototype of top five clusters)	– avg. RMSD from the experimentally known structure (cluster) – RMSD from the experimentally known structure (single structure)
HCPM [18]	– agglomerative strategy with the average link – cluster representative: the structure that is the closest to the the average distance map of the cluster – initial screening based on energy and gyration radius	slower than k-means-based methods but performs better than those	RMSD
UQlust [1]	– projection of 3D coordinates into a suitable 1D structural profile – geometric consensus ranking – largest cluster: best cluster	– time and memory efficient than ClusCo, Sk-means – better than ClusCo and Pleiades	MaxSub Score
bcl-Cluster [5]	– relies upon pre-calculated pairwise distances – accommodates a variety of proximity measures – offers a pre-clustering step	offers a wide range of proximity measures to choose from	several proximity measures
Parallel Ward Clustering [14]	the nearest neighbor chain algorithm for hierarchical Ward clustering of protein structures	– faster than ClusCo – clustering results comparable to Clusco	TM-Score (for single structure selection)
NMF-MAD [2]	domain-specific feature-based non-negative matrix factorization	better than the methods mentioned before this one in this table	RMSD (for single structure selection)
SNMF-DS [24]	symmetric non-negative matrix factorization on RMSD-based distance matrix	better than NMF-MAD	RMSD (for single structure selection)
NTF-REL [23]	utilizes tensors to capture multiview of protein tertiary structures	better than all of the other approaches mentioned in this table in terms of quality assessment	multiple proximity measures

Factorization-Based Methods: NMF-MAD [2] opens a new avenue by exploring the matrix factorization route to identify tertiary structures of biologically active proteins using domain-specific features. SNMF-DS [24] dives further in this direction to demonstrate a feature-free, nonparametric method based on the symmetric non-negative matrix factorization. Furthermore, NTF-REL [23] shifts from matrix-factorization to tensor-factorization to serve as a quality assessment method for protein tertiary structures in addition to grouping the same. Other mentionable techniques for unsupervised learning of molecular structures include uniform time clustering, regular space clustering along with Markov state modeling (MSM) offered by *PyEMMA* [34], *ICON* [37], granular clustering based on growing local similarities [19], clustering based on the distribution of local minima [27], conserved residue clustering [35], score-based clustering using ligand RMSD. Table 2 captures the key properties as well as provides a brief comparison of the discussed methods.

3.4 Evaluation Metrics

Two sets of metrics are typically employed in unsupervised learning literature. The first set (consisting of external metrics) is designated here as application-specific. The second set contains internal metrics, termed here as application-agnostic. It is worth noting that unsupervised learning research for molecular structure data presents several adaptations of both external and internal metrics as inspired by domain insights.

Application-Specific Metrics: While most of these works assume that there is no external information about what states the clusters capture (that is, no ground truth), they often leverage the availability of known experimental structure(s) for a protein of interest in the PDB. Many external metrics take this into account. SCAR computes the average RMSD of the structures (w.r.t. experimentally known structure) in a cluster to compare the quality of the cluster [10]. A similar process is followed in [7,21,28,44,45](with TM-score in [30,31]). *Purity* is used with the graph-based techniques [4,25]. For a given cluster, it computes the fraction of structures that are very similar to the experimentally known structure in the cluster over the total number of structures contained by the cluster [4].

Application-Agnostic Metrics: The works that employ these metrics don't take into account the availability of the experimentally known structure or other external information. *ClusCo* selects the best cluster by $\mathtt{min}(\frac{\langle R \rangle}{f})$, where f denotes the fraction of elements in a particular cluster and $\langle R \rangle$ denotes the average *RMSD* between cluster elements. And the cluster center of the best cluster is selected as the best structure. *ClusPro* [13] takes into account the center of the most populated cluster as the best structure. To determine the cluster quality, *SPICKER* considers normalized structure cluster density, D defined as, $D = \frac{M}{\langle RMSD \rangle M_t}$, where M is the multiplicity of structures in the cluster, M_t is the total number of structures to be clustered and $\langle RMSD \rangle$ denotes the average RMSD of the structures in the cluster. *ICON* [37] measures the cluster quality in terms of cluster concentration, $CC_j = \frac{N_j}{\overline{RMSD_j}}$; where N_j is the number of structures in the cluster and $\overline{RMSD_j}$ denotes the average RMSD of the cluster. Besides, *UQlust* [1] reports the centroids of the five largest clusters as the top scoring structures. *HCPM* [17,18] considers *RMSD* distribution of the intra-cluster structures and inter-cluster's centroids to assess the clusters' quality. On the other hand, network properties e.g., average degree, clustering coefficient, and size of the communities are taken into account by Zhou et al. [46] to analyze the communities/clusters obtained from amino acid network-based graph representation of the structures. The Clustering Coefficient, C can be computed as, $C_i = \frac{2E_i}{m(m-1)}$ where m is the degree of vertex i and E_i denotes the number of connections from all neighbors of i. The clustering coefficient C of a cluster is the average of C_i. To capture the size and internal similarity of a cluster, Li et al. [29] compute a confidence score for each cluster, $CS = \frac{\sum_{i \in c} \sum_{j \in c} sim(i,j)}{\sum_{i=1}^{n} \sum_{j=1}^{n} sim(i,j)}$ where n is the total number of structures, c is the cluster under observation and $sim(i, j)$ denotes the corresponding entry in the similarity matrix. The cluster center with the maximum confidence score contributes to the selection of the best structure.

4 Discussion

Even though significant works have been conducted (most being application-specific), there are ample open problems and ways forward such as:

i. How to pick a structure that represents a cluster (beyond traditional ways of doing that)? For instance, Akhter et al. [2] consider the potential energy of a structure and employ a density-based structure weighting scheme to do so.

ii. The design of a proper distance function will remain an interesting research direction. For instance, how to design a meaningful distance function (beyond the L1 norm) for the dihedral angle representation and consider it for clustering?

iii. Are the structures themselves sufficient for finding patterns or consideration of additional properties of molecular structures (e.g., energy-based features [2]) is needed to improve the clustering results?

iv. Subspace clustering presents an unleveraged direction at the moment. Besides, one can also consider dimension weighting and reduce the problem to learning the weights of the different dimensions for optimizing an objective function that corresponds to the quality of the clustering.

5 Conclusion

This article presents an effort to organize the landscape of research on the clustering of molecular structures of proteins. As highlighted, direct comparison of existing methods is sometimes difficult due to the lack of standards regarding benchmark datasets and metrics used. As identified above, several directions of research may improve current efforts to reveal the underlying organization of molecular structures to elucidate structural states as an exploratory step toward understanding the behavior of a dynamic molecule.

References

1. Adamczak, R., Meller, J.: Uqlust: combining profile hashing with linear-time ranking for efficient clustering and analysis of big macromolecular data. BMC Bioinform. **17**(1), 546 (2016)
2. Akhter, N., et al.: Improved protein decoy selection via non-negative matrix factorization. IEEE/ACM Trans. Comput. Bio. Bioinfom. **19**(3), 1670–1682 (2021)
3. Akhter, N., Chennupati, G., Kabir, K.L., Djidjev, H., Shehu, A.: Unsupervised and supervised learning over the energy landscape for protein decoy selection. Biomolecules **9**(10), 607 (2019)
4. Akhter, N., Shehu, A.: From extraction of local structures of protein energy landscapes to improved decoy selection in template-free protein structure prediction. Molecules **23**(1) (2018)
5. Alexander, N., et al.: bcl:: Cluster: a method for clustering biological molecules coupled with visualization in the pymol molecular graphics system. In: The 1st International Conference on Computational Advances in Bio and Medical Sciences (ICCABS), pp. 13–18 (2011)
6. Álvarez, Ó., et al.: Principal component analysis in protein tertiary structure prediction. J. Bioinform. Computat. Biol. **16**(02) (2018)
7. Berenger, F., Shrestha, R., Zhou, Y., Simoncini, D., Zhang, K.Y.: Durandal: fast exact clustering of protein decoys. J. Comput. Chem. **33**(4), 471–474 (2012)
8. Berenger, F., Zhou, Y., Shrestha, R., Zhang, K.Y.: Entropy-accelerated exact clustering of protein decoys. Bioinformatics **27**(7), 939–945 (2011)
9. Berman, H.M., Bourne, P.E., Westbrook, J., Zardecki, C.: The protein data bank. In: Protein Structure, pp. 394–410. CRC Press (2003)
10. Betancourt, M.R., Skolnick, J.: Finding the needle in a haystack: educing native folds from ambiguous ab initio protein structure predictions. J. Comput. Chem. **22**(3), 339–353 (2001)
11. Boutonnet, N.S., Rooman, M.J., Wodak, S.J.: Automatic analysis of protein conformational changes by multiple linkage clustering. J. Mol. Biol. **253**(4), 633–647 (1995)
12. Cazals, F., Dreyfus, T.: The structural bioinformatics library: modeling in biomolecular science and beyond. Bioinformatics **33**(7), 997–1004 (2017)
13. Comeau, S.R., et al.: Cluspro: an automated docking and discrimination method for the prediction of protein complexes. Bioinformatics **20**(1), 45–50 (2004)
14. Dang, H.V., Schmidt, B., Hildebrandt, A., Tran, T.T., Hildebrandt, A.K.: CUDA-enabled hierarchical ward clustering of protein structures based on the nearest neighbour chain algorithm. Int. J. High Perform. Comput. Appl. **30**(2), 200–211 (2016)
15. Estrada, T., Armen, R., Taufer, M.: Automatic selection of near-native protein-ligand conformations using a hierarchical clustering and volunteer computing. In: The 1st ACM International Conference on Bioinformatics and Computational Biology, pp. 204–213 (2010)
16. Ferone, A., Maratea, A.: Decoy clustering through graded possibilistic c-medoids. In: International Conference on Fuzzy Systems, pp. 1–6. IEEE (2017)
17. Gront, D., Hansmann, U.H., Kolinski, A.: Exploring protein energy landscapes with hierarchical clustering. Int. J. Quant. Chem. **105**(6), 826–830 (2005)
18. Gront, D., Kolinski, A.: Hcpm—program for hierarchical clustering of protein models. Bioinformatics **21**(14), 3179–3180 (2005)

19. Guzenko, D., Strelkov, S.V.: Granular clustering of de novo protein models. Bioinformatics **33**(3), 390–396 (2017)
20. Han, X., Li, L., Lu, Y.: Selecting near-native protein structures from predicted decoy sets using ordered graphlet degree similarity. Genes **10**(2), 132 (2019)
21. Harder, T., Borg, M., Boomsma, W., Røgen, P., Hamelryck, T.: Fast large-scale clustering of protein structures using gauss integrals. Bioinformatics **28**(4), 510–515 (2012)
22. Jamroz, M., Kolinski, A.: Clusco: clustering and comparison of protein models. BMC Bioinform. **14**(1), 62 (2013)
23. Kabir, K.L., Bhattarai, M., Alexandrov, B.S., Shehu, A.: Single model quality estimation of protein structures via non-negative tensor factorization. In: The 11th International Conference on Computational Advances in Bio and Medical Sciences (ICCABS), pp. 3–15 (2021)
24. Kabir, K.L., Chennupati, G., Vangara, R., Djidjev, H., Alexandrov, B.S., Shehu, A.: Decoy selection in protein structure determination via symmetric non-negative matrix factorization. In: International Conference on Bioinformatics and Biomedicine (BIBM), pp. 23–28. IEEE (2020)
25. Kabir, K.L., Hassan, L., Rajabi, Z., Akhter, N., Shehu, A.: Graph-based community detection for decoy selection in template-free protein structure prediction. Molecules **24**(5), 854 (2019)
26. Kabir, K.L., Ma, B., Nussinov, R., Shehu, A.: Fewer dimensions, more structures for improved discrete models of dynamics of free versus antigen-bound antibody. Biomolecules **12**(7) (2022)
27. Li, H.: A model of local-minima distribution on conformational space and its application to PSP. Proteins: Struct. Function Bioinform. **64**(4), 985–991 (2006)
28. Li, H., Zhou, Y.: Scud: fast structure clustering of decoys using reference state to remove overall rotation. J. Comput. Chem. **26**(11), 1189–1192 (2005)
29. Li, L., Yan, H., Lu, Y.: Selecting near-native protein structures from ab initio models using ensemble clustering. Quant Biol. **6**(4), 307–312 (2018)
30. Li, S.C., Bu, D., Li, M.: Clustering 100,000 protein structure decoys in minutes. IEEE/ACM Trans. Comput. Biol. Bioinform. **9**(3), 765–773 (2011)
31. Li, S.C., Ng, Y.K.: Calibur: a tool for clustering large numbers of protein decoys. BMC Bioinform. **11**(1), 25 (2010)
32. Mereghetti, P., Ganadu, M.L., Papaleo, E., Fantucci, P., De Gioia, L.: Validation of protein models by a neural network approach. BMC Bioinform. **9**(1), 66 (2008)
33. Moll, M., Schwarz, D., Kavraki, L.E.: Roadmap methods for protein folding. Protein Struct. Predict. 219–239 (2008)
34. Scherer, M.K., et al.: Pyemma 2: a software package for estimation, validation, and analysis of Markov models. J. Chem. Theory Comput. **11**(11), 5525–5542 (2015)
35. Schueler-Furman, O., Baker, D.: Conserved residue clustering and protein structure prediction. Proteins: Struct. Function Bioinform. **52**(2), 225–235 (2003)
36. Shortle, D., Simons, K.T., Baker, D.: Clustering of low-energy conformations near the native structures of small proteins. Natl. Acad. Sci. **95**(19), 11158–11162 (1998)
37. Subramani, A., DiMaggio, P.A., Jr., Floudas, C.A.: Selecting high quality protein structures from diverse conformational ensembles. Biophys. J . **97**(6), 1728–1736 (2009)
38. Tribello, G.A., Gasparotto, P.: Using dimensionality reduction to analyze protein trajectories. Front. Mol. Biosci. **6**, 46 (2019)
39. Wang, Q., Shang, Y., Xu, D.: A new clustering-based method for protein structure selection. In: International Joint Conference on Neural Networks, pp. 2891–2898. IEEE (2008)

40. Wu, H., Li, H., Jiang, M., Chen, C., Lv, Q., Wu, C.: Identify high-quality protein structural models by enhanced k-means. BioMed Res. (2017)
41. Wu, S., Skolnick, J., Zhang, Y.: Ab initio modeling of small proteins by iterative tasser simulations. BMC Biol. **5**(1), 17 (2007)
42. Zaman, A.B., Kamranfar, P., Domeniconi, C., Shehu, A.: Reducing ensembles of protein tertiary structures generated de novo via clustering. Molecules **25**(9), 2228 (2020)
43. Zemla, A.: LGA: a method for finding 3d similarities in protein structures. Nucleic Acids Res. **31**(13), 3370–3374 (2003)
44. Zhang, J., Xu, D.: Fast algorithm for clustering a large number of protein structural decoys. In: International Conference on Bioinformatics and Biomedicine, pp. 30–36. IEEE (2011)
45. Zhang, Y., Skolnick, J.: Spicker: a clustering approach to identify near-native protein folds. J. Comput. Chem. **25**(6), 865–871 (2004)
46. Zhou, J., Yan, W., Hu, G., Shen, B.: Amino acid network for the discrimination of native protein structures from decoys. Protein Peptide Sci. **15**(6), 522–528 (2014)

Fast and Succinct Compression of k-mer Sets with Plain Text Representation of Colored de Bruijn Graphs

Enrico Rossignolo and Matteo Comin(✉)

Department of Information Engineering, University of Padua, Padua, Italy
{enrico.rossignolo,matteo.comin}@unipd.it

Abstract. A fundamental operation in computational genomics is the reduction of input sequences into their constituent k-mers. Designing space-efficient ways to represent a k-mer collection is essential to improve the scalability of bioinformatics analyses. A widely used approach involves converting the k-mer set into a de Bruijn graph and then producing a compact plain text representation by identifying the minimum path cover. In this article, we present USTAR-CR, a novel algorithm for compressing multiple k-mer sets. USTAR-CR leverages node connectivity principles in the colored de Bruijn graph for a more compact plain text representation, combined with an efficient encoding of k-mers colors. We tested USTAR-CR on real read datasets and compared it with the state-of-the-art GGCAT. USTAR-CR demonstrated superior performance in terms of compression, requiring less memory and being significantly faster (up to 51x) https://github.com/enricorox/USTAR-CR.

Keywords: k-mer Sets · Compression · Colored Bruijn graphs · Plain Text Representation

1 Introduction

k-mer algorithms have emerged as top-performing tools in bioinformatics, excelling across various analyses by working with sets of k-mer substrings instead of processing reads or alignments directly. These methods have shown exceptional performance across numerous applications. In genome assembly, for instance, tools like Spades [4] use k-mer-based strategies to reconstruct entire genomes with remarkable accuracy. In metagenomics, many methods [3,6,23,30,32] leverage k-mers for classifying microorganisms in complex samples. Similarly, in genotyping, several tools [11,18,19,31] utilize k-mers rather than alignments to identify genetic variations across individuals and populations. Additionally, numerous k-mer-based methods [12,17] have been introduced to enable efficient sequence searching in large databases.

Supplementary Information The online version contains supplementary material available at https://doi.org/10.1007/978-3-032-02489-3_5.

M. Alser et al. (Eds.): ICCABS 2025, LNCS 15599, pp. 54–65, 2026.
https://doi.org/10.1007/978-3-032-02489-3_5

k-mer methods are appreciated for their adaptability and scalability, often handling datasets containing billions of k-mers. One central challenge in k-mer-based approaches is the efficient storage and querying of the massive k-mer sets generated from sequencing data. As dataset sizes increase, reducing the storage footprint and query times for k-mer sets has become a vital research focus. Conway and Bromage [9] showed that at least $\log \binom{4^k}{n}$ bits are required for the worst-case lossless storage of n k-mers. However, sequencing datasets often exhibit a spectrum-like property, where redundancies are common, allowing data structures to optimize storage requirements well below this theoretical bound [7,8].

A common approach to reduce redundancy in a k-mer set K is to represent it as a set of maximal unitigs. A unitig is a non-branching path in the de Bruijn graph, where nodes correspond to the k-mers of K and edges represent overlaps between k-mers. Each unitig u can be expressed as a string called spell(u) with length $|u| + k - 1$, where $|u|$ is the number of k-mer substrings in spell(u). For example, the unitig (AAC, ACG, CGT) is represented as the string $AACGT$. This technique allows $|u|$ k-mers to be stored using $|u| + k - 1$ characters instead of $k \times |u|$, significantly reducing space when unitigs are long, as in real datasets. The entire set K can thus be represented by the set of maximal unitigs U, with the property that $x \in K$ if and only if x is a substring of spell(u) for some $u \in U$.

A notable variation of the de Bruijn graph is the colored de Bruijn graph, initially developed for de novo assembly and variant genotyping [2,13]. This type of graph is built from a collection of datasets, such as different sequencing datasets or genome sequences, and associates each k-mer with the identifiers (or "colors") of the datasets in which it appears. Colored de Bruijn graphs provide a compressed representation of k-mers across multiple datasets while retaining information (i.e., color) for each k-mer to identify its dataset origins. This model has since been applied in pangenomics [33], bacterial genome querying [16], among other areas.

The state-of-the-art tool for the compression of k-mer sets is GGCAT [10], which uses a colored de Bruijn graph to compress the input k-mers into a set of unitigs with colors. GGCAT proved to be a several times improvement over Cuttlefish 2 [14] and a two orders of magnitude improvement over BiFrost [16] for colored construction and querying. However, for large datasets, GGCAT still requires several hours of computation and a large amount of memory. For example, to compress 649K bacterial genomes, GGCAT takes at least 21.48GB of RAM and more than 11 h.

In this paper, we present USTAR-CR, an algorithm for k-mer sets compression based on a plain text representation of a colored de Bruijn graph. USTAR-CR is able to efficiently compress large sets of k-mers in minute space, with a small memory and time footprint.

1.1 Related Works

In the realm of k-mer set compression, plain text representation has become a highly practical approach. Formally, this representation is a set of strings that includes all k-mers derived from the input sequences (considering both forward and reverse-complemented forms) while excluding any extraneous k-mers. This structure is referred to as a spectrum-preserving string set (SPSS). Note that this definition differs from that of Rahman and Medvedev [24], who specify an additional constraint ensuring that each k-mer appears no more than once.

Independently, Rahman and Medvedev [24] and Břinda, Baym, and Kucherov [5] introduced the concept of representing k-mers in plain text without repetition for a more compact storage. They respectively termed this representation the "Spectrum-Preserving String Set" (SPSS) [24] and "simplitigs" [5]. To avoid confusion with the newer SPSS definition, we will use the term simplitigs here.

Both Rahman and Medvedev with UST and Břinda, Baym, and Kucherov with ProphAsm present algorithms that use greedy approaches to merge consecutive unitigs into this type of representation. These heuristic methods notably reduce both the number of strings (string count, SC) and the cumulative length (cumulative length, CL) necessary for representing a k-mer set. Lowering CL directly decreases memory needed for string storage, while reducing SC minimizes the size of the index structure for storing the strings, thereby improving overall storage efficiency.

Recently, a heuristic called USTAR (Unitig STitch Advanced constRuction) was introduced [26,27], which leverages graph connectivity to more efficiently traverse de Bruijn graphs. By capitalizing on graph density and node connectivity, USTAR improves path selection for constructing the path cover, achieving better compression results, especially in denser graphs.

In a recent study [29], researchers presented the first algorithm to find an SPSS of minimum cumulative length (CL) that permits repeated k-mers, achieving significantly improved compression over simplitigs. This optimal algorithm, called Matchtigs, utilizes a many-to-many min-cost path query and a min-cost perfect matching approach, and it was shown to be polynomially solvable. However, its complexity of $O(n^3m)$, where n is the number of nodes and m is the number of arcs in the de Bruijn graph, makes it impractical for large datasets. To improve efficiency, the authors also proposed a heuristic approach, Greedy Matchtigs, which sacrifices the optimal matching to produce a compact SPSS more quickly. USTAR2 [28] addresses the path cover problem on a de Bruijn graph to create an optimal k-mer set representation, aiming to minimize the cumulative length of compressed data. It achieves this by strategically choosing paths based on node connectivity and reusing previously traversed nodes, which leads to compression ratios that exceed those of tools like UST and USTAR, that do not allow repeated k-mers, and match the efficiency of Greedy Matchtigs while requiring significantly less time and memory.

The above methods focus on single k-mer set compression. For compressing multiple k-mer sets, GGCAT is the leading tool for constructing compacted, optionally colored de Bruijn graphs. GGCAT combines k-mer counting with unitig construction by adding "context" information, enabling the computation of valid global unitigs within each input bucket. GGCAT also implements a method inspired by BiFrost [16] that produces compact color maps with greatly improved build times. Unlike BiFrost, which uses an individually compressed color bitmap for each k-mer, GGCAT maps each color set to an index. To store each color set efficiently, GGCAT computes differences between consecutive colors and applies run-length encoding. When written to disk, the color set indices of consecutive k-mers in each unitig are also run-length encoded. GGCAT can also be combined with Greedy Matchtigs to further optimize unitig compression.

In the next sections, we introduce USTAR-CR, a faster and memory-efficient greedy heuristic designed to generate a compact SPSS for large colored de Bruijn graphs. USTAR-CR is based on the USTAR2 paradigm [28], but it enables the simultaneous compression of multiple k-mer sets, with the efficient storage of colors.

2 Method

2.1 Preliminaries

In this paper, we consider a string composed of characters from the alphabet $\Sigma = \{A, C, T, G\}$. A string of length k is termed a "k-mer", and its "reverse complement" is denoted as $rc(\cdot)$. Since the DNA strand of origin is generally unknown, we treat a k-mer and its reverse complement as equivalent. A k-mer set K can be compressed by creating a representation S composed of strings of arbitrary lengths, such that the set of all k-mers extracted from S precisely matches the original set K. The "spectrum" of a set of strings S is defined as the collection of all k-mers and their reverse complements that appear in at least one string $s \in S$, formally: $spec_k(S) = \{t \in \Sigma^k \mid \exists s \in S : t \text{ or } rc(t) \text{ is a substring of } s\}$. Given a set of input k-mers K, our goal is to find a minimal collection of strings that preserves this spectrum. Formally, we define:

Definition 1. *A Spectrum Preserving String Set (SPSS) for the input set of k-mers K is a collection of strings, denoted as S, where each string in S has a length of at least k, and such that $spec_k(K) = spec_k(S)$.*

The essential feature of the SPSS is its ability to contain the same collection of k-mers as the original set K, including both the k-mers and their reverse complements. A straightforward way to evaluate the size of a string set S is to compute its *cumulative length*, defined as the sum of all string lengths: $CL(S) = \sum_{s \in S} |s|$, where $|s|$ is the length of string s.

Problem 1. Given a k-mer set K, find the Spectrum Preserving String Set S with minimum $CL(S)$.

USTAR2 (Unitig STitch Advanced constRuction) [28] offers an efficient heuristic for compressing large k-mer datasets. USTAR2 is a linear-time algorithm that utilizes the connectivity properties of the de Bruijn graph (dBG) to ensure effective compression while navigating the graph efficiently. It approximates a minimum SPSS by leveraging the compacted de Bruijn graph generated by BCALM2 [8]. The USTAR2 method begins by selecting a seed node and attempts to build a path by linking adjacent nodes until further extension is unfeasible. This process repeats with new seed nodes until all nodes are covered by paths. USTAR2's effectiveness depends on two key operations: selecting the optimal seed node and extending the path through available connections. Specifically, USTAR2 chooses the most imbalanced node as the starting seed and extends each path by selecting nodes with fewer connections, preserving highly connected nodes for future paths. This strategy reduces the cumulative length (CL) by creating fewer, longer strings and minimizing isolated nodes.

2.2 USTAR-CR: SPSS with Colors Reordering

USTAR2 can compress individual k-mer sets and generate a plain text representation of the de Bruijn graph. In USTAR-C, we extend this approach to compress multiple k-mer sets, where each set is associated with a specific color. Encoding colors for each k-mer in a de Bruijn graph involves two primary challenges: (1) efficiently tracking all colors associated with each k-mer and (2) storing these colors in a way that minimizes both space and processing time. For colored graphs, we extend our algorithm with an approach inspired by GGCAT. Instead of using a separate (compressed) color bitmap for each k-mer, GGCAT groups colors into color set indices, a method similar to [1]. Additionally, GGCAT encodes each color set by calculating differences between consecutive colors and applying RLE (Run-Length Encoding). When saving data to disk, the color set indices of consecutive k-mers within each unitig are also run-length encoded. This strategy is highly efficient, as unitigs are typically "variation-free" and therefore have only a small number of color set indices associated with their k-mers.

Suppose that USTAR-C computed a representation of a set of colored k-mers with $k = 3$. In Table 1, we report a minimal example of a SPSS representation made of five sequences of length 6 with related color set indices. The colors are grouped into sets represented by an index (see Appendix Table ??). Each sequence in the representation can not be further merged and compressed, because they do not share an overlap or, equivalently, they are not connected in the de Bruijn graph. Instead, each number can be compressed individually using RLE, obtaining two runs per sequence with a total of 10 runs, detailed in the last row of Table 1.

Table 1. An example of a set of sequences with associated color set indices, $k = 3$. Each index represents a set of colors (see Appendix Table ??). The plain representation of colors can be compressed using RLE.

sequences	AATAGA	ACTTCG	CCAGGC	CCTCTG	CTTGAA
colors set indices (plain)	2 2 1 1	4 4 3 3	2 1 1 1	3 3 3 2	3 2 2 2
colors set indices (RLE)	2-2 1-2	4-2 3-2	2-1 1-3	3-3 2-1	3-1 2-3

However, we note that some sequences share the same colors. We can exploit these similarities to further compress the colors, with a better run-lenght encoding. More precisely, if two sequences that share the same end-point color set index are one next to the other, we can merge the two encodings of the color into a single run-length encoding. Thus, the encoding of colors, like in the previous example, can be improved by reordering the sequences. Following this idea, we further extend our algorithm with an optimal reordering of sequences and colors, and named this implementation USTAR-CR.

The algorithm adds to the USTAR-C pipeline the construction of an end-point weight graph (*ewG*). Like a de Bruijn graph, in an *ewG* each sequence is represented by a node u with two sides, labeled with the first and last color set associated with the sequence. Arcs connect the node sides that share the same color set, allowing us to merge two runs for each link in any path in the graph.

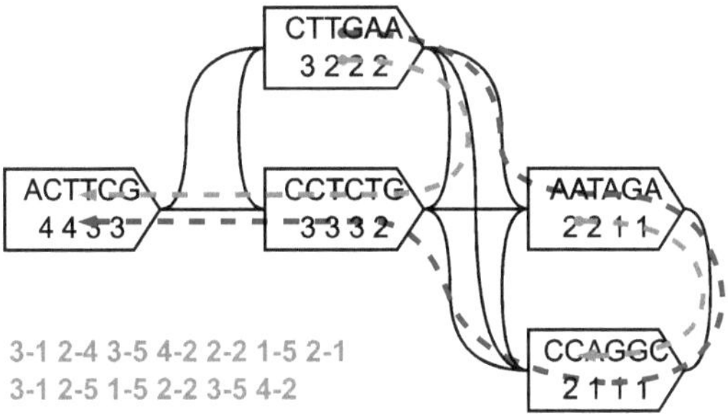

Fig. 1. An end-point weight graph (*ewG*) constructed with the sequences in Table 1. In red, an arbitrary path cover that produced 7 runs. In green, a path cover constructed using the node connectivity that generates 6 runs.

This setup allows us to follow paths in the *ewG* to reorder the sequences and their colors, facilitating the merging of adjacent runs. To maximize the number of merged runs and ensure each sequence is represented exactly once, it is essential to generate long, disjoint paths. Consequently, minimizing the number of runs becomes equivalent to finding a minimum vertex-disjoint path cover. An exact solution for a similar problem is provided in [21], however, it involves additional preprocessing steps that can slow down the pipeline. Instead, we can efficiently find an approximate solution using a greedy strategy, as implemented in USTAR [27], that constructs a path cover by selecting the less connected nodes.

Consider the *ewG* example in Fig. 1, constructed using the sequences from Table 1. An arbitrary path cover, highlighted in red, may consist of two paths with a total of 7 runs. In contrast, the green path cover is built by extending to the node with the minimum degree at each step. Using this strategy, we are able to obtain a single long path that visits every nodes, generating only 6 runs. In conclusion, leveraging the *ewG* significantly improved color compression. Starting with 10 initial runs, the *ewG* reduced the runs count to 7 with an arbitrary path cover, and further down to 6 runs using the path cover computed by USTAR-CR.

In the next section, we present the evaluation of USTAR-CR on real sequencing data, comparing its compression and performance against other state-of-the-art tools.

3 Results

In this section, we present the comparative analysis of our methods against the state-of-the-art tool for colored k-mer compression, GGCAT. In particular, we evaluated GGCAT (maximal unitigs) and GGCAT GM (greedy matchtigs) against USTAR-C, with just the plain RLE encoding of colors, and USTAR-CR, with the optimal reordering of color runs.

We used as benchmark a dataset of 20 files containing sequencing reads (see Appendix, Table ??) also used in other studies [5,8,15,20,25]. These datasets varied in terms of read type (paired or single-ended), read length, and coverage, providing a diverse testbed to assess the robustness and performance under different conditions. We merged all datasets together associating to each k-mer one or more colors based on the files where it appears. In Appendix Table ?? are reported the number of distinct k-mers to compress, ranging from 400 millions to 2 billions depending on the k-mer length.

To compare the performance of these tools, we considered three key metrics: Cumulative Length (CL), Sequence Count (SC), and the number of runs. The Cumulative Length (CL) gives an indication of how well the compression algorithm reduces data size. The Sequence Count (SC) helps assess the fragmentation of the k-mers during compression. The number of runs ($\#runs$) provides insights into the compression efficiency for repetitive elements. These metrics offer a comprehensive view of the efficiency and quality of the uncompressed k-mers representation produced by each tool.

To evaluate the effectiveness of each tool after compression, we analyzed the total compressed size, which included the sequence file compressed using MFCompress [22], the color table, and, specifically for USTAR-C and USTAR-CR, the color file compressed with bzip3. In the following, we analyze the performance of all tools in terms of compression, execution times, and memory usage.

3.1 Compression of k-mer Sets

In the first experiment, we considered the common k-mer length $k = 31$. In Table 2 are reported the results before compression.

Table 2. Tools comparison before compression, using Cumulative Length (CL), Sequence Count (SC) and number of runs ($\#runs$). GGCAT GM provided the best values for CL and SC while USTAR-CR produced the minimum $\#runs$.

k=31	GGCAT	GGCAT GM	USTAR-C	USTAR-CR
CL	6,266,509,634	**3,290,519,704**	3,681,600,490	3,681,600,490
SC	135,191,765	**41,341,022**	43,520,096	43,520,096
#runs	468,952,986	78,730,727	92,771,770	**54,026,538**

As expected, the worst tool is GGCAT which computed maximal unitigs and obtained the highest CL, SC, and number of runs. There is still a high level of redundancy between the sequences and the associated colors that can be lowered. USTAR-C and USTAR-CR greatly reduced CL and SC while GGCAT GM obtained the smallest values for the same metrics. On the other hand, USTAR-CR obtained the lowest $\#runs$ showing the effectiveness of the RLE optimization through sequence reordering.

Table 3. Tools comparison: considering the sequences file compressed with MFCompress (*sequences*), the colors file compressed with bzip3 (*colors*), the colors table, and the total compression. All the measures are expressed in bytes. USTAR-CR excelled in all the metrics, followed by USTAR-C and GGCAT GM.

k = 31	GGCAT	GGCAT GM	USTAR-C	USTAR-CR
sequences	3,198,171,033	1,028,520,445	915,769,659	908,641,367
colors	–	–	98,742,763	50,959,415
colors table	50,281	50,318	50,281	50,281
total compression	3,198,221,314	1,028,570,763	1,014,562,703	**959,651,063**

The compression values for each methods are shown in Table 3. As expected, GGCAT produced the largest FASTA file due to the presence of unitig linkage information. GGCAT GM shrank the file size of the sequences by nearly one-third while USTAR-C and USTAR-CR further reduced the metric by relocating color information to an external file and compressing them separately from the sequences. As expected, given the lower $\#runs$ of USTAR-CR, the color file is better compressed, producing the best total compression. Overall, the total file size produced by USTAR-CR is 70% smaller than GGCAT and 6.7% compared to GGCAT GM.

3.2 Compression: Different k-mer Lengths

In the following paragraph, we explore the effects of using various k-mer lengths on colored k-mer compression. A different value of k impacts the number of k-mers and the number of links between them, generally making the graph more dense as k decreases (see Appendix, Table ??).

The tools are executed for $k \in \{15, 21, 31, 41\}$ and the final compression is displayed in Fig. 2. The weak compression of GGCAT is confirmed for all values of k. For $k = 15$, the best result is obtained by USTAR-CR, followed by USTAR-C and GGCAT GM. Again, for $k = 21$, USTAR-CR reached the minimum compression, followed by GGCAT GM and USTAR-C. However, the gap between USTAR-CR and USTAR-C is much wider, meaning that RLE reordering optimization is necessary to minimize the compression. Similarly, for $k = 41$, the best value is obtained by USTAR-CR followed by USTAR-C and GGCAT GM. In this last case, we observe the largest gap between USTAR-CR

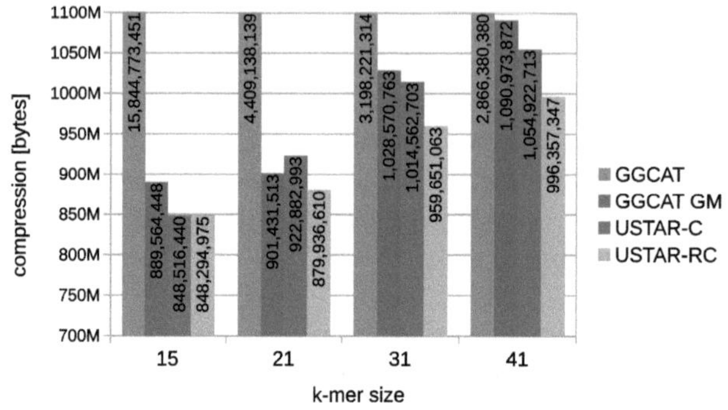

Fig. 2. Total compression varying the k-mer size.

and GGCAT GM, where the former improves the compression by 8.67% w.r.t. to the latter.

3.3 Compression: Different Number of Colors

In the previous section, we show that GGCAT GM is the main competitor of USTAR-CR. Next, we study the performance while varying the number of colors, i.e. the number of files considered. We subsampled the datasets considering the number of colors $c \in \{10, 15, 20\}$ and $k = 31$ (see Appendix Table ??). Note that with the increase in c, the number of k-mers grows and the color subsets exponentially rise, challenging the compression of colored k-mers. The results are detailed in Table 4.

Table 4. Number of runs (#runs) and total compression varying the number of colors (c).

k = 31	c = 10		c = 15		c = 20	
	GGCAT GM	USTAR-CR	GGCAT GM	USTAR-CR	GGCAT GM	USTAR-CR
#runs	36,587,632	26,418,625	40,832,766	29,116,036	78,730,727	54,026,538
total compression	433,708,310	**410,570,603**	495,357,773	**464,890,012**	1,028,570,763	**959,651,063**

USTAR-CR systematically obtained the smallest #runs, leading to the best total compression for all the number of colors taken into account. The largest difference between GGCAT GM and USTAR-CR is obtained for $c = 20$, where the *#runs* is 31.4% smaller and the total compression improves by 6.7%.

3.4 Time and Memory Usage

In this section, we compare the execution time and memory usage of the tools that have shown the best compression, namely GGCAT GM and USTAR-CR.

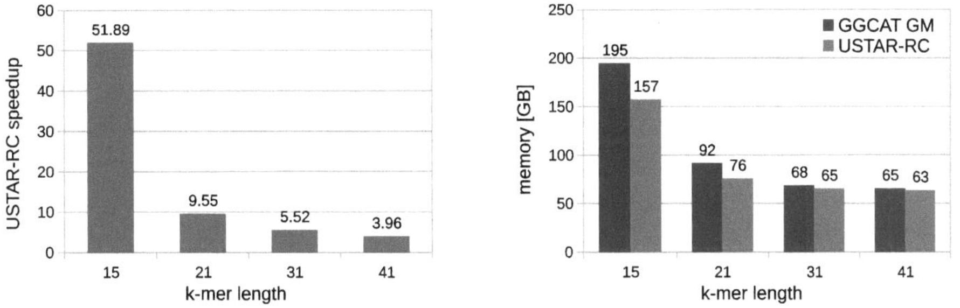

(a) Speedup varying k-mer length.

(b) Memory varying k-mer length.

Fig. 3. Speed up and memory requirement of USTAR-CR with respect to GGCAT GM.

In Fig. 3 we analyze the execution performances while varying the k-mer lengths used for compression. The speedup of USTAR-CR with respect to GGCAT GM ranges between 3.92× and 51.89×. For $k = 15$, we have the maximum speedup of 51.89× for USTAR-CR w.r.t. GGCAT GM, with a memory requirement of $157GB$; 19.2% less than GGCAT GM. A similar trend can be observed for memory usage, where USTAR-CR always required less resouces compared to GGCAT GM.

Overall, our tool consistently demonstrates faster performance and lower resource consumption. Additionally, USTAR-CR is currently implemented as a single-threaded application, while GGCAT GM utilizes multithreading, indicating a potential for further improvement in our tool.

4 Conclusions

In this article, we introduce USTAR-CR, a new algorithm designed to compress multiple k-mer sets efficiently. USTAR-CR utilizes node connectivity within the colored de Bruijn graph to achieve a more compact plain-text representation and incorporates an optimized encoding method for k-mer colors.

The comparative evaluation of USTAR-CR against GGCAT and GGCAT GM for colored k-mer compression shows clear advantages in both compression efficiency and use of resources. For a common k-mer length of 31, USTAR-CR achieved the lowest number of runs ($\#runs$) thanks to the color reordering, substantially reducing color redundancy and resulting in a compressed file size 70% smaller than GGCAT's and 6.7% smaller than GGCAT GM's. Across varying k-mer lengths, USTAR-CR continued to lead, especially at lower k values where graph density increased. With increasing numbers of colors, USTAR-CR consistently produced the fewest $\#runs$ and achieved the highest compression efficiency.

In terms of speed and memory usage, USTAR-CR proved highly efficient, achieving up to a 51.89× speedup over GGCAT GM at $k = 15$ and maintaining significant advantages across all tested k values. Memory usage was also lower for USTAR-CR, especially at smaller k values, establishing it as a resource-conscious tool. In summary, USTAR-CR surpasses current state-of-the-art tools, offering fast, memory-efficient, and highly compressed representations of colored k-mer sets.

Acknowledgments. Authors are supported by the Project funded under the National Recovery and Resilience Plan (NRRP), Mission 4 Component 2 Investment 1.4 - Call for tender No. 3138 of 16 December 2021, rectified by Decree n.3175 of 18 December 2021 of Italian Ministry of University and Research funded by the European Union NextGenerationEU.

References

1. Almodaresi, F., Pandey, P., Patro, R.: Rainbowfish: a succinct colored de Bruijn graph representation. In: Schwartz, R., Reinert, K. (eds.) 17th International Workshop on Algorithms in Bioinformatics (WABI 2017). Leibniz International Proceedings in Informatics (LIPIcs), vol. 88, pp. 18:1–18:15. Schloss Dagstuhl – Leibniz-Zentrum für Informatik, Dagstuhl, Germany (2017)
2. Andreace, F., Lechat, P., Dufresne, Y., Chikhi, R.: Comparing methods for constructing and representing human pangenome graphs. Genome Biol. **24** (2023)
3. Andreace, F., Pizzi, C., Comin, M.: Metaprob 2: metagenomic reads binning based on assembly using minimizers and k-mers statistics. J. Comput. Biol. **28**(11), 1052–1062 (2021)
4. Bankevich, A., et al.: Spades: a new genome assembly algorithm and its applications to single-cell sequencing. J. Comput. Biol. **19**(5), 455–477 (2012)
5. Břinda, K., Baym, M., Kucherov, G.: Simplitigs as an efficient and scalable representation of de Bruijn graphs. Genome Biol. **22**(1), 1–24 (2021)
6. Cavattoni, M., Comin, M.: Classgraph: improving metagenomic read classification with overlap graphs. J. Comput. Biol. **30**(6), 633–647 (2023). pMID: 37023405
7. Chikhi, R., Holub, J., Medvedev, P.: Data structures to represent a set of k-long DNA sequences. ACM Comput. Surv. (CSUR) **54**(1), 1–22 (2021)
8. Chikhi, R., Limasset, A., Medvedev, P.: Compacting de Bruijn graphs from sequencing data quickly and in low memory. Bioinformatics **32**(12), i201–i208 (2016)
9. Conway, T.C., Bromage, A.J.: Succinct data structures for assembling large genomes. Bioinformatics **27**(4), 479–486 (2011)
10. Cracco, A., Tomescu, A.: Extremely fast construction and querying of compacted and colored de Bruijn graphs with GGCAT. Genome Res. (2023)
11. Denti, L., Previtali, M., Bernardini, G., Schönhuth, A., Bonizzoni, P.: Malva: genotyping by mapping-free allele detection of known variants. IScience **18**, 20–27 (2019)
12. Harris, R.S., Medvedev, P.: Improved representation of sequence bloom trees. Bioinformatics **36**(3), 721–727 (2020)
13. Iqbal, Z., Caccamo, M., Turner, I., Flicek, P., McVean, G.: De novo assembly and genotyping of variants using colored de Bruijn graphs. Nat. Genet. **44**, 226–32 (2012)

14. Khan, J., Kokot, M., Deorowicz, S., Patro, R.: Scalable, ultra-fast, and low-memory construction of compacted de Bruijn graphs with cuttlefish 2 (2021)
15. Kokot, M., Długosz, M., Deorowicz, S.: KMC 3: counting and manipulating k-mer statistics. Bioinformatics **33**(17), 2759–2761 (2017)
16. Luhmann, N., Holley, G., Achtman, M.: Blastfrost: fast querying of 100,000s of bacterial genomes in bifrost graphs. Genome Biol. **22** (2021)
17. Marchet, C., Iqbal, Z., Gautheret, D., Salson, M., Chikhi, R.: Reindeer: efficient indexing of k-mer presence and abundance in sequencing datasets. Bioinformatics **36**(Supplement_1), i177–i185 (2020)
18. Marcolin, M., Andreace, F., Comin, M.: Efficient k-mer indexing with application to mapping-free SNP genotyping. In: Lorenz, R., Fred, A.L.N., Gamboa, H. (eds.) Proceedings of the 15th International Joint Conference on Biomedical Engineering Systems and Technologies, BIOSTEC 2022, Volume 3: BIOINFORMATICS, 9–11 February 2022, pp. 62–70 (2022)
19. Monsu, M., Comin, M.: Fast alignment of reads to a variation graph with application to SNP detection. J. Integr. Bioinform. **18**(4), 20210032 (2021)
20. Pandey, P., Bender, M.A., Johnson, R., Patro, R.: Squeakr: an exact and approximate k-mer counting system. Bioinformatics **34**(4), 568–575 (2018)
21. Pibiri, G.E.: On weighted k-mer dictionaries. Algorithms Mol. Biol. **18**(1), 3 (2023)
22. Pinho, A.J., Pratas, D.: Mfcompress: a compression tool for fasta and multi-fasta data. Bioinformatics **30**(1), 117–118 (2014)
23. Qian, J., Comin, M.: Metacon: unsupervised clustering of metagenomic contigs with probabilistic k-mers statistics and coverage. BMC Bioinform. **20**(367) (2019)
24. Rahman, A., Medvedev, P.: Representation of k-mer sets using spectrum-preserving string sets. In: Schwartz, R. (ed.) RECOMB 2020. LNCS, vol. 12074, pp. 152–168. Springer, Cham (2020). https://doi.org/10.1007/978-3-030-45257-5_10
25. Rizk, G., Lavenier, D., Chikhi, R.: Dsk: k-mer counting with very low memory usage. Bioinformatics **29**(5), 652–653 (2013)
26. Rossignolo, E., Comin, M.: USTAR: Improved compression of k-mer sets with counters using de Bruijn graphs. In: Guo, X., Mangul, S., Patterson, M., Zelikovsky, A. (eds.) ISBRA 2023. LNCS, vol. 14248, pp. 202–213. Springer, Singapore (2023). https://doi.org/10.1007/978-981-99-7074-2_16
27. Rossignolo, E., Comin, M.: Enhanced compression of k-mer sets with counters via de Bruijn graphs. J. Comput. Biol. **31**(6), 524–538 (2024)
28. Rossignolo, E., Comin, M.: Ustar2: fast and succinct representation of k-mer sets using de Bruijn graphs. In: Proceedings of the 17th International Joint Conference on Biomedical Engineering Systems and Technologies - Volume 1: BIOINFORMATICS, pp. 368–378. INSTICC, SciTePress (2024)
29. Schmidt, S., Khan, S., Alanko, J.N., Pibiri, G.E., Tomescu, A.I.: Matchtigs: Minimum plain text representation of k-mer sets. Genome Biology (Online) **24** (2023)
30. Storato, D., Comin, M.: K2mem: discovering discriminative k-mers from sequencing data for metagenomic reads classification. IEEE/ACM Trans. Comput. Biol. Bioinf. **19**(1), 220–229 (2022)
31. Sun, C., Medvedev, P.: Toward fast and accurate SNP genotyping from whole genome sequencing data for bedside diagnostics. Bioinformatics **35**(3), 415–420 (2019)
32. Wood, D.E., Salzberg, S.L.: Kraken: ultrafast metagenomic sequence classification using exact alignments. Genome Biol. **15**(3), 1–12 (2014)
33. Zekic, T., Holley, G., Stoye, J.: Pan-Genome Storage and Analysis Techniques, pp. 29–53. Springer, New York (2018)

Enhancing Protein Side Chain Packing Using Rotamer Clustering and Machine Learning

Mohammed Alamri[1], Mohammad Al Sallal[2], Kamal Al Nasr[1(✉)], Muhammad Akbar[1], and Ahmad Jad Allah[1]

[1] Tennessee State University, Nashville, TN 37209, USA
kalnasr@tnstate.edu
[2] HCA Healthcare, Nashville, TN 37203, USA

Abstract. One of the challenges and a significant part of a protein structure's prediction in three-dimensional space is a side chain prediction/packing. This area of research has a large importance, due to its various applications in protein design. In recent years, many methodologies and techniques have been crafted for side chain prediction such as DLPacker, FASPR, SCWRL4 and OPUS-Rota4. In this research, we address the problem from a different perspective. We employed a machine learning model to predict the side chain packing of protein molecules given only the Cα trace. We analyzed 32,000 protein molecules to extract important geometrical features that can distinguish between different orientations of side chain rotamers. We designed and implemented a Random Forest model to tackle this problem. Given the accuracy of existing state-of-the-art approaches, our model represents an improvement. The results of our experiment show that Random Forest is highly effective, achieving a total average accuracy of 73.7% for proteins and 73.3% for individual amino acids.

Keywords: Protein Structure · Side Chain Prediction · Protein Cα Trace · Side Chain Rotamer · Side Chain Packing

1 Introduction

1.1 Problem Background

Protein is a complex molecule that plays a fundamental function in our bodies. Proteins are composed of chains and molecules known as amino acids. (a.k.a. residues). In addition, all proteins consist of 20 varieties of amino acids which are made by carboxyl group (COOH), amine group (NH_2), and side chain (R-group) [1]. These groups are molecules made of atoms. The carboxyl group and amine group form the backbone of the amino acids. Amino acids have the same backbone. What distinguishes one amino acid from another is the side chain. Each amino acid has the same atoms that form its side chain. However, side chain structural configuration can be different in orientation based on many factors. Each possible configuration is called rotamer. Rotamers can be defined using the dihedral angles, called chi angles, between the bonds formed by its

M. Alser et al. (Eds.): ICCABS 2025, LNCS 15599, pp. 66–77, 2026.
https://doi.org/10.1007/978-3-032-02489-3_6

atoms. For instance, chi1 is the dihedral angle around the bond Cα-Cβ and formed by the atoms: N - Cα - Cβ – Cγ. The size of the side chain determines the number of chi angles defined for each rotamer. Some amino acids have no chi angles such as ALA and GLY and some others have chi angles ranging from 1 to 4. Figure 1 shows three different configurations/rotamers for the side chain of amino acid ARG in protein ID: 135L. For ARG 14, the four chi angles are: -68.5, 177.97, -99.71, and 120.57 respectively. For ARG 68, the four chi angles are: 58.22, 166.56, 161.29, and 112.9 respectively. Finally, for ARG 128, the four chi angles are: -71.46, -55.74, 104.76, and -145.36 respectively.

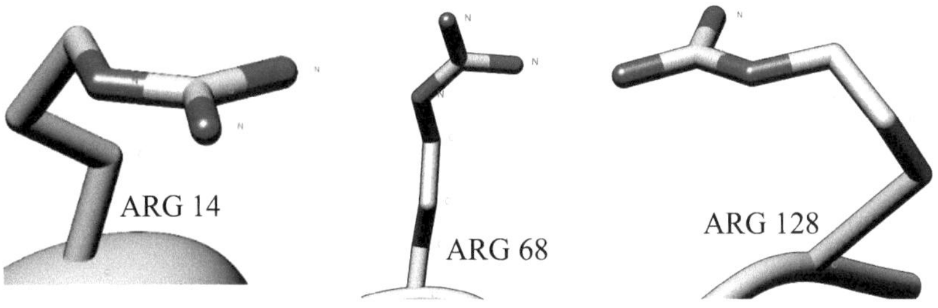

Fig. 1. Different rotamer configurations for amino acid ARG in protein ID: 135L

The protein side chain is closely related to biological function [2], and therefore, an accurate structural determination of side chains is essential to serve the biological function. Predicting protein side chains is crucial because it gives an insight to the protein function [3]. Predicting side chain can be significant to serve several applications such as homology design, and protein modeling. These applications depend on protein side chain conformations prediction from its backbone structure and amino acid sequence (also called side chain packing) [4, 5].

1.2 Literature Review

Side chain prediction is usually completed by searching for possible side chain conformation and evaluating every backbone structure by using some scoring function. If we assume that a target protein's side chain is approximately similar, the search space can be significantly reduced. The accurate and fast side chain prediction is significant for protein prediction and design, either for ab initio protein structure or homology modeling.

Recently, there are many methods and modeling techniques that have been developed, such as AlphaFold [6], AlphaFold2 [7], DLPACKER [8], SCWRL [9], OPUS-Rota4 [10], FASPR [11], and AttnPacker [12]. However, protein side chain prediction remains a difficult challenge. Most of these methods place side chains in a fixed backbone, whether generated from simulations or from a parent structure. More accurate and faster methods for a side chain prediction of protein are still required.

For the past 50 years, protein structural 3-D prediction has been a difficult and challenging task. Recently, some applications depended on AlphaFold. AlphaFold exceeds other techniques, especially at the 14th protein structure prediction Critical Assessment with 95% Cα deviation residue for 87 proteins from 0.96Å [6, 13–16]. In addition,

AlphaFold latest version supports machine learning and integrate biological and physical knowledge, which is helpful for deep learning algorithms to solve the problem of protein modeling [6]. Nonetheless, the performance of AlphaFold prediction is perfect for protein backbone, but not clear for side chains [17, 18].

DLPacker [8] uses a Neural Network model to predict the side chain in three steps: an input generation, Neural Network model, and the side chain reconstruction. DLPacker brings the data entries from protein data bank (PDB), which are grouped together based on their similarity at 50% threshold. From all groups, DLPacker only selects a single structure that has the highest resolution and then reconstructs utilizing a PDB-redo algorithm [19]. DLPacker discards any groups with a resolution of less than 2.5Å. After defining an input box, each atom is packed on a network and divided into 28 channels. The channels are five channels (one channel for C, one channel for N, one channel for O, one channel for S, and one channel for other elements), 21 channels for amino acid types, one channel for a partial charge, and one channel for the label. The improvement is achieved with most of the amino acids. For instance, hydrophobic amino acids obtained the most improvement percentage, close to 50%. Other amino acids received about 20% improvement.

SCWRL [9] is a method used to determine side chains of residues given the backbone. SCWRL is easy to use for seven reasons: 1) rotamer library for a new backbone. 2) averaging through conformations samples for positions in a library. 3) hydrogen bonding function. 4) interaction of van der Waals forces between atomic potentials. 5) fast detection. 6) algorithm of tree decomposition. 7) all parameters optimization through determining interaction graph. Moreover, there are many versions of SCWRL, and the popular version is SCWRL 3 and SCWRL 4. In addition, SCWRL 4 improves prediction accuracy.

OPUS-Rota [10] is an open-source tool which is considered an important method for side chains prediction. The first module is OPUS-RotaNN2. The second module is OPUS-RotaCM, where it calculates the orientation and distance between various residue pair's side chains. The third module is OPUS-Fold2, which guides side chain modeling. The results of OPUS-Rota4 on side chain predictions are closer to native residues (i.e., RMSD 0.588 and 0.472) than AlphaFold2, while OPUS-Rota4 prediction was at RMSD values 0.535 and 0.407.

FASPR [11] is one of the new methods used for predicting side chains. In FASPR, an input is the backbone of the protein and an optional amino acid sequence to superimpose with the backbone. When comparing FASPR performance with other methods (i.e., SCWRL4, RASP, CISRR, and SCATD) on a dataset of 379 backbones, this method outperforms SCWRL 4, and CISRR. The prediction accuracy for FASPR was 69.1% for each side chain.

AttnPacker [12] is a recent deep learning method that directly predicts the coordinates of side chain atoms. Unlike others, AttnPacker directly incorporates backbone 3D geometry to simultaneously compute all side-chain coordinates without delegating to a discrete rotamer library or performing expensive conformational search and sampling steps improving computation efficiency and decreasing inference time.

In this research, we are addressing the problem using Cα trace only. The advantage of using the Cα trace instead of using the full backbone atoms is the robustness and tolerance to the missing information. Many protein molecules are missing one or more atomic structural information. 30–40% of the determined protein models are missing at least one atom's structural information or more [20]. Therefore, the accuracy of prediction methods that use all atoms will be negatively impacted.

2 Methodology

This research's main goal is to build a machine learning (ML) model to predict protein's side chain configuration using protein's Cα trace only. The basic idea is to develop one model for each amino acid type and use these models collaboratively to predict protein's side chains. Figure 2 depicts the framework of our methodology we used to build our approach.

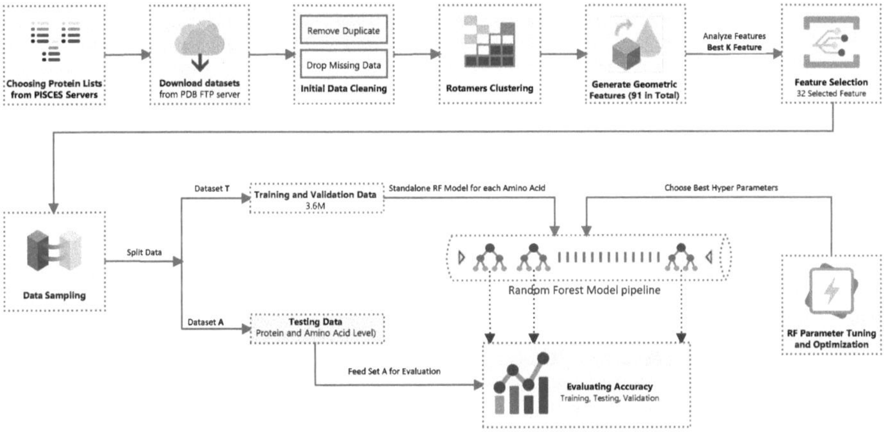

Fig. 2. The framework of our model

2.1 Rotamers Clustering

In this research, we use backbone dependent library for training purposes for our model [21]. In backbone dependent library, the total number of rotamers available for each amino acid are different. For instance, ARG has 110,889 rotamers while ASP has 12,321 rotamers. These rotamers are divided into groups based on statistical bins. A bin for a given phi and psi angle values are used to decide a group of rotamers that are common for such backbone configuration. There are 1,369 bins. 37 bins for phi against and 37 bins for psi ranges from -180 to 180 at 10 degrees step. For instance, if the value of phi is 85 and psi is −22, the chosen bin is (90, −20). Each bin recommends a group or common rotamers for this backbone configuration. Further, the number of rotamers in each bin's group is different. The numbers range from 3 to 80 rotamers. When our initial machine learning model was developed, we used the number of the rotamer in each group as a label to train the model. However, this numbering/labeling scheme confused our model.

For example, in bin x of the amino acid ARG, there are 11 unique rotamers, which are numbered from 1 to 11. Similarly, for bin y, of the same amino acid, there are 15 rotamers labeled from 1 to 15. The rotamer 5 (for example) in bin x is not the same as rotamer 5 in bin y. On the other hand, when we analyzed the rotamers and their structures, we found that many of them are geometrically/structurally similar. Therefore, we applied a clustering approach to unify the labeling and group all similar rotamers in one label. After clustering, label i for any amino acid is the same across all bins.

Table 1. The table shows our amino acids list, number of chi angles, total number of rotamers, and total number of clusters for each amino acid.

AA	Chi	Rotamers	Clusters	AA	Chi	Rotamers	Clusters
ALA	0	0	0	LYS	4	110,889	82
ARG	4	110,889	117	MET	3	36,963	13
ASN	2	24,642	11	PHE	2	82,14	17
ASP	2	12,321	5	PRO	2	2,738	3
CYS	1	4,107	6	SER	1	4,107	5
GLN	3	49,284	40	THR	1	4,107	3
GLU	3	36,963	31	TRP	2	12,321	10
GLY	0	0	0	TYR	2	8,214	16
HIS	2	12,321	9	VAL	1	4,107	4
ILE	2	12,321	6	LYS	4	110,889	17
LEU	2	12,321	16	MET	3	36,963	13

To unify labels for our machine learning model, we grouped our rotamers for each amino acid into clusters of rotamers that are structurally similar. For each amino acid, clustering is performed to group all similar roamers that are within a given arbitrary root mean square deviation (RMSD) threshold value. The given threshold is different for each amino acid based on the size of that amino acid. For instance, the threshold we used for amino acid ARG was 0.8 and LEU was 0.3. The clustering method we used to create clusters of rotamers for each amino acid is based on the frequently utilized mean-shift algorithm in the field of machine learning. We start with one random rotamer from the library and create the first cluster. This rotamer is considered the mean/centroid of the cluster. For every new rotamer, we calculate the RMSD between this rotamer and the centroids of existing clusters. We add it to the closest cluster if the RMSD is within the threshold. Otherwise, a new cluster will be created and this rotamer will be added to it as its centroid. Every time a new rotamer is added to a cluster, the new centroid is re-calculated. Table 1 shows the list of amino acids, number of chi angles in each amino acid, total of rotamers for each amino acid, and the number of clusters after applying our clustering approach. Note that clusters of the same amino acids may have different sizes (i.e., number of rotamers). Figure 3 shows a sample consisting of 136 rotamers for GLN in one cluster that contains 1,356 rotamers at 0.3 RMSD cutoff (left), a sample of 137 rotamers for ARG in one cluster that contains 1,369 rotamers at 0.8 RMSD cutoff (middle), and a sample of 134 rotamers for PHE in one cluster that contains 1,337 rotamers at 0.35 RMSD cutoff. From the figure we can see that some rotamers are structurally aligned well when overlapped.

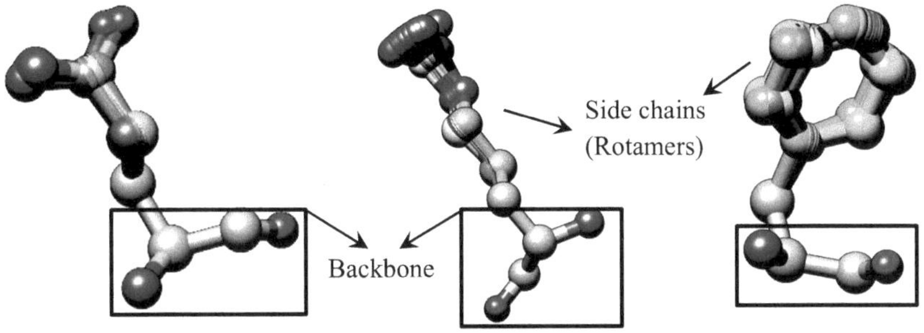

Fig. 3. Samples of rotamers from clusters for residues GLN (left), ARG (middle), and PHE (right). The boxed part is the backbone. All rotamers are overlapped to show their similarities.

2.2 Data Sets

We collected data sets from PDB. We utilized three protein lists from [22] (data not shown). Our data set consists of 32,000 PDB protein lists came with the maximum 0.286 R - factor, 3.0 Å resolution or better collected from the three data sets such that only non-duplicate protein chains were extracted from the sets. A total residue number of around 9.3 million (before data cleaning).

Cleaning Data. In the process of preparing our dataset, we accurately cleaned the data to ensure its quality and reliability. This involved the removal of any redundant protein chains or chains with missing structural information, specifically focusing on Secondary Structure Elements (SSEs) and Cα coordinates. By eliminating these instances, we enhanced the dataset's integrity and consistency. After this data-cleaning process, we randomly picked approximately 3.7 million residues to form our final dataset, referred to as set D. This carefully culled step was essential in guaranteeing the robustness and suitability of our dataset for subsequent analyses and research endeavors.

Dividing Dataset. We divided set D into two sets: the first set was used for training (called set T), and the second set was used for testing to report the accuracy of the model (we named this set A). Set A includes two subsets, set A1 and set A2. Set A1 contains 20 proteins which have 3,436 residues, we already removed it from the original set (D) and utilized it for testing. Set A2 contains 500 randomly chosen rotamers/side chain for each amino acid type. The total of A2 is 18,000 (ALA and GLY have no rotamers). Finally, the set T was chosen from sets D-A. Set T consists of around 3.6 residues. Moreover, Set T was divided randomly into 20% for validation (called Tv), and 80% for training (called Tt).

2.3 Geometric Features

Our ML model and algorithms are founded on geometric features that were gathered from protein structures, particularly Cα traces. We extracted the geometric features that we thought could distinguish between residues and side chains for every amino acid. Some of these features are inspired by [23–25]. The tentative total of features we had was 91 features. The work on features continued to be added, removed, and updated as needed.

Feature Analysis. We analyzed all 91 features to select the best and most important features for our machine-learning model process to make the process more accurate. We utilized a "K-Means" algorithm to analyze our features, related to amino acids and their rotamers [26].

We used the K-Means clustering technique to accurately divide and arrange the data. By using a systematic methodology, we were able to precisely identify the 32 features that had the most influence. These features are crucial to our dataset, since they have a significant impact on forming our results and insights. Their significance is substantial, since they provide a comprehensive comprehension of the underlying patterns and trends inherent in our data.

Best Feature Selection. We divided the 32 features into five categories. Each group provides a unique geometric perspective related to the side chain of protein.

The first group set is torsion angles which is the intersection angle between two surfaces. Every surface has three Cα. A torsion angle (Tα) is a kind of dihedral angle. Moreover, the torsion angle describes the connection between two molecular segments through the link. The torsion angle is essential to understanding geometric conformation. We used the torsion angle for side chain prediction, and we calculated the torsion angle by utilizing four consecutive coordinates of Cα. The second group set is a set of triangular angles (Rα), which include three angle values. The third group is vector angles (Vα), which is an angle created between two vectors in Cα coordinates. The fourth group is Axis distances (Dα), the distance between two points' projections on a virtual axis. We used (Dα) for residue (i) of interest, such as residues distance (i - 1) and (i - 2) on the axis projection formed by the axis connecting residues (i-2) and ($i+2$).

2.4 Random Forest and Tuning

We developed 18 Random Forest (RF) models to predict the best side chain for each amino acid given a Cα backbone trace of a protein. Each amino acid type has its own model, and these models work together to predict side chains when work on an entire protein. Performance metrics across different amino acids were analyzed to determine the effectiveness of our RF model. The model was trained on a training set Tt, which constituted 80% of the data, and evaluated on a test set Tv, which constituted the remaining 20%.

A RF algorithm is a strong ML technique, and it is applied for regression tasks and classification. RF algorithm belongs to a learning family group, where multiple individual decision trees are included to form the powerful prediction model. RF models are familiar for their efficiency and ability to handle complex and big datasets, and it has high accuracy and flexibility against overfitting [27].

Parameter Tuning. We began by fine-tuning select hyperparameters within our model. This optimization aimed to identify the best set for our ML model to ensure optimal performance. Techniques such as grid search and random search were employed to systematically traverse through a range of values for each hyperparameter.

Hyperparameter Tuning. In the realm of ML, hyperparameter tuning is an essential step to optimize the performance of algorithms. The parameters that define the model architecture, unlike the internal parameters learned during training, need to be set before training starts. For this study, a systematic approach was adopted to fine-tune the hyperparameters of the RF classifier.

For the RF classifier, after an exhaustive search, the optimal parameters were identified as follows: a maximum tree depth of 40, a criterion of "gini," and a maximum in features set into the square root of the total features. The search space for hyperparameters was defined in two modes: a comprehensive set for production and a limited set for testing purposes, ensuring a balance between computational efficiency and model performance.

Hyperparameter Optimization. In the process of model selection and optimization, a systematic hyperparameter tuning was conducted for the RF algorithm. Utilizing a 5-fold cross-validation approach, the RF model was subjected to nine distinct hyperparameter sets, leading to a total of 45 individual training runs. This rigorous tuning process was instrumental in identifying the most optimal model configuration, ensuring robust and reliable performance in subsequent evaluations.

Table 2. Random Forest model testing accuracy percentage for 20 proteins (Set A1).

Protein ID	Accuracy	Protein ID	Accuracy	Protein ID	Accuracy	Protein ID	Accuracy
6XRR	87.4%	6QTB	76.6%	5KWM	73.1%	4NZR	67.5%
2VN6	82.8%	2G6B	74.8%	3KRU	72.7%	3H9M	66.2%
6WUD	81.7%	3EBV	70.4%	7RMN	72.3%	6I5S	65.3%
6HC1	79.7%	8ERM	74.6%	4KH8	72.1%	7QZJ	64.9%
3BM7	79.7%	3PAS	74.1%	6PBM	68.6%	1XRE	69.7%

3 Experimental Results

We evaluated the efficiency of our model by employing Set A. Set A is composed of two subsets, namely Set A1 and Set A2. Set A1 comprises 20 proteins, totaling 3,436 amino acid residues. Set A2 comprises 500 randomly selected amino acids for each amino acid type, amounting to 18,000 residues. To assess the correctness of our model, we conducted a comparison between the label of the predicted rotamer/cluster and the labeled rotamer in the native structure. A match between the anticipated native rotamer and the labeled rotamer was classified as a hit, whereas any other outcome was classified as a miss. Prior to implementing our model, the ground truth side chains were excluded from the native structures of the testing sets.

The RF model had a significant overall average accuracy, highlighting its efficacy in predicting rotamers. When evaluated on set A1 of 20 proteins, the RF algorithm achieved an average accuracy rate of 73.7% (see Table 2). The RF method has proven to be effective for this prediction task, exhibiting a notable level of overall accuracy. Notable proteins with high testing accuracy include 6XRR (87.4%), 2VN6 (82.8%), 6WUD (81.7%), 6HC1 (79.7%), and 3BM7 (79.7%). Moreover, conducting a more detailed analysis of specific components of chosen proteins unveiled diverse levels of precision. In the case of protein 6XRR, specific amino acids like THR, ILE and PRO demonstrated accuracies of the overall prediction due to the decent performance of our model with these amino acids as shown in Table 3, resulting in an overall accuracy of 87.4%. On the other hand, protein 3H9M dominated by residues such as ARG, GLN, and HIS with accuracies of 51%, 62.4% and 69% respectively, resulting in a total accuracy of 66.2%. These comprehensive assessments offer a sophisticated comprehension of the model's performance, examining it at the protein level. The accuracy of these proteins varies from 64.9% to 87.4%, with a mean accuracy of 73.7%. This suggests a robust performance by the model, especially for specific proteins.

Table 3. The accuracy of the models on individual amino acids (Set A2)

Amino Acid	Accuracy	Amino Acid	Accuracy	Amino Acid	Accuracy
MET	60.8%	ILE	80.4%	LEU	76.5%
THR	88.6%	PRO	84.5%	ASN	77.7%
LYS	70.2%	ASP	80.6%	HIS	69.0%
TRP	72.4%	SER	79.8%	TYR	68.2%
PHE	69.0%	CYS	72.2%	ARG	51.0%
GLN	62.4%	VAL	87.5%	GLU	68.8%

To have a more profound understanding of the model's performance, we analyzed the performance of our models at amino acids level. We assessed the effectiveness of our RF model by analyzing various important metrics, such as accuracy (see Table 3) and average RMSD between the native side chain and the predicted rotamer (see Fig. 4). These metrics offer valuable information on the accuracy and uniformity of our predictions.

The accuracy of predicting 500 individual amino acids varies considerably as listed in Table 3. THR demonstrates the highest accuracy at 88.6%, indicating extremely trustworthy predictions, while MET exhibits a lower accuracy of 60.8%. This was reflected on Table 3 as expected. Amino acids, with high prediction accuracy, positively impact the accuracy of proteins they dominate. On the contrary, if a protein is dominated by low prediction accuracy amino acids, its overall accuracy is lower.

The average RMSD quantifies the average discrepancy between the predicted rotamer of the side chains and their ground truth structures. A smaller RMSD value signifies a higher accuracy in the prediction. The distribution of RMSD as in Fig. 4 illustrates and reflects the performance of the RF model on various amino acids as found in Table 3. For instance, the performance of THR is outstanding (i.e., 88.6% accuracy), with an average RMSD of 0.059 and a standard deviation of 0.048. This suggests that the model accurately and consistently predicts the side chain conformations of THR with high precision. Conversely, the performance of the model on amino acid like MET is lower (60.8%), with an average RMSD of 0.542 and a standard deviation of 0.220. This suggests that there is greater variety in the prediction errors.

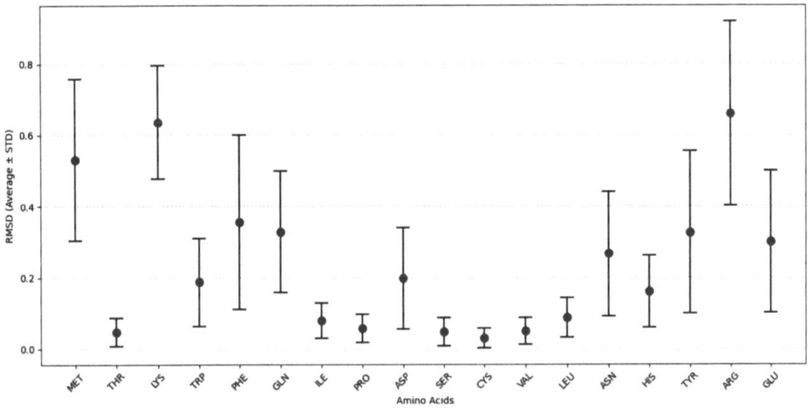

Fig. 4. RMSD Distribution by Amino Acid

4 Conclusion

Proteins are complex molecules that play fundamental functions in our bodies. Proteins are composed of molecule blocks known as amino acids. The structure of amino acids can be divided into two parts: backbone and side chain. Amino acids share the same backbone structure. Amino acids differ because of the uniqueness of their side chains. The side chain of the same amino acid adopts different configurations (called rotamers) based on its location in the protein. Amino acids' Side chains are closely related to biological functions, and therefore, an accurate prediction of the correct side chain is essential to serve the biological function.

To address the problem of predicting the side chains of a given Cα trace of a protein, we conducted a thorough investigation to find geometrical features that can be used to capture the local environment of a given amino acids that impacts the structure of its side chain. We introduced 91 features then analyze them and maintain the most 32 features that play an important role in deciding the correct rotamer of a given amino acid. The features are used to build a Random Forest machine learning model to solve the problem. The model was tested on two different sets. One set consists of 20 proteins and the other set consists of 18,000 individual amino acids. The actual results unambiguously demonstrate that Random Forest regularly attained a high average accuracy. Although Random Forest has demonstrated promising outcomes. The results of this study provide a strong basis for future efforts in predicting protein's side chains. We are confident that using more sophisticated approaches and models can enhance the accuracy and practicality of our predictions.

Acknowledgments. This Research was funded by NSF Projects (2153807) and (2409093).

Disclosure of Interests. The authors have no competing interests to declare that are relevant to the content of this article.

References

1. Ridley, M.: Genome, 1st edn. Harper Perennial, New York (2000)
2. Miao, Z., Cao, Y.: Quantifying side-chain conformational variations in protein structure. Sci. Rep. **6**(1), 37024 (2016)
3. Alberts, B., Hopkin, K., Johnson, A., Morgan, D., Roberts, K.: Essential Cell Biology 6th edn: W. W. Norton & Compan (2023)
4. Mobley, D.L., Dill, K.A.: Binding of small-molecule ligands to proteins: "what you see" is not always "what you get." Structure **17**(4), 489–498 (2009)
5. Al Nasr, K., He, J.: Constrained cyclic coordinate descent for cryo-EM images at medium resolutions: beyond the protein loop closure problem. Robotica **34**(8), 1777–1790 (2016)
6. Jumper, J., et al.: Highly accurate protein structure prediction with AlphaFold. Nature **596**(7873), 583–589 (2021)
7. Skolnick, J., Gao, M., Zhou, H., Singh, S.: AlphaFold 2: why it works and its implications for understanding the relationships of protein sequence, structure, and function. J. Chem. Inf. Model. **61**(10), 4827–4831 (2021)
8. Misiura, M., Shroff, R., Thyer, R., Kolomeisky, A.B.: DLPacker: deep learning for prediction of amino acid side chain conformations in proteins. Proteins: Structure, Function, and Bioinformatics **90**(6), 1278–1290 (2022)
9. Krivov, G.G., Shapovalov, M.V., Dunbrack, R.L.: Improved prediction of protein side-chain conformations with SCWRL4. Proteins: Structure, Function, and Bioinformatics **77**(4), 778–795 (2009)
10. Xu, G., Wang, Q., Ma, J.: OPUS-Rota4: a gradient-based protein side-chain modeling framework assisted by deep learning-based predictors. Briefings in Bioinformatics **23**(1) (2021)
11. Huang, X., Pearce, R., Zhang, Y.: FASPR: an open-source tool for fast and accurate protein side-chain packing. Bioinformatics **36**(12), 3758–3765 (2020)
12. McPartlon, M., Xu, J.: An end-to-end deep learning method for protein side-chain packing and inverse folding. Proc. Natl. Acad. Sci. **120**(23), e2216438120 (2023)
13. Kryshtafovych, A., Schwede, T., Topf, M., Fidelis, K., Moult, J.: Critical assessment of methods of protein structure prediction (CASP)—Round XIV. Proteins: Structure, Function, and Bioinformatics **89**(12), 1607–1617 (2021)
14. Tetchner, S., Kosciolek, T., Jones, D.T.: Opportunities and limitations in applying coevolution-derived contacts to protein structure prediction. Bio-Algorithms and Med-Systems **10**(4), 243–254 (2014)
15. AlQuraishi, M.: AlphaFold at CASP13. Bioinformatics **35**(22), 4862–4865 (2019)
16. Jumper, J., et al.: Applying and improving AlphaFold at CASP14. Proteins: Structure, Function, and Bioinformatics **89**(12), 1711–1721 (2021)
17. Terwilliger, T.C., et al.: AlphaFold predictions are valuable hypotheses and accelerate but do not replace experimental structure determination. Nat. Methods **21**(1), 110–116 (2024)

18. Zhao, H., et al.: Exploring AlphaFold2′s performance on predicting amino acid side-chain conformations and its utility in crystal structure determination of B318L protein. Int. J. Mol. Sci. **24**(3), 2740 (2023)
19. Joosten, R.P., et al.: PDB_REDO: automated re-refinement of X-ray structure models in the PDB. J. Appl. Crystallogr. **42**(3), 376–384 (2009)
20. Law, S.M., Frank, A.T., Brooks, C.L., III.: PCASSO: a fast and efficient cα-based method for accurately assigning protein secondary structure elements. J. Comput. Chem. **35**(24), 1757–1761 (2014)
21. Dunbrack, R.L., Jr., Karplus, M.: Backbone-dependent rotamer library for proteins application to side-chain prediction. J. Mol. Biol. **230**(2), 543–574 (1993)
22. Wang, G., Dunbrack, R.L., Jr.: PISCES: a protein sequence culling server. Bioinformatics **19**(12), 1589–1591 (2003)
23. Al Sallal, M., Chen, W., Al Nasr, K.: Machine learning approach to assign protein secondary structure elements from cα trace. In: 2020 IEEE International Conference on Bioinformatics and Biomedicine (BIBM): 16–19 Dec. 2020, pp. 35–41 (2020)
24. Al Nasr, K., Sekmen, A., Bilgin, B., Jones, C., Koku, A.B.: Deep learning for assignment of protein secondary structure elements from Cα coordinates. In: 2021 IEEE International Conference on Bioinformatics and Biomedicine (BIBM): 9–12 Dec. 2021, pp. 2546–2552 (2021)
25. Sekmen, A., Al Nasr, K., Bilgin, B., Koku, A.B., Jones, C.: Mathematical and machine learning approaches for classification of protein secondary structure elements from Cα coordinates. Biomolecules **13**(6), 923 (2023)
26. Ran, X., Zhou, X., Lei, M., Tepsan, W., Deng, W.: A novel K-means clustering algorithm with a noise algorithm for capturing urban hotspots. Appl. Sci. **11**(23), 11202 (2021)
27. Schonlau, M., Zou, R.Y.: The random forest algorithm for statistical learning. Stand. Genomic Sci. **20**(1), 3–29 (2020)

Can Language Models Reason About ICD Codes to Guide the Generation of Clinical Notes?

Ivan Makohon(✉), Jian Wu, Bintao Feng, and Yaohang Li

Old Dominion University, Norfolk, VA 23529, USA
imako001@odu.edu

Abstract. In the past decade a surge in the amount of electronic health record (EHR) data in the United States, attributed to a favorable policy environment created by the Health Information Technology for Economic and Clinical Health (HITECH) Act of 2009 and the 21st Century Cures Act of 2016. Clinical notes for patients' assessments, diagnoses, and treatments are captured in these EHRs in free-form text by physicians, who spend a considerable amount of time entering them. Manually writing clinical notes may take considerable amount of time, increasing the patient's waiting time and could possibly delay diagnoses. Large language models (LLMs), such as GPT-3 possess the ability to generate news articles that closely resemble human-written ones. We investigate the usage of Chain-of-Thought (CoT) prompt engineering to improve the LLM's response in clinical note generation. In our prompts, we incorporate International Classification of Diseases (ICD) codes and basic patient information along with similar clinical case examples to investigate how LLMs can effectively formulate clinical notes. We tested our CoT prompt technique on six clinical cases from the CodiEsp test dataset using GPT-4 as our LLM and our results show that it outperformed the standard zero-shot prompt.

Keywords: Large language models · generative AI · chain-of-thought (CoT) · natural language processing · information retrieval · clinical note generation · International Classification of Diseases (ICD) codes

1 Introduction

In the past decade, there has been a surge in the amount of electronic health record (EHR) data in the United States. In 2008, only 42% of office-based physicians had access to an EHR. This figure has now risen to 88% as reported in 2021 [1]. This increase can be attributed to a favorable policy environment created by the Health Information Technology for Economic and Clinical Health (HITECH) Act of 2009 [2] and the 21st Century Cures Act of 2016 [3].

Clinical notes for patients' assessments, diagnoses, and treatments are captured in these EHRs in free-form text by physicians, who spend a considerable amount of time entering them into computers. These notes offer valuable insights based on real-time observed data, which have shown to enhance the predictive capabilities of medical decision-making models [4]. Despite the rich information contained in these notes, it

M. Alser et al. (Eds.): ICCABS 2025, LNCS 15599, pp. 78–89, 2026.
https://doi.org/10.1007/978-3-032-02489-3_7

is likely some details are excluded from publicly available LLMs due to restrictions on access to their content, a consequence of the Health Insurance Portability and Accountability Act (HIPAA) of 1996 [5]. HIPAA plays a crucial role in safeguarding the privacy and security of patients' protected health information (PHI) in the context of clinical notes.

Despite the rapid growth of medical advancements, the quality of healthcare has unfortunately fallen behind [6–8]. One significant contributing factor to this decline is physician burnout. Physicians experience emotional exhaustion, demotivation, and detachment from their patients caused by the demanding and stressful nature of their work. A primary culprit for this burnout is the inconvenient and inefficient structure of EHRs, which requires excessive data entry and clinical note-taking [4, 6–8].

The medical scribe industry has emerged to handle the burdensome documentation tasks behind the scenes [9], but relying on non-professional scribes poses challenges, because they often lack the necessary medical expertise. To address physician burnout challenge, we focus our attention to large language models (LLMs) given that a remarkable progress has been made in recent years with some observations suggesting that they exhibit more powerful reasoning abilities as the model size increases [10, 11].

Our paper makes the following research contributions:

1. We evaluate GPT-4's performance in generating patient current history of present illness (HPI) based on a task instruction, using diagnosis codes and relevant patient information as input.
2. We explore and apply Chain-of-Thought (CoT) prompting, using clinical cases as examples to guide GPT-4 in generating clinical notes.

2 Related Works

The rapid advancements in LLMs have greatly enhanced their ability to comprehend patterns and relationships between words and phrases more effectively by developing a general understanding of grammar, syntax, and semantic relationships to generate text, bringing their output closer to human-level quality in areas of news compositions, story generation, and code generation [12]. LLMs like GPT-3 [10] and GPT-4 [11] have demonstrated impressive performance on downstream NLP tasks, even in zero-shot and few-shot settings. With its substantial capacity, it possesses the ability to generate news articles that closely resemble human-written ones, making it difficult to distinguish between the two [10]. This poses a particular challenge in detecting LLM-generated text, which is crucial for ensuring responsible AI governance [12]. GPT-4 is said to adhere more closely to guardrails, ensuring a higher level of responsible text generation.

Prompt engineering (or In-Context Prompting) [13, 14] emerged as a recent field focused on crafting and refining prompts to effectively harness techniques aimed at interacting with LLM to guide its behavior towards specific goals, without making changes to the model weights. Since the recent releases of LLMs, Google researchers recently revolutionized a prompting strategy in solving word problems across five different LLMs [15]. Several prompt engineering techniques [16, 17] have emerged and significantly improves the performance of LLMs on many natural language generation tasks. Recent studies, such as CoT [14, 15], Tree-of-Thought (ToT) [18, 19], and

Graph-of-Thought (GoT) [20, 21] have shown to improve the reasoning and accuracy performance of LLMs by providing rationales for a given word or phrase [14, 15, 18–21]. Although self-verification [22] and self-consistency [23] have enhanced performance in CoT prompting, recent prompting techniques such as ToT [18, 19] and GoT [20, 21] have shown improvement, though their effectiveness is still being assessed. CoT [24] has demonstrated that LLMs are capable of reasoning through multiple-choice questions on medical board exams. For our purposes, can it reason about ICD codes along with some patient information to generate clinical notes?

Previous endeavors have demonstrated that employing an attention mechanism in a multi-label classification task can effectively yield ICD codes from clinical notes [25] and shows that numerous prior research endeavors have revolved around classifying ICD codes using clinical notes as their primary input data. Our work in this paper is to reverse this process by generating comprehensive clinical notes, guided by provided ICD codes and supplemented basic patient information using instructional prompting techniques. In a recent study [26], LLMs were investigated using zero-shot prompting to predict ICD-10 codes. ICD-10 codes were provided in their prompt: "Predict these ICD-10 codes to the best of your ability" without any patient information or a clear instruction task to generate clinical notes. Based on their outputs, the LLMs outputs predicted just the ICD codes titles, not actual patient clinical notes.

Additionally, recent studies have explored the use of LLMs for generating clinical notes with the use of prompt engineering. These include leveraging LLMs to convert transcribed interactions into structured notes through structured prompting and integration of supplementary data for improved quality [27], developing a specialized medical LLM to understand and summarize medical conversations using zero-shot prompt for note generation [28], and providing rapid access to medical information via a chatbot that utilizes a predefined system prompt to perform contextual searches and Retrieval-Augmented Generation (RAG) techniques [29]. Some of the challenges in clinical note generation through use of LLMs are captured in this study [30], which highlights the feasibility of training efficient open-source LLMs for clinical note generation, with opportunities for further exploration in domain adaptation, data selection, and reinforcement learning from human feedback (RLHF). RLHF helps align LLMs with human preferences and can be applied in two ways: outcome-supervised, which focuses on improving the overall quality of the text, and process-supervised, which provides more detailed guidance on specific text components, such as reasoning steps, as seen in approaches like InstructGPT [16].

We conduct experiments on the closed-source GPT-4 using semantic searches and the CoT prompting technique to query similar clinical cases based on the given ICD codes or text references. To our knowledge, we are the first to perform experiments of this kind using diagnosis codes (ICD codes) as input along with basic patient information to generate clinical notes using LLMs and CoT prompting instructions. We seek to answer our research question: Can LLMs reason about ICD codes to guide the generation of clinical notes using instruction prompting?

3 Methodology

This paper explores a method for guiding the generation of clinical notes by using an LLM (GPT-4) while providing a task instruction, ICD codes and patient information utilizing CoT instruction prompting as rationale prompts with examples of clinical notes diagnosis with similar ICD codes.

3.1 Semantic Search and Clinical Cases

CodiEsp, introduced during the CodiEsp track for CLEF eHealth 2020, is recognized as a gold-standard annotated data source [31]. The dataset is comprised of 1000 clinical cases, where the clinical notes are translated from Spanish to English. It encompasses both ICD-10 CM and PCS codes, distributed across three randomly sampled datasets: the Training set contains 500 clinical cases. The Development (validation) and the Test set each contains 250 clinical cases. The text-reference column consists of text used during the annotated process using the Brat visualization tool [32]. Hereafter, we will refer to text-reference column as the Text Reference.

We combine CodiEsp's training and validation datasets (750) for our semantic search embedding query, while reserving the test dataset (250) for selecting six clinical cases samples for evaluating ground-truth against generative text. We converted the texts in the combined dataset of 750 clinical cases into numerical vector representations with OpenAI's text embedding model (text-embedding-3-small). Our objective is to leverage the embedding-driven retrieval to tap into the rich semantic features present in other clinical cases. This is achieved through the use of query ICD codes or text references, which facilitates the provision of clinical case examples for use as "thoughts" in our CoT prompt. As we will demonstrate, these semantic searches provide an efficient approach to identifying examples resembling the examples in the prompts. The embedded query (ICD code or text reference) is used to pinpoint the most relevant clinical cases by assessing their proximity within the embedding space, utilizing document similarity to rank and present the top-n most suitable clinical cases. For each query, the cosine similarity is used to identify the top-n most similar clinical cases. We randomly selected six clinical cases with less than 200 words from the test dataset that contain 1 or more ICD codes or text references. The breakdown of the ICD codes, text references and word count for each clinical case is shown in Table 1.

Table 1. Clinical Case Samples (counts).

Clinical Case	CodiEsp ArticleID	ICD Code	Text Reference	Clinical Note
A	S0213–12852003000600002-1	2	2	99
B	S1130–05582017000100031-1	1	1	113
C	S1130–01082008001000008-1	4	5	128
D	S1130–01082009000500011-1	9	9	107
E	S1130–01082008000100009-1	10	11	107
F	S1130–01082006001000017-1	9	9	90

3.2 Prompt Format

Our standard prompt, referred to as the baseline, is formatted as task instructions to generate the HPI clinical notes based on the given ICD codes with zero-shot prompt. Our CoT semantic search (CoT prompting) is formatted as instructions to guide the output of language model by controlling its generated text. In the CoT instruction prompt, each experiment contains the ground-truth clinical case's ICD codes along with basic patient information. In addition, the top-10 similar clinical cases are provided by semantic search query (based on the ground-truth ICD codes or text references), which uses contextualized word embeddings and the cosine similarity function to find related clinical cases, are provided as prompts. These inputs act as rationales, enabling the LLM to learn and generate the intended clinical notes based on the provided ICD codes.

Our CoT prompting takes inputs, such as:

- Task instruction.
- ICD codes for the diagnosis and/or procedure.
- Examples of similar clinical cases using the semantic search (ICD codes or text references) query.
- Basic patient information (age and gender).

3.3 Metrics

Cosine distance is the complement of cosine similarity, which measures the angular difference between two vector representations in a multi-dimensional space. It is a mathematical function that quantifies the degree of dissimilarity between two vectors based on their orientation rather than their magnitude. The cosine distance formula is defined as:

$$Cosine\,Distance = 1 - \frac{A \cdot B}{\|A\|\,\|B\|} = \frac{\sum_{i=1}^{n} A_i B_i}{\sqrt{\sum_{i=1}^{n} A_i^2}\sqrt{\sum_{i=1}^{n} B_i^2}} \tag{1}$$

where, $A \cdot B$ is the dot product of the sentence vectors A and B, $\|A\|$ and $\|B\|$ are the magnitudes of the vectors, and the result gives a measure of the angular distance between the vectors.

Transformer-based models, like Bidirectional Encoder Representations from Transformers (BERT) [33], capture both syntactic and semantic relationships between words by generating contextualized word embeddings. To assess the similarity between machine-generated and ground-truth documents, we use BERT. Both documents are processed through the BERT model (bert-large-cased) to obtain embeddings, which are then used to calculate sentence similarity:

- Using the special "classification" [CLS] token of each sentence. The [CLS] token in BERT serves as a means to gather a holistic representation of the input sequence. The output of [CLS] is inferred by all other words in this sentence. This implies that the [CLS] contains all information in other words, which makes [CLS] a representation for sentence-level classification.
- Calculating the MEAN of the sentence embeddings provides a way quantify the overall cosine distance between the sentences based on semantic meaning.

3.4 Experiment

For our experiment, we use OpenAI's closed-source GPT-4 (gpt-4) model as the foundational LLM for all experiments, with the following parameters: *seed* (123), *temperature* (0), *top_p* (0.000001), *frequency_penalty* (0), and *presence_penalty* (0). We establish a baseline for our results using standard zero-shot prompt and compare it against the results from our CoT prompts, which utilize a semantic search query based on the provided ICD codes or text references from ground-truth clinical case samples. Additionally, basic patient information from the ground-truth data is provided as supplementary prompts. Our semantic search query introduces an extra prompt, which includes the top-10 most similar clinical cases based on the given ICD codes or text references.

For each clinical case sample, we collect results from 100 API calls to GPT-4, with each call treated as an independent interaction. This ensures there is no memory or history from previous interactions, making each response independent. The clinical case's top-10 relatedness scores from the semantic search query are presented in Table 2. These scores are calculated using cosine distance, which evaluates spatial proximity to identify the top-10 most similar clinical cases based on the provided ICD codes or text references.

Table 2. Semantic Search Query (Top-10 Relatedness Scores).

Clinical Case	ICD Code Relatedness	Text Reference Relatedness
A	0.762, 0.754, 0.729, 0.724, 0.718, 0.715, 0.708, 0.695, 0.687, 0.683	0.563, 0.483, 0.478, 0.469, 0.462, 0.458, 0.433, 0.431, 0.429, 0.417
B	0.720, 0.717, 0.711, 0.701, 0.677, 0.672, 0.653, 0.644, 0.635, 0.630	0.469, 0.434, 0.411, 0.387, 0.380, 0.361, 0.359, 0.338, 0.336, 0.334
C	0.812, 0.780, 0.775, 0.768, 0.768, 0.760, 0.759, 0.754, 0.754, 0.751	0.601, 0.553, 0.545, 0.522, 0.521, 0.520, 0.520, 0.517, 0.512, 0.511
D	0.803, 0.802, 0.798, 0.796, 0.787, 0.783, 0.783, 0.782, 0.782, 0.782	0.630, 0.613, 0.606, 0.568, 0.565, 0.551, 0.547, 0.537, 0.524, 0.522
F	0.846, 0.807, 0.805, 0.795, 0.786, 0.774, 0.770, 0.769, 0.767, 0.757	0.637, 0.632, 0.623, 0.613, 0.610, 0.609, 0.601, 0.598, 0.595, 0.594
G	0.797, 0.775, 0.764, 0.763, 0.759, 0.752, 0.744, 0.741, 0.739, 0.739	0.654, 0.646, 0.645, 0.629, 0.627, 0.625, 0.624, 0.617, 0.615, 0.610

4 Results and Discussions

We evaluate our prompting technique using cosine distance as the primary metric, comparing generated text to ground truth. The results are visualized with a Kernel Density Estimation (KDE) plot, which includes Bootstrap Confidence Intervals (BCIs) for both sentence-level [CLS] and Mean scores. This comparison contrasts out CoT prompting against the baseline zero-shot prompt. The distribution analysis of CoT prompting, which incorporates semantic search queries (such as ICD codes and text references) for similar clinical cases, consistently shows that our method enhances the language model's ability to capture the underlying patterns and reasoning of the ICD codes and text references, as well as basic patient information, compared to the baseline zero-shot prompt.

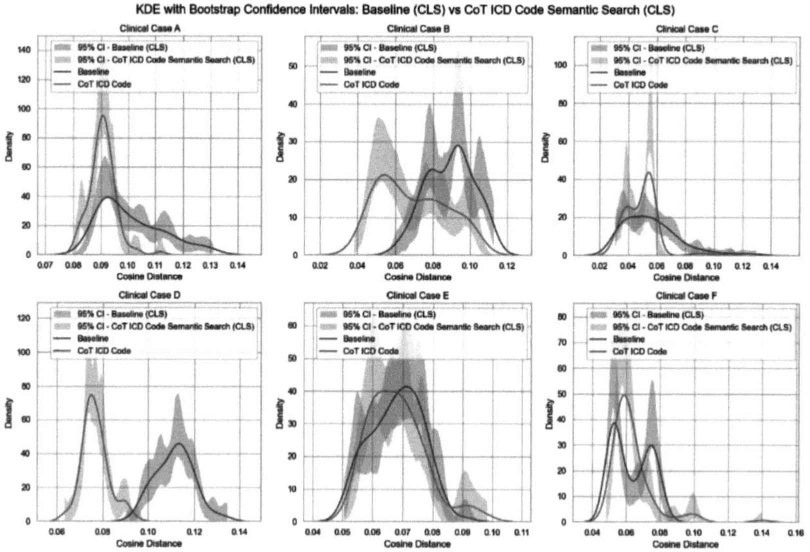

Fig. 1. Illustration of the six clinical cases using KDE with BCIs to compare the Baseline and CoT ICD code semantic search, based on cosine distance score of sentence-level [CLS].

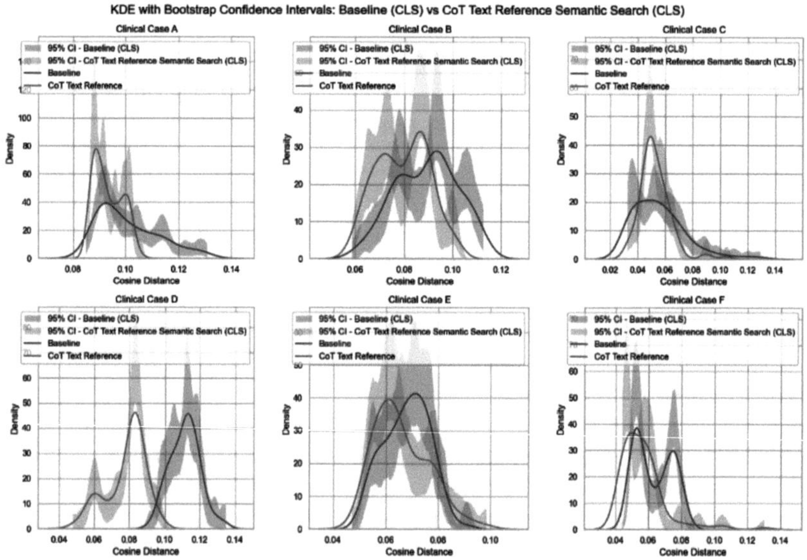

Fig. 2. Illustration of the six clinical cases using KDE with BCIs to compare the Baseline and CoT text reference semantic search, based on cosine distance score of sentence-level [CLS].

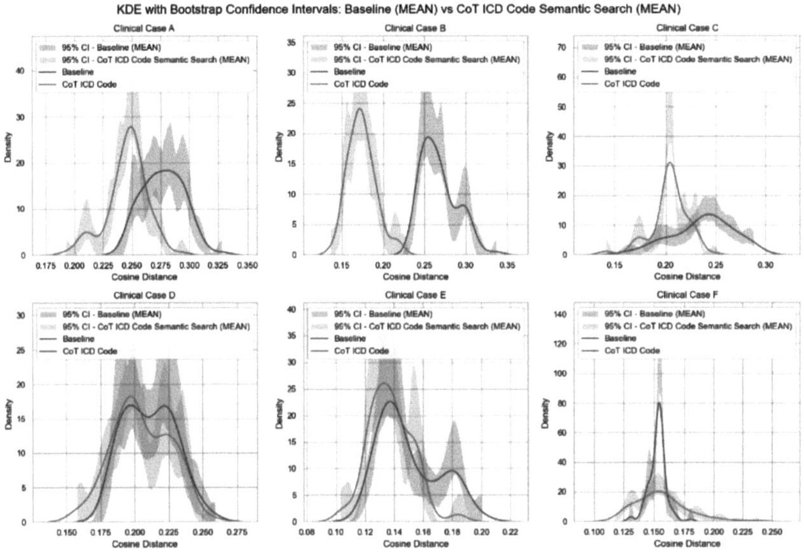

Fig. 3. Illustration of the six clinical cases using KDE with BCIs to compare the Baseline and CoT ICD code semantic search, based on cosine distance score of sentence-level MEAN.

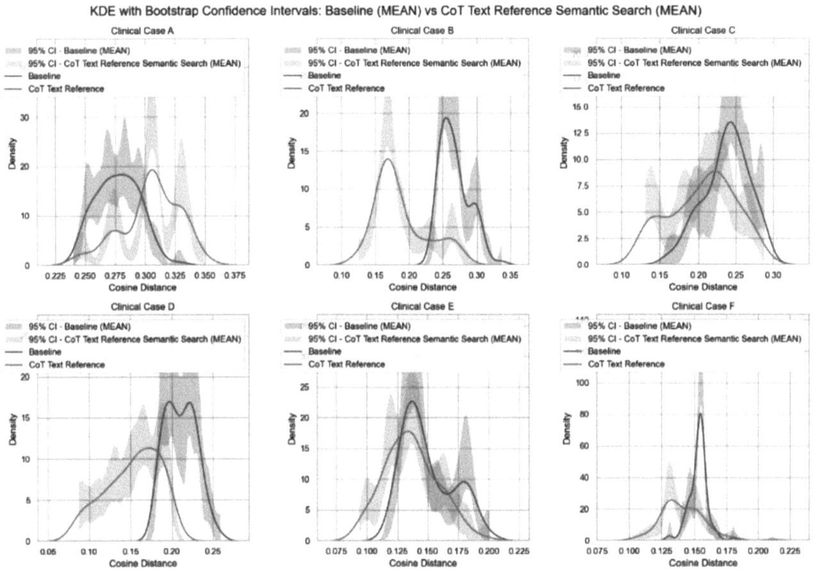

Fig. 4. Illustration of the six clinical cases using KDE with BCIs to compare the baseline and CoT text reference semantic search, based on cosine distance score of sentence-level MEAN.

The KDE with BCI plots (Figs. 1, 2, 3 and 4) reveal a leftward shift in the peaks for the CoT semantic search prompting technique, indicating that its distribution has lower values compared to the baseline prompts. This shift highlights notable differences in semantic alignment with ground-truth clinical cases. The inclusion of BCIs provides statistical validation, enhancing the robustness and interpretability of these findings. However, as shown in Fig. 4, we observe that in clinical case A, the baseline prompt outperforms the CoT text reference prompt. This may be due to the presence of two text references (pain, toothache). In contrast, in Fig. 3, the two ICD codes (K08.89, R52) clinical case A perform better than the baseline prompt. The word "pain" for the text reference could be too general or nonspecific to accurately capture the clinical details needed for generating the HPI or guiding the model's output (Fig. 5).

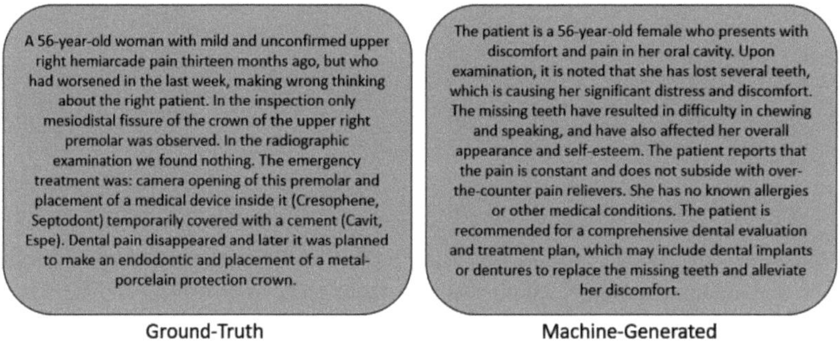

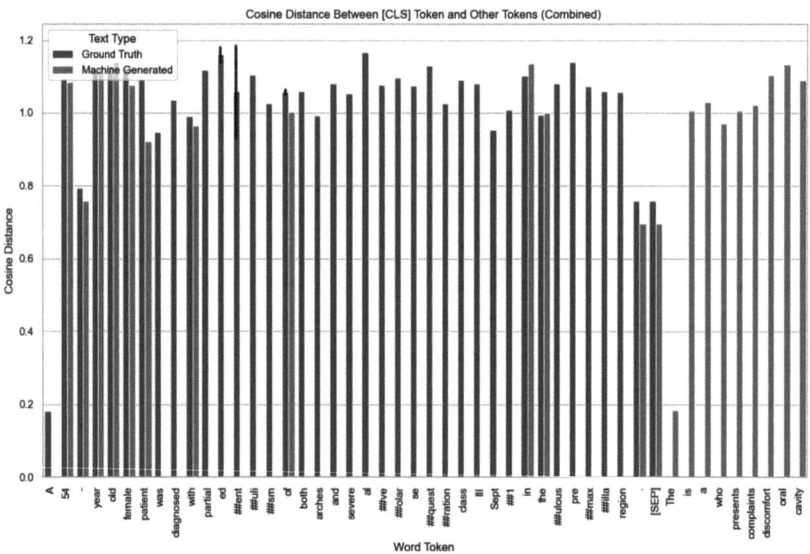

Fig. 5. A sample from Clinical Case A showing ground-truth and the machine-generated text.

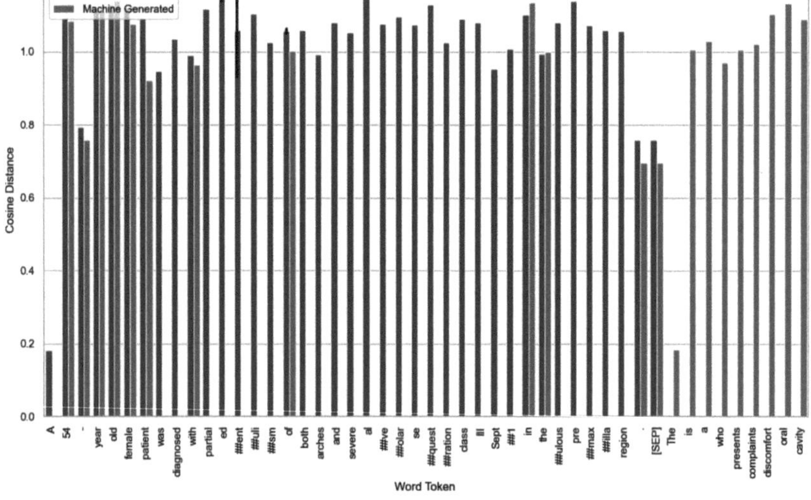

Fig. 6. Illustration of a sample from Clinical Case A, showing the cosine distance scores of the [CLS] tokens for both ground-truth and machine-generated text.

Figure 6 shows that some words are present in the ground-truth but missing from the machine-generated text. It also reveals that ICD-10 codes K08.89 (pain) and R52 (toothache) are linked to terms like "oral," "cavity," and "discomfort," which appear in the generated text, all related to oral health. Both texts include "A 54-year-old female," confirming alignment in basic patient information. However, the cosine distance between the [CLS] tokens is slightly over 1.0, likely due to contextual differences in the surrounding text. As seen in Fig. 4, the ground-truth [CLS] embedding captures a more specific medical context, while the generated text is broader, which could explain the higher cosine distance.

Overall, by using visual comparisons, such as overlaying KDE plots for all clinical cases and prompting techniques, we were able to assess the extent of the observed shifts. These plots, along with the cosine distance measurements, confirm that the differences in semantic similarity performance between the techniques are both statistically significant and practically meaningful. For instance, clinical cases A and B exhibit a significant leftward shift in the peaks (Figs. 1 and 3) for the ICD code semantic search, indicating that their distributions have lower values compared with the baseline. The ICD code prompting technique appears to outperform the Text Reference, likely due to higher relatedness scores for the clinical case examples (Table 2). From observation, these visual insights combined with the precision offered by the BCIs show that both CoT prompting technique results distributions differ from the baseline standard prompt, which suggest that this technique can guide the generation of clinical notes through instruction prompt using similar clinical cases.

5 Conclusion and Future Work

Our study constructs the HPI clinical notes using CoT prompting with ICD codes, clinical case examples, and basic patient information. Experiments were conducted across various clinical cases (clinical cases with 1 ICD code, 2 ICD codes, and several cases with multiple ICD codes), comparing results obtained from a baseline zero-shot prompt and two CoT prompting templates. Through cosine distance analysis, we compared the generated text with ground-truth text, addressing whether LLMs can effectively reason about ICD codes to produce clinical notes using CoT prompting.

Our analysis concludes that the GPT-4 LLM is capable of reasoning about ICD codes using our CoT semantic search prompting techniques over the baseline zero-shot prompt to produce clinical notes. Also, comparing an EHR to a single ground-truth may not be effective, as different doctors write EHRs differently. Thus, human evaluation should be considered to ensure alignment with automatic metrics.

5.1 Future Work

To enhance the prompting techniques, we propose some follow up studies not only in areas of using other instruction prompting techniques, but in these areas as well to enrich LLMs reasoning:

1. CoT Prompting using Patient's Past Medical History

The Medical Information Mart for Intensive Care (MIMIC-III) [34] offers clinical data on 30-day ICU readmissions, allowing past medical history and admission notes to guide model predictions for future visits.

2. Fine-Tune an LLM to become more biased towards the Physician's output
LLMs are prone to biases from training data, but they can be fine-tuned for individual physicians using personalized data. This adjustment, achieved through instruction prompting, allows the model to better meet specific needs. Physician notes can be extracted for this fine-tuning from the MIMIC-III dataset.
3. Use Retrieval Augmented Generation (RAG) in conjunction with CoT prompting
RAG enables retrieval of relevant information, such as patient data from medical databases, to inform the instruction prompting process. Our semantic search embeddings identify the most relevant documents based on query similarity (e.g., patients with similar ICD codes). This helps guide LLM text generation, while RAG minimizes "hallucinations" by feeding relevant facts into the model, improving the accuracy and relevance of clinical note generation.

References

1. Office of the National Coordinator for Health Information Technology. "Office-based Physician Electronic Health Record Adoption," Health IT Quick-Stat #50. https://www.healthit.gov/data/quickstats/office-based-physician-electronic-health-record-adoption
2. Burde, H.: Health Law the HITECH ACT - An Overview. In: the Virtual Mentor, **13**(3), 172–175 (2011)
3. U.S. Food & Drug. 21st Century Cures Act. https://www.fda.gov/regulatory-information/selected-amendments-fdc-act/21st-century-cures-act (2020)
4. Nelson, D.D.: Copying and pasting patient treatment notes. In: American Medical Association Journal of Ethics. March 2011 **13**(3), 144–147 (2011)
5. Moore, W., Frye, S.A.: Review of HIPAA, part 1: history, protected health information, and privacy and security rules. The Journal of Nuclear Medicine Technology **47**, 269–272 (2019)
6. Arndt, B.G., et al.: Tethered to the EHR: primary care physician workload assessment using EHR Event log data and time-motion observations. The Annals of Family Medicine **15**, 419–426 (2017)
7. Gellert, G.A., Ramirez, R., Webster, S.L.: The rise of the medical scribe industry: implications for the advancement of electronic health records. Journal of the American Medical Association (JAMA) **313**(13), 1315–1316 (2015)
8. Kroth, P.J., et al.: Association of electronic health record design and use factors with clinician stress and burnout. In: Journal of the American Medical Association (JAMA) Network Open **2**(8), e199609 (2019)
9. Liu, P.J.: Learning to Write Notes in Electronic Health Records (2018)
10. Brown, T.B., et al.: Language models are few-shot learners. In: the 34th Conference on Neural Information Processing Systems (NeurIPS), Vancouver, Canada (2020)
11. OpenAI. GPT-4 Technical Report (2023)
12. Wu, J., Yang, S., Zhan, R., Yuan, Y., Wong, D.F., Chao, L.S.: A Survey on LLM-Generated Text Detection: Necessity, Methods, and Future Directions. In: ArXiv (2023)
13. Ye, S., Hwang, et al.: In-Context Instruction Learning. In: ArXiv (2023)
14. Wei, J., et al.: Chain of Thought Prompting Elicits Reasoning in Large Language Models. in ArXiv (2022)

15. Zhang, Z., Zhang, A., Li, M., Zhao, H., Karypis, G., Smola, A.J.: Multimodal Chain-of-Thought Reasoning in Language Models. in ArXiv, (2023)
16. Li, J., Tang, T., Zhao, W.X., Nie, J., Wen, J.: Pre-trained language models for text generation: a survey. ACM Comput. Surv. **56**, 1–39 (2022)
17. Sahoo, P., et al.: A Systematic Survey of Prompt Engineering in Large Language Models: Techniques and Applications. in ArXiv, (2024)
18. Yao, S., et al.: Tree of thoughts: deliberate problem solving with large language models. In the 37th Conference on Neural Information Processing Systems (NeurIPS) (2023)
19. Long, J.: Large Language Model Guided Tree-of-Thought. in ArXiv (2023)
20. Besta, M., et al.: Graph of thoughts: solving elaborate problems with large language models. In the 38th AAAI Conference on Artificial Intelligence (AAAI-24) (2024)
21. Yao, Y., Li, Z., Zhao, H.: Beyond Chain-of-Thought, Effective Graph-of-Thought Reasoning in Large Language Models. in ArXiv, (2023)
22. Weng, Y., et al.: Large language models are better reasoners with self-verification. in the Findings of the Association for Computational Linguistics: EMNLP 2023, pp. 2550–2575, December 6–10 (2023)
23. Wang, X., Wei, J., Schuurmans, D., Le, Q., Chi, E.H., Zhou, D.: Self-Consistency Improves Chain of Thought Reasoning in Language Models (2022)
24. Li'evin, V., Hother, C.E., Winther, O.: Can large language models reason about medical questions?. in Patterns **5** (2022)
25. Makohon, I., Li, Y.: Multi-label classification of ICD-10 coding & clinical notes using MIMIC & CodiEsp. in the 2021 IEEE EMBS International Conference on Biomedical and Health Informatics (BHI), pp. 1–4 (2021)
26. Lee, S.A., Lindsey, T.: Do Large Language Models understand Medical Codes? in arXiv, June 6 (2024)
27. Biswas, A., Talukdar, W.: Intelligent Clinical Documentation: Harnessing Generative AI for Patient-Centric Clinical Note Generation. in ArXiv, (2024)
28. Yuan, D., et al.: A continued pretrained LLM approach for automatic medical note generation. In the North American Chapter of the Association for Computational Linguistics (2024)
29. Leong, H.Y., et al.: A GEN AI Framework for Medical Note Generation (2024)
30. Wang, H., et al.: Adapting Open-Source Large Language Models for Cost-Effective, Expert-Level Clinical Note Generation with On-Policy Reinforcement Learning. in ArXiv (2024)
31. Miranda-Escalada, A., et al.: Overview of Automatic Clinical Coding: Annotations, Guidelines, and Solutions for non-English Clinical Cases at CodiEsp Track of CLEF eHealth 2020. In the Conference and Labs of the Evaluation Forum (2020)
32. Stenetorp, P., et al.: brat: a Web-based tool for NLP-assisted text annotation. In the Conf. of the European Chapter of the Association for Computational Linguistics (2012)
33. Devlin, J., Chang, M., Lee, K., Toutanova, K.: BERT: pre-training of deep bidirectional transformers for language understanding. In the Proceedings of the 2019 Conference of the North American Chapter of the Association for Computational Linguistics: Human Language Technologies, **1**, pp. 4171–4186, Minneapolis, Minnesota (2019)
34. Johnson, A.E., et al.: MIMIC-III, a freely accessible critical care database. In Scientific Data 3, 160035 (2016)

Link Prediction in Disease-Disease Interactions Network Using a Hybrid Deep Learning Model

Ashwag Altayyar(✉) and Li Liao

University of Delaware, Newark, DE 19716, USA
{ashwag,liliao}@udel.edu

Abstract. Discovering disease-disease association based on the underlying biological mechanisms is an essential biomedical task in modern biology as understanding these relationships will assist biologists in discovering the pathogenesis, diagnosis, and intervention of human diseases. Recently, deep learning on graph and graph neural networks have achieved promising performance in modeling complex biological structures and learning compact representations of interconnected data. Inspired by the success of graph neural networks in learning subgraph representations, we propose a novel framework, SNN-VGA, designed to predict potential disease comorbid pairs. We first model disease-associated genes as subgraphs in the protein-protein interactions network and learn disentangled disease module representations using a subgraph neural network model. The learned embeddings are leveraged by the variational graph auto-encoder to predict disease comorbidity in the disease-disease interactions network. Empirical results from a benchmark dataset demonstrate that our method performs competitively compared with the state-of-the-art model, with an AUROC of 0.96.

Keywords: Association Prediction · Comorbidity · Disease · Graph Convolution Network · Subgraph Neural Networks · Variational Graph Auto-Encoder

1 Introduction

In cells, the majority of cellular components exert their functions through the interactions with other cellular components [1]. Cellular functions are regulated by a complex network of molecular interactions, known as the interactome, which involves physical and functional interactions between various biological macromolecules such as proteins [2]. Since protein–protein interactions (PPIs) are intrinsic to most of the complex biological processes, any disruption of these interactions may cause malfunction and potentially lead to diseases. It has been shown that, the analysis of PPIs is important for understanding the molecular mechanisms of diseases, which can improve the prognostics and treatment for human disorders [3]. Often, the interconnectivity of the PPIs

M. Alser et al. (Eds.): ICCABS 2025, LNCS 15599, pp. 90–102, 2026.
https://doi.org/10.1007/978-3-032-02489-3_8

network allows genetic abnormalities to propagate through the network connections and indirectly influence the activity of other gene products [1]. Therefore, perturbations in the PPIs network can lead to the simultaneous presence of two or more diseases in the same individual, a phenomenon referred to as comorbidity [4]. The etiology of disease comorbidities involves several mechanisms. Previous studies have identified the comorbidity patterns through shared associated genes between diseases [5, 6]. Beyond genetic overlap, network-based structure has made substantial contributions to the advancement of biological systems [7], which allows for the exploration of cellular-level connections encoded by PPIs to reveal the underling mechanism of comorbidity. Therefore, direct interactions between causative proteins of two diseases were analyzed to uncover the molecular mechanisms driving disease comorbidity [8]. Other studies have suggested that diseases may co-occur because they are co-regulated by high-level biological factors, such as shared cellular processes and biological pathways [9, 10]. The random walk algorithm was proposed to explore unexplained disease similarity by analyzing the connections between disease-related genes in the PPIs network [8, 11].

Most of the approaches described above are based on analyzing biological factors and local network structures underlying the development of comorbid diseases. However, the PPIs network is large and complex, which requires more advanced methods to reveal intricate relationships to explain or predict disease comorbidity. Indeed, studies have been developed to consider the disease module theory to quantify the associations between diseases [6, 12]. In recent years embedding representation has been applied to disease biology. LINE [13] was used to map each gene in the PPIs network into a low-dimensional vector space to capture the intricate similarities between diseases [14]. CoGO is a model that used graph convolutional network (GCN) to measure disease similarity according to the structure of gene ontology and the gene interaction network [15]. Another work employed isometric feature mapping (Isomap), an extension of multidimensional scaling (MDS) that applied geodesic distance on the PPIs network for identifying disease comorbidities. In this approach, the nodes' coordinates were derived by preserving the shortest path distances between node pairs through eigenvalue decomposition and double centering of the distance matrix [16]. Despite these advances, the mentioned studies have certain limitations when inferring disease associations including:

1. Many comorbid disease pairs remain undiscovered in the medical literature. As a result, negative samples, which represent disease pairs that do not co-occur more frequently than expected by random chance, are sparse leading to imbalanced training data.
2. Disease modules contain rich higher-order connectivity patterns, both internally among member genes and externally through interactions with the rest of the network. Most of the previous work elucidates disease associations depending mainly on learning the representation of each gene associated with each disease separately without considering the interconnections of genes related to each disease module.

3. Some of the afore-mentioned methods rely on the location of disease modules within the PPIs network to predict disease relationships. They assumed that gene products associated with a disease segregate in the same neighborhood. In reality, many of the disease modules can be localized in one region of the network or distributed across multiple local neighborhoods, each with non-trivial internal topology.

By analyzing a benchmark PPIs dataset used for comorbidity prediction with disease module separation [6] and PPIs network Isomap embedding [16], we made the following observations:

- All disease modules form multiple disjoint components in the PPIs network. The minimum number of connected components observed across all disease modules is three. However, the maximum number of connected components observed across all disease modules is 276, which represents a high degree of fragmentation within a particular disease subgraph.
- The largest connected component (LCC) across all disease modules contains 85% of the proteins belonging to a specific subgraph. Despite the presence of large and connected components, the majority of the largest connected components, approximately 93%, include less than half of the proteins within the disease module.

These observations indicate the fragmented nature of disease-related genes throughout PPIs network, as they are not organized into cohesive clusters but rather existed in isolated groups. Motivated by the above analyses, we propose a deep learning framework, Subgraph Neural Networks-Variational Graph Auto-encoder (SNN-VGA), as depicted in Fig. 1. It consists of two models Subgraph Neural Networks (SUBGNN) [17] and Variational Graph Auto-Encoder (VGAE) [18]. SUBGNN is used to generate meaningful representations for disease subgraphs on the PPIs network while considering the fragmented topology of each disease subgraph. These learned representations by SUBGNN are further leveraged during the construction of a disease-disease interactions network (DDIs) to denote the features associated with diseases in the network. Then, we formulate disease comorbidity prediction using the constructed disease graph as a link prediction problem and exploit the advancement of VGAE to determine whether there is a missing link between two diseases in the DDIs network.

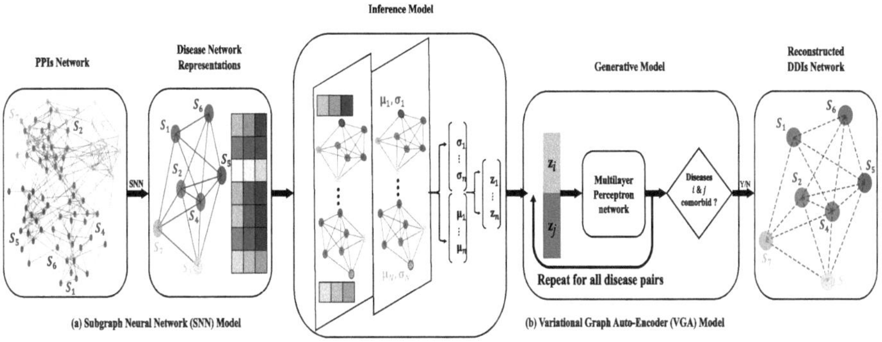

Fig. 1. The Hybrid Deep Learning Architecture of SNN-VGA Combining Subgraph Neural Network and Graph Variational Auto-Encoder. **(a) SNN Model for Disease Representations:** This step takes as input a protein-protein interactions (PPIs) network, in which we identify disease modules via known disease-gene associations. Each disease module corresponds to a subgraph comprised of nodes representing genes associated with the disease. The output of this step is a disease network (much like a condensation graph of these disease modules), along with a feature vector for each disease module, characterizing its positional and structural relationships with other disease modules learned by SNN from the underlying PPIs network. **(b) VGA Model for Predicting Disease Comorbidities: Inference Model:** It consists of a two-layer Graph Convolutional Network that processes the disease feature vectors and the structure of the disease network to infer the latent disease embedding used to predict interactions between diseases. The mean ($\boldsymbol{\mu}$) and variance ($\boldsymbol{\sigma}$) vectors represent the parameters of the learned probabilistic distribution for each node in the latent space. These vectors are used to sample the latent space to create an embedding vector **Z** for each disease node, capturing the essential features of each disease node in a lower-dimensional space. **Generative Model:** For each pair of diseases i and j (colored yellow and green respectively), their **Z** vectors are concatenated into a single representation, which is then processed by a Multilayer Perceptron network to determine the comorbidity between these two diseases: if yes, an edge is added to connect these two disease nodes; otherwise, they remain unconnected. This process is repeated for all disease pairs. **Reconstructed Disease-Disease Interactions Network:** The output network generated by the model aims to predict disease comorbidities by leveraging the latent embedding and incorporating DDIs ground truth to enhance prediction accuracy.

2 Materials and Methods

2.1 Biological Data

Protein-Protein Interactions (PPIs). The PPIs interactome describes the interactions between proteins within the cell. Our PPIs data is derived from [6, 16], contains 13,460 proteins and 141,296 interactions, including regulatory, binary, literature-curated, metabolic enzyme-coupled, protein complexes, kinase-substrate pairs, and signaling interactions. We model the PPIs interactome as a graph $G_{PPIs} = (V, E)$ that contains two main elements $V = 1, \ldots, n$ is the set of nodes representing proteins, and $E \subseteq V \times V$ is the set of undirected edges that indicate the interactions between the proteins. The largest connected component in this graph includes 13,329 nodes and 141,150 edges, covering more than 99% of the nodes and edges in the dataset used for this study. We focus on the LCC because it represents the most biologically relevant interactions, where the involved proteins frequently participate in significant cellular processes. It is generally believed that small connected components (many of them are singletons) in the current incomplete PPIs network are a result of missing edges, which correspond to interactions yet discovered, conceivably due to their minor/obscure roles, and that those small connected components, with missing edges once detected, will be connected to form into a larger component or merge to the LCC. Therefore, it has been a common practice adopted in similar and related studies to focus on LCC for PPIs networks [16, 17].

Disease Data. The disease-gene associations dataset is obtained from [6, 16]. The dataset contains a list of 299 diseases, and each disease has a set of genes that are known to be associated with the disease.

2.2 Disease Modules in the PPIs Network

Disease-gene associations and the interactions between them can be modeled in a PPIs network as subgraphs consisting of both known human diseases and disease-related genes. Each subgraph represents a disease module that contains a set of proteins which collectively contribute to a cellular function within the PPIs network and are implicated in causing the disease. In this work, we have constructed 299 disease modules as subgraphs, each consisting of gene products related to a specific disease.

2.3 Disease Comorbidity

To validate our proposed method, we utilize a Medicare dataset of disease history that includes 10,743 disease pairs [6, 16]. In order to quantify the comorbidity for each disease pair, the relative risk (RR) of observing a pair of diseases d_i and d_j, affecting the same patient, is computed using the following equation:

$$\mathrm{RR}_{ij} = \frac{C_{ij}N}{P_i P_j} \tag{1}$$

where C_{ij} is the number of patients affected by both diseases, N is the total number of patients in the population, and P_i and P_j are the prevalence of diseases i and j respectively. The prevalence of a disease refers to the proportion of the total population that is affected by a given disease. When the RR exceeds a specific threshold, it indicates that the co-occurrence of two diseases is more frequent than would be expected by a random chance. In this study, we set the threshold for the RR at two different values: 0 and 1 to investigate how it may affect the learning and performance of the model. When the threshold on RR is set at 1, the data contains 6,269 comorbid disease pairs, whereas setting the RR value to zero gives rise to 8,874 disease pairs, which are used to construct DDIs network, as described in the following sections.

2.4 Disease Network Representation

Given the dataset of disease-associated genes and the PPIs network, we adopt a subnetwork embedding model called SUBGNN that captures the topology of disease subgraphs.

It creates representations for all disease modules, which have varying sizes and multiple distributed connected components throughout the graph, as shown in Fig. 1(a).

Subgraph Representations. Given a PPIs network as a graph $G_{PPIs} = (V, E)$, where $V = 1, \ldots, n$ consists of a set of nodes denote the proteins, and edges $E \subseteq V \times V$ represent the interactions between them. $S = (V', E')$ is a disease subgraph of G_{PPIs} if $V' \subseteq V$ and $E' \subseteq E$ where nodes in each disease subgraph denote the product of genes associated with the disease, and the edges indicate the interactions between them. Each subgraph has a unique label y_S defines distinct disease and may include multiple connected components $S^{(c)}$. Given disease subgraphs $\mathcal{S} = \{S_1, S_2, \ldots, S_n\}$, SUBGNN is designed to identify the unique structure of subgraphs via three property-aware channels, each designated to explore a different aspect of subgraph topology which are position, neighborhood, and structure described in Table 1. SUBGNN specifies a mechanism that propagates neural messages at the subgraph level, between the subgraph components and randomly sampled anchor patches . Anchor patches $\mathcal{A}_X = \left\{A_X^{(1)}, \ldots, A_X^{(nA)}\right\}$ are subgraphs that are randomly sampled from the underlying graph G_{PPIs} in a channel-specific manner, where each anchor patch corresponds to one of the SUBGNN's channels, defined as $\mathcal{A}_P$, $\mathcal{A}_N$ and $\mathcal{A}_S$. Each propagated message conveys information about the relationship between a specific anchor patch and a subgraph component as follows:

$$\mathrm{MSG}_X^{A \to S} = \gamma_X(S^{(c)}, A_X).\,\mathbf{a}_X \tag{2}$$

where X is the channel, γ_X is a similarity function between the component $S^{(c)}$ and the anchor patch A_X, and $\mathbf{a}_X$ is the learned embedding of A_X. There are three types of similarity functions that determine the relative weighting of each anchor patch in building the subgraph component representations. For the position channel, the similarity function is defined as follow:

$$\gamma_P\left(S^{(c)}, A_P\right) = \frac{1}{\left(d_{sp}\left(S^{(c)}, A_P\right) + 1\right)} \tag{3}$$

where d_{sp} represents the average shortest path (SP) on the graph between the connected component $S^{(c)}$ and the anchor patch A_P specified for position channel. In contrast, for the neighborhood channel, the similarity function is $\gamma_N\left(S^{(c)}, A_N\right) = 1$ in the case of an internal neighborhood and $\gamma_N\left(S^{(c)}, A_N\right) \leq K$ for a border neighborhood that includes the subset of neighbor nodes within a k-hop distance from the connected component nodes. For the structure channel, the similarity function is given by:

$$\gamma_S\left(S^{(c)}, A_S\right) = \frac{1}{\left(\mathrm{DTW}\left(d_{S^{(c)}}, d_{A_S}\right) + 1\right)} \tag{4}$$

here, $d_{S^{(c)}}$ and d_{A_S} are the ordered degree sequences for the subgraph component and anchor patch, respectively, which are compared by the normalized dynamic time warping (DTW) measure [19], a similarity measure that calculates the optimal alignment between two sequences by minimizing the cumulative distance. The messages are then transformed into an order-invariant hidden representation $\mathbf{h}_{X,c}$ for the subgraph component $S^{(c)}$, as follows:

$$\mathbf{g}_{X,c} = \mathrm{AGG}_M\left(\left\{\mathrm{MSG}_X^{A_X \to S^{(c)}} \forall A_X \in \mathcal{A}_X\right\}\right) \tag{5}$$

$$\mathbf{h}_{\mathrm{X},c} \leftarrow \sigma\left(\mathbf{W}_{\mathrm{X}} \cdot [\mathbf{g}_{\mathrm{X},c}; \mathbf{h}_{\mathrm{X},c}]\right) \tag{6}$$

The outcome of applying these equations is a channel specific hidden representation $\mathbf{h}_{\mathrm{X},c}$ for each connected component $S^{(c)}$ of subgraph S and channel X, where $\mathbf{W}_{\mathrm{X}}$ is a layer-wise learnable weight matrix for channel X, σ is a non-linear activation function, AGG_M is a function that aggregates messages received from anchor patches, and $\mathbf{h}_{\mathrm{X},c}$ is the representation of the connected component at the previous layer, which gets updated and passed to the next layer of the model. The model is designed as such to learn a d_s-dimensional subgraph representation $\mathbf{z}_S \in \mathbb{R}^{d_s}$ for each disease subgraph $S \in \mathcal{S}$. This representation encapsulates the collective properties of all subgraph components using three channels across all layers, which can be then used for comorbidity prediction.

Table 1. Six properties of subgraph topology in Subgraph Neural Network.

Position	Internal	The distances between S_i's components
	Border	The distances between S_i and the rest of G_{PPIs}
Neighborhood	Internal	Defines a set of internal nodes of S_i
	Border	Defines a set of border nodes of S_i
Structure	Internal	The internal connectivity of $S^{(c)}$ within S_i
	Border	The border connectivity of $S^{(c)}$ within S_i

2.5 Disease-Disease Interaction Prediction

We address disease comorbidity as a task of predicting potential edges between diseases in a network as shown in Fig. 1(b). We consider a Graph $G_{disease} = (V, E)$, where $V = 1, \ldots, n$ represents the set of nodes each denoting a disease, and $E\ E \subseteq V \times V$ is a set of edges that capture the interactions between diseases. The adjacency matrix of $G_{disease}$ denoted by $\mathbf{A} \in \mathbb{R}^{n \times n}$ satisfies $A_{ij} \neq 0$ if and only if $\left(v_i, v_j\right) \in E$ suggesting the existence of a relationship between disease pairs. Specifically, with the RR threshold set to 1, only disease pairs with an RR value of 1 or higher are considered connected by positive edges in the disease graph, indicating their comorbidities. Conversely, assigning a more relaxed threshold at RR = 0 allows disease pairs with an RR value of 0 or higher to be connected by positive edges. Additionally, each node in the graph is associated with a d-dimensional feature vector generated by SUBGNN model. All disease feature vectors are stored in the disease feature matrix $\mathbf{X} \in \mathbb{R}^{n \times d}$.

Variational Graph Auto-Encoder (VGAE). VGAE is a framework for unsupervised learning specifically designed for graph-structured data. It combines the power of GCN with probabilistic modeling to learn low-dimensional latent representations of nodes in a graph. In particular, the latent representations for an undirected graph are learned by leveraging the graph structure represented by an adjacency matrix $\mathbf{A}$ and observed node attributes $\mathbf{X}$ to encode the graph structure and produce a posterior approximation $q_\phi(\mathbf{Z} \mid \mathbf{X}, \mathbf{A})$ over the latent variables $\mathbf{Z}$. Subsequently, the decoder reconstructs the original graph structure from these latent variables that consist of a compressed representation of the graph's structure and features. We introduce a component for disease comorbidity prediction based on the VGAE model in our designed formwork, as illustrated in Fig. 1(b). To the best of our knowledge, our model is the first attempt to implement VGAE for comorbidity prediction.

Inference Model. The inference model aims to compute latent representations $\mathbf{Z}$ via multiple graph convolution layers to capture the structural similarities between diseases. We initially adopt two convolutional layers of a GCN to learn more informative representations of diseases. We then embed these representations into a low-dimensional latent space. The encoder model is defined as:

$$q(\mathbf{Z} \mid \mathbf{X}, \mathbf{A}) = \prod_{i=1}^{\mathbf{N}} q(\mathbf{z}_i \mid \mathbf{X}, \mathbf{A}) \tag{7}$$

$$q(\mathbf{z}_i \mid \mathbf{X}, \mathbf{A}) = \mathcal{N}(\mathbf{z}_i \mid \mu_i, \mathrm{diag}(\sigma_i^2)) \tag{8}$$

where $\mu = \mathrm{GCN}_{\mu}(\mathbf{X}, \mathbf{A})$ and $\log\sigma = \mathrm{GCN}_{\sigma}(\mathbf{X}, \mathbf{A})$ are the matrices of μ_i and $\log\boldsymbol{\sigma}_i$ representing the parameters of the learned distribution that describes the latent variables $\mathbf{Z}$. $\mathrm{GCN}_{\mu}(\mathbf{X}, \mathbf{A})$ and $\mathrm{GCN}_{\sigma}(\mathbf{X}, \mathbf{A})$ denote a two-layer GCN defined as $\mathrm{GCN}(\mathbf{X}, \mathbf{A}) = \widetilde{\mathbf{A}}\,\mathrm{ReLU}(\widetilde{\mathbf{A}}\mathbf{X}\mathbf{W}_0)\mathbf{W}_1$, where $\mathbf{W}_i$ are the weight matrices. The symmetrically normalized adjacency matrix $\widetilde{\mathbf{A}}$ is given by $\widetilde{\mathbf{A}}=\mathbf{D}^{-\frac{1}{2}}\mathbf{A}\mathbf{D}^{-\frac{1}{2}}$, where $\mathbf{D}$ is the degree matrix. The Rectified Linear Unit function is defined as $\mathrm{ReLU}(\cdot) = \max(0, \cdot)$.

Generative Model. The Generative model maps disease feature vectors from the latent space generated by the encoder into the original disease graph. The structure of the decoder component influences the model's flexibility and ability to capture the expressiveness of the learned features. Therefore, to enhance these aspects, a multilayer perceptron (MLP) neural network is employed to predict the probability of links between diseases in the network, as illustrated in Fig. 1(b). The latent representations corresponding to each disease pair are concatenated and fed into a MLP neural network to predict the likelihood of edges in the disease network. We propose the following decoder network to reconstruct the original disease graph $\mathbf{A}$:

$$p(\mathbf{A} \mid \mathbf{Z}) = \prod_{i=1}^{N} \prod_{j=1}^{N} p(A_{ij} \mid \mathbf{z}_i, \mathbf{z}_j) \tag{9}$$

For each pair of nodes i and j in the disease network, the probability of the existence of an edge between them is calculated using MLP as the following expression:

$$p(A_{ij} = 1 \mid \mathbf{z}_i, \mathbf{z}_j) = \sigma\left(\mathbf{W}_2(\mathrm{ReLU}(\mathbf{W}_1\mathbf{Z}_{ij} + \mathbf{b}_1)) + \mathbf{b}_2\right) \tag{10}$$

where $\mathbf{Z}_{ij} = [\mathbf{z}_i, \mathbf{z}_j]$ represents the concatenated latent representations corresponding to diseases i and j, and the parameters $\mathbf{W}_i$ and $\mathbf{b}_i$ are the decoder weight matrix and bias vectors, respectively. $\sigma(\cdot)$ is defined as the logistic sigmoid function. The final output determines the predicted probability of a link between diseases i and j.

Training Objective. We optimize the variational lower bound $\mathcal{L}$ w.r.t. the variational parameters $\mathbf{W}_i$, given by:

$$\mathcal{L} = \mathbb{E}_{q(\mathbf{Z}|\mathbf{X}, \mathbf{A})}\left[\log p(\mathbf{A}|\mathbf{Z})\right] - \mathrm{KL}\left[q(\mathbf{Z}|\mathbf{X}, \mathbf{A}) \,||\, p(\mathbf{Z})\right] \tag{11}$$

here, $\mathrm{KL}\left[q(\cdot) \,||\, p(\cdot)\right]$ denotes the Kullback-Leibler divergence between $q(\cdot)$ and $p(\cdot)$. We assume a Gaussian prior for $p(\mathbf{Z})$, expressed as $p(\mathbf{Z}) = \prod_i p(\mathbf{z}_i) = \prod_i \mathcal{N}(\mathbf{z}_i \mid 0, \mathbf{I})$.

3 Experiments

3.1 Datasets

Our experiment is conducted on benchmark datasets for PPIs and disease-associated genes [6, 16], which form an underlying base graph including subgraphs with their associated labels as known diseases. Because of the use of LCC in our method, genes that are not in the LCC will be dropped from our experiments, which may cause some loss of information and make the prediction task more challenging. On the other hand, use of LCC allows us to focus on genes that are on the LCC and hence more informative in terms of the degree of interconnectivity and interactions with other genes. The statistics of the datasets are summarized in Table 2.

Table 2. Statistics of the benchmark datasets.

Dataset	#Nodes	#Edges
Protein–Protein Interactions	13,460	141,296
Disease–Disease Interactions (RR = 0)	299	8,874
Disease–Disease Interactions (RR = 1)	299	6,269

3.2 Experimental Setup and Evaluation Methods

We build our implementation of SNN-VGA by leveraging two distinct platforms: the Facebook machine learning library "PyTorch" [20–22], and the scikit-learn machine learning library [23]. We detail the experimental setups and evaluation methods for disease module representations and comorbidity prediction, respectively.

3.3 Disease Module Representations

We use the experimental setups proposed by SUBGNN. Initially the model is trained using Graph Isomorphism Network (GIN) [24] on link prediction to generate node and meta node embeddings for each node within the subgraph of the PPIs network. Subsequently, these trainable nodes embeddings are utilized to implement SUBGNN model, which generates feature vectors for each disease module.

3.4 Predicting Comorbidities Between Disease Modules

For the experimental settings of VGAE, we employ a transductive link prediction split in which the same graph structure is partitioned into the training, validation, and test sets. From the entire graph, 70% of the edges are designated as positive samples for the training set. Additionally, we sample 20% of the edges for validation and 10% for testing, which serve as positive samples, i.e., node pairs that are connected with an edge. Concurrently, for the training, validation, and test sets, we also randomly sample an equal number of negative samples, i.e., node pairs that are unconnected.

Parameters Selection for VGAE Architecture. The architecture of VGAE significantly influences the prediction performance of the model. Accordingly, we empirically set the dimensions of both the hidden layer and latent variables to 128 and 64, respectively. These values were selected based on validation set performance to balance model complexity and generalization. Additionally, we initialize the weights as described in reference [25]. We train the model for 50 epochs using Adam optimizer [26] with a learning rate of 0.001.

Evaluation Measures. We apply a nested cross-validation procedure [27], for model assessment and selection. Our model is trained using a 10-fold-within-5-fold nested-CV procedure to obtain an unbiased estimate of model performance while simultaneously optimizing the parameters. We calculate the reconstruction probability of the test edges to evaluate the ability of the model to classify comorbid versus non-comorbid disease pairs. We employ common evaluation metrics to measure the prediction performance of the SNN-VGA model, which include accuracy, precision, recall, F-measure (F1), average precision (AP), and the receiver operating characteristic (ROC) curve score.

4 Results

The averaged model performance for the comorbidity prediction task is reported in Table 3. We evaluate our method's performance by setting the comorbidity RR threshold values at 0 and 1. A threshold of 1 emphasizes stronger disease associations, while a threshold of 0 incorporates a wider range of associations, thus increasing edges between diseases and enhancing both graph connectivity and model training. Moreover, it enables the model to capture more complex relationships between the diseases and learn meaningful representations. As illustrated in Table 3, with an RR threshold of 0, SNN-VGA achieves remarkably high scores with an area under the ROC curve (AUROC) of 0.96 and an AP of 0.95. At the stricter RR threshold of 1, although there is a slight decline in the performance, SNN-VGA still yields strong results with an AUROC of 0.94 and an AP of 0.92. The superior performance of our method, particularly at the RR = 0 threshold can be attributed to its ability to effectively leverage the increased connectivity within the disease network, which in turn leads to more comprehensive analysis of potential disease associations. Figure 2 represents the ROC curves and their related areas under the curves that exhibit the performance of SNN-VGA across distinct test sets. For each test set, we run the model with different random initialization, and we then obtain the mean result and standard error derived from 10 runs that further emphasize the model consistency and statistical reliability under different conditions.

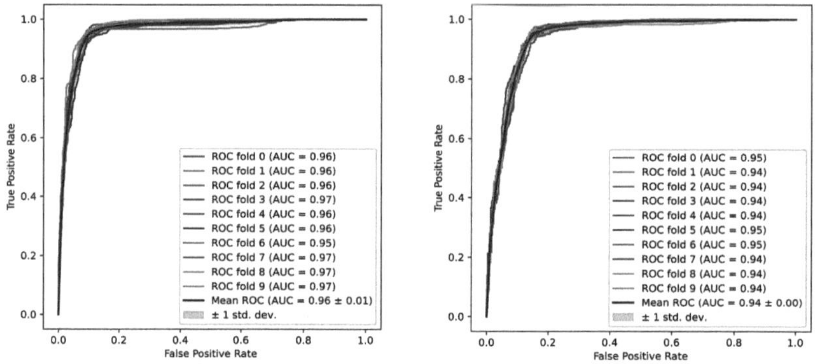

(a) RR = 0 **(b) RR = 1**

Fig. 2. ROC curves for each fold of the 10-fold cross-validation, along with the mean result and standard error for the test sets across different RR values.

Table 3. Comparison of averaged model performance using our method and state-of-the-art method for thresholds RR = 0 and RR = 1.

Model		Indicators					
		AUROC	Accuracy	Precision	Recall	F1	AP
SNN-VGA (Ours)	RR = 0	0.96 ± 0.01	0.92 ± 0.01	0.91 ± 0.01	0.94 ± 0.01	0.92 ± 0.01	0.95 ± 0.00
	RR = 1	0.94 ± 0.00	0.89 ± 0.01	0.87 ± 0.00	0.92 ± 0.01	0.89 ± 0.01	0.92 ± 0.00
Weighted Geometric Embedding [* MERGEFORMAT 16]	RR = 0	0.90	0.90	0.90	0.90	0.89	-
	RR = 1	0.76	0.70	0.70	0.70	0.69	-

In our comparison, we include the recent state-of-the-art method "Weighted Geometric Embedding" [16] for predicting comorbid diseases. This method mapped the PPIs network into a low-dimensional geometric space using the MDS technique. Each disease module was characterized by features derived from its projection in the geometric space, which were subsequently used to train support vector machine and random forest classifiers for comorbidity classification. Table 3 presents the results of this performance comparison. It can be observed that our SNN-VGA model achieves outstanding performance in the disease comorbidity prediction task as compared with the "Weighted Geometric Embedding" method. In particular, our approach enhances the AUROC score and accuracy by 6% and 2%, respectively.

5 Conclusion

In this study, we introduce SNN-VGA, a novel computational approach that integrates biological data into a single network and employs graph deep learning paradigms to predict disease comorbidity. We develop two distinct models. Initially, a SUBGNN is adopted to produce a set of feature vectors, each representing a specific disease module. Then, a model based on VGAE is applied to reconstruct the DDIs network for predicting disease comorbidities. By addressing shortcomings observed in related work, our method significantly outperforms a state-of-the-art method in cross-validation experiments on a benchmark dataset, as measured by common metrics. It demonstrates that our approach, by integrating a network comprised of diseases and a network of PPIs, cross linked via known disease-gene associations, offers a powerful platform for analyzing disease similarities with a unified graph-theoretic framework.

References

1. Barabási, A.-L., Gulbahce, N., Loscalzo, J.: Network medicine: a network-based approach to human disease. Nat. Rev. Genet. **12**(1), 56–68 (2011). https://doi.org/10.1038/nrg2918
2. Luck, K., Sheynkman, G.M., Zhang, I., Vidal, M.: Proteome-scale human interactomics. Trends Biochem. Sci. **42**(5), 342–354 (2017). https://doi.org/10.1016/j.tibs.2017.02.006
3. Gonzalez, M.W., Kann, M.G.: Chapter 4: Protein interactions and disease. PLoS Comput. Biol. **8**(12), e1002819 (2012). https://doi.org/10.1371/journal.pcbi.1002819
4. Feinstein, A.R.: The pre-therapeutic classification of comorbidity in chronic disease. J. Chronic Dis. **23**(7), 455–468 (1970). https://doi.org/10.1016/0021-9681(70)90054-8
5. Goh, K.-I., Cusick, M.E., Valle, D., Childs, B., Vidal, M., Barabási, A.-L.: The human disease network. Proc. Natl. Acad. Sci. U.S.A. **104**(21), 8685–8690 (2007). https://doi.org/10.1073/pnas.0701361104
6. Menche, J., et al.: Uncovering disease-disease relationships through the incomplete interactome. Science **347**(6224), 1257601 (2015). https://doi.org/10.1126/science.1257601
7. Barabási, A.-L., Oltvai, Z.N.: Network biology: understanding the cell's functional organization. Nat. Rev. Genet. **5**(2), 101–113 (2004). https://doi.org/10.1038/nrg1272
8. Ko, Y., Cho, M., Lee, J.-S., Kim, J.: Identification of disease comorbidity through hidden molecular mechanisms. Sci. Rep. **6**(1), 39433 (2016). https://doi.org/10.1038/srep39433
9. Rual, J.-F., et al.: Towards a proteome-scale map of the human protein–protein interaction network. Nature **437**(7062), 1173–1178 (2005). https://doi.org/10.1038/nature04209
10. Rubio-Perez, C., et al.: Genetic and functional characterization of disease associations explains comorbidity. Sci. Rep. **7**(1), 6207 (2017). https://doi.org/10.1038/s41598-017-04939-4
11. Hamaneh, M.B., Yu, Y.-K.: DeCoaD: determining correlations among diseases using protein interaction networks. BMC. Res. Notes **8**(1), 226 (2015). https://doi.org/10.1186/s13104-015-1211-z
12. Ni, P., Wang, J., Zhong, P., Li, Y., Wu, F.-X., Pan, Y.: Constructing disease similarity networks based on disease module theory. IEEE/ACM Trans. Comput. Biol. Bioinf. **17**(3), 906–915 (2020). https://doi.org/10.1109/TCBB.2018.2817624
13. Tang, J., Qu, M., Wang, M., Zhang, M., Yan, J., Mei, Q.: LINE: large-scale information network embedding. In: Proceedings of the 24th International Conference on World Wide Web, pp. 1067–1077. Florence, Italy (2015). https://doi.org/10.1145/2736277.2741093

14. Li, Y., Keqi, W., Wang, G.: Evaluating disease similarity based on gene network reconstruction and representation. Bioinformatics **37**(20), 3579–3587 (2021). https://doi.org/10.1093/bioinformatics/btab252
15. Chen, Y., Hu, Y., Hu, X., Feng, C., Chen, M.: CoGO: a contrastive learning framework to predict disease similarity based on gene network and ontology structure. Bioinformatics **38**(18), 4380–4386 (2022). https://doi.org/10.1093/bioinformatics/btac520
16. Akram, P., Liao, L.: Prediction of comorbid diseases using weighted geometric embedding of human interactome. BMC Med. Genomics **12**(S7), 161 (2019). https://doi.org/10.1186/s12920-019-0605-5
17. Alsentzer, E., Finlayson, S.G., Li, M.M., Zitnik, M.: Subgraph neural networks. In: Advances in Neural Information Processing Systems **33**, pp. 8017–8029. Vancouver, Canada (2020). https://doi.org/10.48550/arXiv.2006.10538
18. Kipf, T.N., Welling, M.: Variational graph auto-encoders. In: NIPS Workshop on Bayesian Deep Learning, pp. 1–3. Barcelona, Spain (2016). https://doi.org/10.48550/arXiv.1611.07308
19. Mueen, A., Keogh, E.: Extracting optimal performance from dynamic time warping. In: Proceedings of the 22nd ACM SIGKDD International Conference on Knowledge Discovery and Data Mining, pp. 2129–2130. San Francisco, USA (2016). https://doi.org/10.1145/2939672.2945383
20. Fey, M., Lenssen, J.E.: Fast graph representation learning with PyTorch Geometric. In: International Conference on Learning Representations Workshop on Representation Learning on Graphs and Manifolds, New Orleans, USA (2019). https://doi.org/10.48550/arXiv.1903.02428
21. Paszke, A., et al.: PyTorch: An imperative style high-performance deep learning library. In: Advances in Neural Information Processing Systems **32**, pp. 8026–8037. Vancouver, Canada (2019). https://doi.org/10.48550/arXiv.1912.01703
22. Falcon, W., et al.: PyTorchLightning/pytorch-lightning: 0.7.6 release. Zenodo, (2020). https://doi.org/10.5281/zenodo.3828935
23. Pedregosa, F., et al.: Scikit-learn: machine learning in Python. J. Mach. Learn. Res. **12**, 2825–2830 (2011). https://doi.org/10.48550/arXiv.1201.0490
24. Xu, K., Hu, W., Leskovec, J., Jegelka, S.: How powerful are graph neural networks? In: International Conference on Learning Representations, New Orleans, USA (2019). https://doi.org/10.48550/arXiv.1810.00826
25. Glorot, X., Bengio, Y.: Understanding the difficulty of training deep feedforward neural networks. In: Proceedings of the 13th International Conference on Artificial Intelligence and Statistics, pp. 249–256. Sardinia, Italy (2010)
26. Kingma, D.P., Ba, J.L.: Adam: A method for stochastic optimization. In: International Conference on Learning Representations, San Diego, USA (2015). https://doi.org/10.48550/arXiv.1412.6980
27. Stone, M.: Cross-validatory choice and assessment of statistical predictions. J. Roy. Stat. Soc. Ser. B Methodol. **36**(2), 111–133 (1974). https://doi.org/10.1111/j.2517-6161.1974.tb00994.x

Model Selection for Sparse Microbial Network Inference Using Variational Approximation

Shibu Yooseph(✉)

Kravis Department of Integrated Sciences, Claremont McKenna College, Claremont, CA 91711, USA
syooseph@cmc.edu

Abstract. Microbial communities are often composed of taxa from different taxonomic groups. The associations among the constituent members in a microbial community play an important role in determining the functional characteristics of the community, and these associations can be modeled using an edge weighted graph (microbial network). A microbial network is typically inferred from a sample-taxa matrix that is obtained by sequencing multiple biological samples and identifying the taxa abundance in each sample. Motivated by microbiome studies that involve a large number of samples collected across a range of study parameters, here we consider the computational problem of identifying the number of microbial networks underlying the observed sample-taxa abundance matrix. Specifically, we consider the problem of determing the number of *sparse* microbial networks in this setting. We use a mixture model framework to address this problem, and present formulations to model both count data and proportion data. We propose several variational approximation based algorithms that allow the incorporation of the sparsity constraint while estimating the number of components in the mixture model. We evaluate these algorithms on a large number of simulated datasets generated using a collection of different graph structures (band, hub, cluster, random, and scale-free).

Keywords: Microbiome · sparse networks · mixture models · variational approximation

1 Introduction

In a microbial community, the presence and abundance of one constituent taxon may either *directly* or *indirectly* influence the abundance of another [11]. We use the term *microbial association* to capture this concept. The structure, organization, and function of a microbial community is determined by its constituent taxa and their associations. Microbial associations are influenced by various environmental parameters. Here, we investigate the computational problem of determining these associations from microbiome data.

M. Alser et al. (Eds.): ICCABS 2025, LNCS 15599, pp. 103–129, 2026.
https://doi.org/10.1007/978-3-032-02489-3_9

Microbial associations can be represented using a weighted graph (*microbial network*), where the edge weights capture the strength of the pairwise associations between microbial taxa, and where the edge weight sign reflects whether the association is positive or negative [15]. This graph representation can be used to model a variety of microbial interactions, including competition and co-operation [17]. Information about associations between microbial taxa can provide important insights into the microbial community ecology. Microbial networks have been used to identify differences in the organization of microbial communities in health and disease settings [10,17,18].

Microbial associations can be inferred from the underlying covariance structure of the community. This structure can be estimated from a *sample-taxa* abundance matrix, which is generated by sequencing biological samples collected from the environment of interest and identifying the abundances (i.e., counts or proportions) of the constituent taxa in each sample. In a microbiome study, a sample-taxa matrix is typically generated by sequencing a taxonomic marker gene (for instance, the 16S ribosomal RNA gene, which is found in all bacteria [31]) or the constituent genomes (for instance, using a shotgun sequencing approach [28]).

Early methods for microbial network reconstruction assumed a *single* covariance structure in a microbial community, when modeling the conditional dependencies of the random variables that represent taxa abundances [3,7,8,14,15]. However, microbiome studies now routinely involve large cohorts and a wide range of (clinical and/or environmental) metadata variables. In this scenario, the microbiome samples in a study may be associated with *more than one* underlying covariance structure (and thus, more than one microbial network). This has motivated the recent development of computational methods to reconstruct *multiple* networks from a sample-taxa matrix [4,25,26]. In particular, we developed an approach where we treat the multiple network inference problem in a mixture model framework based on generative models [25,26], and we solve the following computational problem: given a sample-taxa abundace matrix generated by a mixture model with K component distributions, estimate the mixing coefficients and the parameters of the K component distributions.

In a mixture model, the component distributions model the observed taxa abundances. We have previously proposed two mixture model frameworks (MixGGM and MixMPLN), based on the type of abundance data in the sample-taxa matrix. In MixGGM [26], we use a Gaussian Graphical Model framework [13] to model the conditional dependencies; we apply a *log-ratio transformation* to the count or proportion data, and model the transformed data using a mixture of multivariate Gaussian distributions. In MixMPLN [25], we assume that the taxa counts are generated by a mixture of Multivariate Poisson Log-Normal (MPLN) distributions [1,12]. In both mixture models, the adjacency matrix representing a microbial network corresponds to the *precision* matrix (i.e. inverse covariance matrix) of the underlying component distribution. A zero entry in the precision matrix indicates conditional independence of the two taxa.

In practice, we do not have *a priori* knowledge of the value of K - the number of component distributions. The *model selection* problem for mixture modeling involves the determination of K. We recently proposed an algorithm to solve the model selection problem for the MixMPLN framework [33]. A solution for the general MixGGM framework has been proposed previously [5]. Both algorithms take a similar approach, where the mixture model is reformulated as a latent variable model, with only the mixing coefficients being treated as parameters, while all other variables, including the means and precision matrices of the component distributions, being treated as latent variables. An evidence lower bound (ELBO) function is defined using suitable factor distributions involving the latent variables, and a variational Expectation Maximization algorithm is used to estimate the parameters of these factor distributions in order to maximize the ELBO function with respect to the mixing coefficients (and thereby approximate the true marginal likelihood).

For a microbial network with d nodes (taxa), while there are $\frac{d(d-1)}{2}$ edge weights that need to be determined, the number of available samples n in a microbiome study is often not large enough, with the result that the system of equations to determine all pairs of taxa associations is under-determined. This issue is typically handled by assuming that the microbial network is sparse (i.e. the number of edges is $O(d)$; alternatively, the number of non-zero entries in the precision matrix is $O(d)$). The sparsity assumption also has an additional natural motivation: while a microbial community may have several hundreds of taxa, environmental barriers (both physical and chemical) can restrict microbial interactions, with the result that any given microbial taxon may only be associated with a small number of other constituent taxa. We have previously proposed algorithms for *sparse* precision matrix estimation for MixGGM [26] and MixMPLN [25] when the number of components, K, is known.

In this paper, we focus on the *model selection* problem for inferring *sparse* microbial networks for the MixGGM and MixMPLN formulations. We denote the model selection problems as MS_MixGGM and MS_MixMPLN, respectively. We provide several algorithms, based on our earlier variational approximation inference framework, that approximate the true marginal likelihood using *sparse* precision matrices for the underlying component distributions. We assess the performance of the proposed algorithms using multiple evaluation criteria, and on a large collection of simulated datasets for the scenario of sample sizes which are typical of microbiome studies. We also apply this framework to analyze a microbiome dataset from a previously published malaria study [32].

2 Methods

We first summarize the previously described variational approximation framework for model selection for MixGGM, and then describe the methods contribution of this paper (in Sect. 2.3) for MS_MixGGM to incorporate sparsity while estimating the precision matrices in the course of approximating the marginal likelihood. The solution for MS_MixMPLN follows a similar approach and is described later in this section.

Notation: Given a matrix X, we use $X_{:i}$ to denote its ith column, $X_{j:}$ to denote its jth row, and x_{ji} to denote its entry in row j and column i. We use n to denote the number of samples, d to denote the number of taxa, and K to denote the number of mixture components. Unless otherwise specified, all vectors are assumed to be column vectors.

2.1 Mixture of Multivariate Gaussian Distributions (MixGGM)

Let $Z_{d \times n}$ denote a Centered Log Ratio (CLR) transformed sample-taxa abundance matrix with d taxa and n samples that is generated by mixture of K multivariate Gaussian distributions and a mixing coefficient vector $\phi = (\phi_1, \phi_2, .., \phi_K)$ [26]. Then, the probability of the observed matrix Z, given the parameters of the mixture model, can be written as

$$p(Z|\phi, \mu_{(1)}, \Omega_{(1)}, .., \mu_{(K)}, \Omega_{(K)}) = \prod_{i=1}^{n} \sum_{l=1}^{K} \phi_l p(Z_{:i}|\mu_{(l)}, \Omega_{(l)}) = \prod_{i=1}^{n} \sum_{l=1}^{K} \phi_l \mathbb{N}_d(Z_{:i}|\mu_{(l)}, \Omega_{(l)}) \quad (1)$$

where $\sum_{l=1}^{K} \phi_l = 1$. The lth component distribution $p(Z_{:i}|\mu_{(l)}, \Omega_{(l)}) = \mathbb{N}_d(Z_{:i}|\mu_{(l)}, \Omega_{(l)})$, where $\mathbb{N}_d(Z_{:i}|\mu_{(l)}, \Omega_{(l)})$ denotes a d-dimensional multivariate Gaussian distribution with mean $\mu_{(l)}$ and precision matrix $\Omega_{(l)}$.

The Latent Variable Model for MS_MixGGM: The mixture model given in Eq. (1) can be reformulated as a latent variable model [2], in which we treat only the mixing coefficients ϕ_l's as parameters while all other variables, including $\mu_{(l)}$, and $\Omega_{(l)}$, where $1 \leq l \leq K$, are treated as latent variables. Let $\Theta = \mathcal{M} \cup \mathcal{T} \cup \mathcal{S}$ denote the set of all latent variables in our model, where $\mathcal{M} = \{\mu_{(l)}|1 \leq l \leq K\}$, $\mathcal{T} = \{\Omega_{(l)}|1 \leq l \leq K\}$, and $\mathcal{S} = \{S_i|1 \leq i \leq n\}$. The set $\mathcal{S}$ denotes component membership information for samples, where $S_i = (s_{i1}, ..., s_{iK})^T$ is a K-dimensional binary vector, also called a 1-of-K binary vector, that is associated with sample $Z_{:i}$. This vector has the property that if $Z_{:i}$ was generated by component r then $s_{ir} = 1$, and that $s_{il} = 0$, for all $l \neq r$.

We now describe the different parts of the generative model. Each S_i is drawn from a multinomial distribution; that is, $S_i \sim$ Multinomial$(1, \phi)$. We have that

$$p(\mathcal{S}|\phi) = \prod_{i=1}^{n} p(S_i|\phi) = \prod_{i=1}^{n} \prod_{l=1}^{K} \phi_l^{s_{il}}$$

Conditional on $\mathcal{S}$, each sample is assumed to be independently drawn from a multivariate Gaussian distribution with parameters $\mu_{(l)}$ and $\Omega_{(l)}$. Thus,

$$p(Z|\mathcal{M}, \mathcal{T}, \mathcal{S}) = \prod_{i=1}^{n} \prod_{l=1}^{K} p(Z_{:i}|\mu_{(l)}, \Omega_{(l)})^{\ s_{il}}$$

Marginalizing the function $p(Z|\mathcal{M}, \mathcal{T}, \mathcal{S}) \times p(\mathcal{S}|\phi)$ over $\mathcal{S}$ results in Eq. (1). We also introduce conjugate priors over each $\mu_{(l)}$ and $\Omega_{(l)}$. We assume that

$\mu_{(l)} \sim \mathbb{N}_d(0, \beta I)$ where I is the $d \times d$ identity matrix, and β is a fixed parameter. We also assume that $\Omega_l \sim \mathbb{W}(\nu, V)$, where $\mathbb{W}(\nu, V)$ is the Wishart distribution [30] with fixed degrees of distribution ν and fixed scale matrix V. The density function for the Wishart distribution is given as

$$\mathbb{W}(\Omega|\nu, V) = \frac{|V|^{\nu/2}\ |\Omega|^{(\nu-d-1)/2}}{2^{\nu d/2}\ \pi^{d(d-1)/4}\ \prod_{j=1}^{d} \Gamma\left(\frac{\nu+1-j}{2}\right)}\ \exp\left(-\frac{1}{2}\mathrm{Tr}(V\Omega)\right)$$

where Tr(.) and $\Gamma(.)$ denote the matrix trace function and Gamma function respectively.

Finally, we set $p(\mathcal{M}) = \prod_{l=1}^{K} p(\mu_{(l)})$ and $p(\mathcal{T}) = \prod_{l=1}^{K} p(\Omega_{(l)})$. Taken together, these specifications allow us to describe the joint distribution of Z and all latent variables, conditioned on the mixing coefficients, as

$$\begin{aligned} p(Z, \Theta \mid \phi) &= p(Z, \mathcal{M}, \mathcal{T}, \mathcal{S} \mid \phi) \\ &= p(Z|\ \mathcal{M}, \mathcal{T}, \mathcal{S}) \times p(\mathcal{S} \mid \phi) \times p(\mathcal{M}) \times p(\mathcal{T}) \end{aligned} \tag{2}$$

The Evidence Lower Bound Function: The marginal likelihood function $p(Z|\phi)$ can be obtained by integrating Eq. (2) over all latent variables in Θ. However, since this approach is not analytically tractable, we use instead a variational approximation method involving a lower bound on the marginal log-likelihood function. This lower bound, called the Evidence Lower Bound (ELBO) function [2,27], will then be maximized with respect to the mixing coefficients. The ELBO function $Q(\Theta)$ is defined as

$$ELBO(Q) = \int Q(\Theta)\ \log\left[\frac{p(Z, \Theta|\phi)}{Q(\Theta)}\right]\ \mathrm{d}\Theta = \ \langle \log p(Z, \Theta|\phi)\rangle_\Theta\ -\ \langle \log Q(\Theta)\rangle_\Theta \tag{3}$$

where $\langle . \rangle_\Theta$ denotes the expectation over the distribution $Q(\Theta)$. For any function $Q(\Theta)$, the following identity holds [2]:

$$ELBO(Q) \leq \log\left(\int p(Z, \Theta|\phi)\ \mathrm{d}\Theta\right) = \log p(Z|\phi)$$

Our goal is to maximize $ELBO(Q)$ using some choice of $Q(\Theta)$. The difference between $ELBO(Q)$ and $\log p(Z|\phi)$ can be shown to be the Kullback-Leibler distance (KL) between $Q(\Theta)$ and the posterior distribution $p(\Theta|Z, \phi)$. Thus, $ELBO(Q)$ is maximum when $Q(\Theta)$ is equal to the posterior [2]. Let $\Theta = \mathcal{M} \cup \mathcal{T} \cup \mathcal{S} = \{\theta_i\}$. We assume that $Q(\Theta) = \prod_t q(\theta_t)$, that is, $Q(\Theta)$ is the product of independent factor distributions $q(\theta_t)$. With this assumption [21], the form of the optimized factor distributions that minimize the KL distance can be computed [2]. For each t, the optimized distribution $q(\theta_t)$ can be shown to have the form

$$q(\theta_t) = \frac{\exp(\langle \log p(Z, \Theta|\phi)\rangle_{\theta_v \neq \theta_t})}{\int \exp(\langle \log p(Z, \Theta|\phi)\rangle_{\theta_v \neq \theta_t})\ \mathrm{d}\theta_t} \tag{4}$$

The optimized factor distributions $q(.)$ for the latent variables S_i, $\mu_{(l)}$, and $\Omega_{(l)}$ have the same functional forms as their respective priors $p(S_i|\phi)$, $p(\mu_{(l)})$, and $p(\Omega_{(l)})$. Specifically,
$q(S_i)$ = Multinomial$(S_i|1, \alpha_i)$, with parameter $\alpha_i = (\alpha_{i1}, \alpha_{i2}, .., \alpha_{iK})$ where $\sum_{l=1}^{K} \alpha_{il} = 1$,
$q(\mu_{(l)}) = \mathbb{N}_d(\mu_{(l)}|m_{(l)}, T_{(l)})$, a multivariate Gaussian with d-dimensional mean vector $m_{(l)}$ and $d \times d$ precision matrix $T_{(l)}$,
$q(\Omega_{(l)}) = \mathbb{W}(\Omega_{(l)}|\eta_{(l)}, C_{(l)})$, a Wishart distribution with degrees of freedom $\eta_{(l)}$ and $d \times d$ scale matrix $C_{(l)}$.

Parameter Updates: Equations (3) and (4) can be used to derive update equations for the variational parameters (i.e. parameters of the factor distributions). These update equations are linked, in the sense that the update equation for a variational parameter is a function of other variational parameters. With estimates for the variational parameters, we can compute the ELBO function using an expansion of Eq. (3) as

$$\begin{aligned} ELBO(Q) =& \langle \log p(Z|\ \mathcal{S}, \mathcal{M}, \mathcal{T}) \rangle + \langle \log p(\mathcal{S} \mid \phi) \rangle + \langle \log p(\mathcal{M}) \rangle + \langle \log p(\mathcal{T}) \rangle \\ &- \langle \log q(\mathcal{S}) \rangle - \langle \log q(\mathcal{M}) \rangle - \langle \log q(\mathcal{T}) \rangle \end{aligned}$$

The formulas for calculating the variational parameters and $ELBO(Q)$ are given in Appendix A. Since the ELBO function approximates the true marginal log-likelihood function $\log p(X|\phi)$, after we have cycled through and estimated the variational parameters, we can then maximize the resulting ELBO with respect to the mixing coefficients. It can be shown that $\phi_l = \frac{1}{n} \sum_{i=1}^{n} \alpha_{il}$, for $1 \leq l \leq K$ [2].

2.2 Variational Expectation Maximization (VEM) Algorithm for MS_MixGGM

Input: Matrix $Z_{d \times n}$, number of components K, the prior parameters β, ν, and V.
Output: Values of the mixing coefficients and the variational parameters that maximize the ELBO function, and the maximum ELBO function value.
Initialization: Initialize the mixing coefficient vector $\phi = (\phi_1, \phi_2, .., \phi_K)$ and the variational parameters α_i, $m_{(l)}$, $T_{(l)}$, $\eta_{(l)}$, and $C_{(l)}$, for $1 \leq i \leq n$, $1 \leq l \leq K$.
Repeat until convergence (that is, the ELBO function does not increase any further):
E-step:
Cycle through the variational parameters and update their estimates.
M-step:
Set $\phi_l = \frac{1}{n} \sum_{i=1}^{n} \alpha_{il}$, for $1 \leq l \leq K$.

2.3 Approximating the Marginal Likelihood Using Sparse Precision Matrices

We noted that the factor distribution $q(\Omega_{(l)})$ can take the form of a Wishart distribution $\mathbb{W}(\Omega_{(l)}|\eta_{(l)}, C_{(l)})$. From the update formula in Appendix A, we observe that $\langle\Omega_{(l)}\rangle = \eta_{(l)}C_{(l)}^{-1}$. While maximizing the ELBO function, if we can constrain the inverse scale matrix $C_{(l)}^{-1}$ to be a sparse matrix, then $\langle\Omega_{(l)}\rangle$ will also be sparse. We describe how this can be achieved.

Recall that $\Omega_{(l)} \in \{\theta_v\}$. The part of the $ELBO(Q)$ function in Eq. 3 that involves only terms associated with $q(\Omega_{(l)})$ and Ω_l (and ignoring constant terms) is

$$\int q(\Omega_{(l)})\ \langle \log p(Z, \Theta|\phi)\rangle_{\theta_v \neq \Omega_{(l)}}\ \mathrm{d}\Omega_{(l)} - \int q(\Omega_{(l)})\ \log q(\Omega_{(l)})\ \mathrm{d}\Omega_{(l)}$$

The above integral can be simplified (see Appendix C) and written as a function $elbo(\eta_{(l)}, C_{(l)}^{-1})$, where

$$\begin{aligned} elbo(\eta_{(l)}, C_{(l)}^{-1}) = &\left[\frac{\sum_{i=1}^{n}\langle s_{il}\rangle + \nu}{2}\right] \log|C_{(l)}^{-1}| - \frac{1}{2}\mathrm{Tr}\left(\eta_{(l)}C_{(l)}^{-1}M\right) + \frac{\eta_{(l)}d}{2} \\ &+ \left[\frac{\sum_{i=1}^{n}\langle s_{il}\rangle + \nu - \eta_{(l)}}{2}\right]\left[\sum_{j=1}^{d}\psi\left(\frac{\eta_{(l)}+1-j}{2}\right)\right] \\ &+ \sum_{j=1}^{d}\log\Gamma\left(\frac{\eta_{(l)}+1-j}{2}\right) \end{aligned} \tag{5}$$

In the above equation

$$M = V + \sum_{i=1}^{n} Z_{:i}Z_{:i}^T\langle s_{il}\rangle - \sum_{i=1}^{n} Z_{:i}\langle s_{il}\rangle\langle\mu_{(l)}\rangle^T - \langle\mu_{(l)}\rangle\sum_{i=1}^{n} Z_{:i}^T\langle s_{il}\rangle + \langle\mu_{(l)}\mu_{(l)}^T\rangle\sum_{i=1}^{n}\langle s_{il}\rangle$$

Our goal is to maximize the function $elbo(\eta_{(l)}, C_{(l)}^{-1})$ while constraining $C_{(l)}^{-1}$ to be sparse. We propose several methods to do this:

1. We can obtain a sparse solution for $C_{(l)}^{-1}$ by penalizing the $elbo(.)$ function by subtracting the l_1-norm of $C_{(l)}^{-1}$. This leads to a goal of maximizing the quantity $elbo(\eta_{(l)}, C_{(l)}^{-1}) - \rho_l \left\|C_{(l)}^{-1}\right\|_1$, where ρ_l is a tuning parameter and $\|.\|_1$ denotes the l_1-norm. This can be accomplished by iteratively updating the values for $\eta_{(l)}$ and $C_{(l)}^{-1}$ using a gradient search. Values for η_l can be obtained by setting the derivative of $elbo(.)$ with respect to η_l, to a 0, and then finding the root. The general problem formulation for maximizing with respect to $C_{(l)}^{-1}$ can be written as below, where $\Sigma_{(l)}$ is the empirical covariance matrix

$$\underset{C_{(l)}^{-1}}{argmax}\left\{log(det(C_{(l)}^{-1})) - Tr(\Sigma_{(l)}C_{(l)}^{-1}) - \rho_l\left\|C_{(l)}^{-1}\right\|_1\right\}$$

We consider two approaches to solve this optimization problem, namely, the Graphical Lasso method (GLASSO) [9] and a tuning-insensitive method for estimating Gaussian graphical models (TIGER) [16]. The tuning parameter ρ_l is typically estimated by picking the best value (from a set of tuning values) using a cross-validation (CV) approach [25] or using a model selection criterion like Extended Bayesian Information Criteria (EBIC) [6,35]. We consider both approaches here. This gives rise to four different methods for estimating a sparse $C_{(l)}^{-1}$, namely, *Glasso_EBIC, Glasso_CV, Tiger_EBIC,* and *Tiger_CV*. In our implementation, for the EBIC methods, we select the best ρ_l based on the minimum EBIC score [35]; for the CV methods, this selection is made using three-fold stratified cross-validation [22].

2. In contrast to the above methods where the zeros in $C_{(l)}^{-1}$ are obtained through optimization, an approach proposed by [29] involves determining zeros using Fisher Z-transformed confidence intervals. In our problem setting, this works as follows: First, determine $C_{(l)}^{-1}$ from the update formula for $C_{(l)}$ (Appendix A). Then, compute a partial correlation matrix Υ from $C_{(l)}^{-1}$, where $\upsilon_{ij} = \frac{-c_{ij}}{\sqrt{c_{ii}c_{jj}}}$; here, υ_{ij} and c_{ij} denote the ij^{th} entries in Υ and $C_{(l)}^{-1}$, respectively. Subsequently, the υ_{ij}'s are Fisher Z-transformed (by computing $\frac{1}{2}\log\left(\frac{1+\upsilon_{ij}}{1-\upsilon_{ij}}\right)$) and the resulting approximate normal distribution of the Z-transformed values is used to determine a confidence interval for each υ_{ij}; this confidence interval can be calibrated to trade off between false positives and negatives. In the final step, entry c_{ij} is set to zero if zero is contained in the confidence interval for υ_{ij}. We refer to this method of identifying zeros in $C_{(l)}^{-1}$ as *Pcor*. For the *Pcor* method, the update formula for $\eta_{(l)}$ is the same as given in Appendix A.

The VEM algorithm described in Sect. 2.2 is referred to as *NoSparse* as it does not enforce any sparsity constraints. The methods *Pcor, Glasso_EBIC, Glasso_CV, Tiger_EBIC,* and *Tiger_CV* are modifications of the VEM algorithm, where the update step for the variational parameters $C_{(l)}^{-1}$ and $\eta_{(l)}$ is done using one of the corresponding five methods.

2.4 Mixture of Multivariate Poisson Log-Normal Distributions (MixMPLN)

The Multivariate Poisson Log-Normal (MPLN) distribution [1] can be used to model count data. This distribution has two layers, with the observed sample count vector $A = (a_1, ..., a_d)^T$ being generated by a mixture of independent Poisson distributions whose means are latent (or hidden), and such that the logarithm of the Poisson means follows a multivariate Gaussian distribution. We use $\lambda = (\lambda_1, \lambda_2, ..\lambda_d)^T$ to denote the latent variable vector representing the logarithm of the Poisson means that is associated with the sample A. The probability density function $p(A|\mu, \Omega)$ of the MPLN distribution with parameters μ and Ω

is

$$p(A|\mu, \Omega) = \int_{\mathbb{R}^d} p(A, \lambda|\mu, \Omega) \ \mathrm{d}\lambda \tag{6}$$

where,

$$p(A, \lambda|\mu, \Omega) = \left[\prod_{j=1}^{d} \frac{e^{-e^{\lambda_j}} e^{\lambda_j a_j}}{a_j!}\right] (2\pi)^{-d/2} |\Omega|^{1/2} e^{-\frac{1}{2}[\lambda-\mu]^T \Omega [\lambda-\mu]}$$

with μ the d-dimensional mean vector, $\Omega_{d\times d}$ the precision matrix, and $|\Omega|$ denoting the determinant of Ω. No simplification of the integral in Eq. (6) is known.

Let $X_{d\times n}$ denote a sample-taxa count matrix with d taxa and n samples that is generated by mixture model consisting of K component MPLN distributions and a mixing coefficient vector $\phi = (\phi_1, \phi_2, .., \phi_K)$. Also, let $\Lambda_{(l)}$ denote the $d \times n$ matrix of latent variable vectors of the n samples that is associated with the lth component. That is, column vector $\Lambda_{(l):i}$ is associated with sample $X_{:i}$. We also use λ_{lji} to denote the jth entry in column vector $\Lambda_{(l):i}$. Then, the probability of the observed sample-taxa count matrix X, given the parameters of the mixture model, can be written as

$$\begin{aligned} p(X|\phi, \mu_{(1)}, \Omega_{(1)}, .., \mu_{(K)}, \Omega_{(K)}) &= \prod_{i=1}^{n} \sum_{l=1}^{K} \phi_l p(X_{:i}|\mu_{(l)}, \Omega_{(l)}) \\ &= \prod_{i=1}^{n} \sum_{l=1}^{K} \phi_l \left[\int p(X_{:i}, \Lambda_{(l):i} \mid \mu_{(l)}, \Omega_{(l)}) \ \mathrm{d}\Lambda_{(l):i} \right] \end{aligned}$$

where

$$p(X_{:i}, \Lambda_{(l):i}|\Theta_{(l)}) = \left[\prod_{j=1}^{d} \frac{e^{-e^{\lambda_{lji}}} e^{\lambda_{lji} x_{ji}}}{x_{ji}!}\right] (2\pi)^{-d/2} |\Omega_{(l)}|^{1/2} e^{-\frac{1}{2}[\Lambda_{(l):i} - \mu_{(l)}]^T \Omega_{(l)} [\Lambda_{(l):i} - \mu_{(l)}]}$$

The Latent Variable Model and the ELBO Function For MS_MixMPLN: Briefly, the latent model for MS_mixMPLN [33] is an extension of the latent model for MS_MixGGM, with $\Theta = \mathcal{L} \cup \mathcal{M} \cup \mathcal{T} \cup \mathcal{S}$ denoting the set of all latent variables. Here, $\mathcal{L} = \{\Lambda_{(l)}|1 \leq l \leq K\}$, $\mathcal{M} = \{\mu_{(l)}|1 \leq l \leq K\}$, $\mathcal{T} = \{\Omega_{(l)}|1 \leq l \leq K\}$, and the component membership set $\mathcal{S} = \{S_i|1 \leq i \leq n\}$. Similar to the generative model for MixGGM, we assume that each S_i is drawn from a multinomial distribution. We have that

$$p(X, \mathcal{L}|\mathcal{M}, \mathcal{T}, \mathcal{S}) = \prod_{i=1}^{n} \prod_{l=1}^{K} p(X_{:i}, \Lambda_{(l):i}|\mu_{(l)}, \Omega_{(l)})^{\ s_{il}}$$

Using the same definition for $p(\mathcal{S}|\phi)$ as we did for MixGGM, along with the same conjugate priors for the $\mu_{(l)}$'s and for the $\Omega_{(l)}$'s, the joint distribution of X and all latent variables, conditioned on the mixing coefficients, is

$$\begin{aligned} p(X, \Theta \mid \phi) &= p(X, \mathcal{L}, \mathcal{M}, \mathcal{T}, \mathcal{S} \mid \phi) \\ &= p(X, \mathcal{L} \mid \mathcal{M}, \mathcal{T}, \mathcal{S}) \times p(\mathcal{S} \mid \phi) \times p(\mathcal{M}) \times p(\mathcal{T}) \end{aligned} \tag{7}$$

The $ELBO(Q)$ function for MS_MixMPLN has the form

$$ELBO(Q) = \langle \log p(X, \mathcal{L} \mid \mathcal{S}, \mathcal{M}, \mathcal{T}) \rangle + \langle \log p(\mathcal{S} \mid \phi) \rangle + \langle \log p(\mathcal{M}) \rangle + \langle \log p(\mathcal{T}) \rangle - \langle \log q(\mathcal{S}) \rangle - \langle \log q(\mathcal{M}) \rangle - \langle \log q(\mathcal{T}) \rangle - \langle \log q(\mathcal{L}) \rangle \tag{8}$$

The factor distributions $q(.)$ for the latent variables in $\mathcal{M} \cup \mathcal{T} \cup \mathcal{S}$ have the same form as for MixGGM. For the $\Lambda_{(l):i}$ latent variable vectors, however, the form of optimized $q(\Lambda_{(l):i})$ is quite unwieldy to work with; thus, instead, we define each $q(\Lambda_{(l):i})$ to be a multivariate Gaussian distribution with a *diagonal* precision matrix. Specifically,

$$q(\Lambda_{(l):i}) = \mathbb{N}_d(\Lambda_{(l):i} | \delta_{(l)i}, D_{(l)i}) = \prod_{j=1}^{d} q(\lambda_{lji}) = \prod_{j=1}^{d} \mathbb{N}\left(\lambda_{lji} | a_{lji}, \frac{1}{b_{lji}}\right)$$

with d-dimensional mean vector $\delta_{(l)i}$ and $d \times d$ precision matrix $D_{(l)i}$. Since $D_{(l)i}$ is a diagonal matrix, the multivariate distribution $q(\Lambda_{(l):i})$ can be written as a product of d independent *univariate* Gaussian distributions $q(\lambda_{lji})$, $1 \leq j \leq d$, where $\delta_{(l)i} = (a_{l1i}, a_{l2i}, .., a_{ldi})^T$ and $\text{diag}(D_{(l)i}) = \left(\frac{1}{b_{l1i}}, \frac{1}{b_{l2i}}, .., \frac{1}{b_{ldi}}\right)$. That is, a_{lji} and b_{lji} denote the mean and variance respectively, of the random variable λ_{lji}, and these are estimated using the Newton-Raphson method [33]. The update formulas for the parameters of the factor distributions for MixMPLN are given in Appendix B. The VEM algorithm for MixMPLN model selection is similar to the one in Sect. 2.2, with the only addition being the inclusion of the updates for the variational parameters $\delta_{(l)i}$ and $D_{(l)i}$, for $1 \leq i \leq n$, $1 \leq l \leq K$.

2.5 Approximating the Marginal Likelihood Using Sparse Precision Matrices

The approach of maximizing the ELBO function while constraining the inverse scale matrix $C_{(l)}^{-1}$ to be a sparse matrix, can be adopted here as well. The equation for $elbo(\eta_{(l)}, \Omega_{(l)})$ has a similar form to Eq. (5), except that the quantity M is dependent on the estimate for $\Lambda_{(l)}$. Specifically,

$$M = V + \sum_{i=1}^{n} \langle \Lambda_{(l):i} \Lambda_{(l):i}^T \rangle \langle s_{il} \rangle - \sum_{i=1}^{n} \langle \Lambda_{(l):i} \rangle \langle s_{il} \rangle \langle \mu_{(l)} \rangle^T - \langle \mu_{(l)} \rangle \sum_{i=1}^{n} \langle \Lambda_{(l):i} \rangle^T \langle s_{il} \rangle + \langle \mu_{(l)} \mu_{(l)}^T \rangle \sum_{i=1}^{n} \langle s_{il} \rangle$$

Thus, we evaluate six methods for MixMPLN as well. The methods are *NoSparse, Pcor, Glasso_EBIC, Glasso_CV, Tiger_EBIC,* and *Tiger_CV*.

2.6 Datasets and Benchmarking

The performances of the different model selection algorithms for MS_MixGGM and MS_MixMPLN were evaluated using a collection of synthetic sample-taxa abundance matrices. For each $d \times n$ sample-taxa matrix, the samples were generated using a mixture model with a mixing coefficient vector ϕ and consisting of K component distributions (for MS_MixGGM, each component distribution was a multivariate Gaussian distribution; for MS_MixMPLN, each component distribution was an MPLN distribution). The precision matrices of the component distributions were generated from an underlying graph structure. Five different types of graph structures were considered: *band, cluster, hub, random,* and *scale-free.* The R *huge* package [34] was used to generate the precision matrices associated with each graph structure. Sample-taxa matrices were generated with number of taxa $d = 80$, number of components $K = 2,\ 3,\ 4$, and number of samples $n = sK$, where s is the number of samples per component ($s = d,\ 2d$, and $4d$). The mixing coefficient vectors (ground-truth) for $K = 2,\ 3$, and 4 were $\phi = \left(\frac{1}{2}, \frac{1}{2}\right)$, $\phi = \left(\frac{1}{3}, \frac{1}{3}, \frac{1}{3}\right)$, and $\phi = \left(\frac{1}{4}, \frac{1}{4}, \frac{1}{4}, \frac{1}{4}\right)$ respectively. For each graph type and combination of parameter values, twenty replicates were generated. Thus, a set of 900 sample-taxa matrices ($DS1$ dataset) was generated for evaluating the MS_MixGGM algorithms, and a separate set of 900 sample-taxa matrices ($DS2$ dataset) was generated for evaluating the MS_MixMPLN algorithms.

Evaluation Criteria: The *predicted* number of components for a sample-taxa matrix was determined from the estimated mixing coefficient vector after applying a threshold ($\tau = 0.01$) to the mixing coefficient values; that is, any component with mixing coefficient value $< \tau$ was not counted towards the predicted number of components. Using this approach, we calculated, for each algorithm, the proportion of times the predicted number of components was *equal to*, *greater than*, and *less than*, K, where K is the ground-truth number of components. This approach allowed us to assess each algorithm's *accuracy* (i.e., the proportion of times the predicted value *is equal* to the ground truth value) along with its underclustering and overclustering tendencies on each of the graph types.

Samples were assigned component membership based on their final α_{il} values, with the ith sample being assigned to that component with the maximum α_{il} value. The resulting *predicted* sample partition was compared with the ground truth sample partition (used to generate the sample-taxa matrix) using the Rand Index measure [24]. This measure is calculated as $\frac{a+b}{\binom{n}{2}}$, where a denotes the number sample pairs that are in the same cluster in both partitions, b denotes the number of sample pairs that are in different clusters in both partitions, and n denotes the total number of samples; the Rand Index has a value between 0 (minimum similarity) and 1 (maximum similarity). Finally, the distance between the ground truth and predicted mixing coefficient vectors was assessed using the Hellinger distance measure [20]. This measure calculates the distance between two discrete probability distributions $P = (p_1, p_2, .., p_m)$ and $Q = (q_1, q_2, .., q_m)$, and is equal to $\sqrt{\frac{1}{2}\sum_{i=1}^{m}(\sqrt{p_i} - \sqrt{q_i})^2}$. This measure has a value between 0 (minimum distance) and 1 (maximum distance). Since the predicted and ground truth values for the number of components may not be the same, we pad the shorter length mixing coefficient vector with 0's to make both vectors have the same length. Subsequently, these vectors are sorted, and the Hellinger distance is computed between the two sorted vectors.

Real Dataset: We also applied our model selection framework to a sample-taxa matrix generated by a microbiome study that explored connections between gut microbiome composition and the risk of *Plasmodium falciparum* infection [32]. In this study, stool samples collected from a cohort were assayed by sequencing the 16S ribosomal RNA gene to determine the bacterial communities they contained. This generated a sample-taxa count matrix with 195 samples and 221 bacterial genera.

3 Results

The VEM algorithms (*NoSparse, Pcor, Glasso_EBIC, Glasso_CV, Tiger_EBIC,* and *Tiger_CV*) for MS_MixGGM and MS_MixMPLN were implemented using the R programming language [23]. Our implementations are available at https://github.com/syooseph/YoosephLab/tree/master/MixtureMicrobialNetworks/SparseModelSelection.

The MS_MixGGM and MS_MixMPLN methods were evaluated using datasets $DS1$ and $DS2$, respectively. For each input sample-taxa matrix, each method was run with a larger value for the number of components (5, 6, and 7 respectively for ground-truth $K = 2$, 3, and 4). The values for the priors were as follows: $\beta = 10^{-6}$, $\nu = d$, and V set to a diagonal matrix with all entries equal to $d+1$. Each method was run using 26 different starting points on an input, where 25 starting points were generated using random partitions of the samples in the sample-taxa matrix and one starting point was generated using the K-means algorithm [19] to partition the samples. Each partition was used to initialize the estimates for the mixing coefficients and the variational parameters. For each starting point, the algorithm was allowed to run for 25 iterations (or less, if it reached convergence). The output with the maximum ELBO value was selected. For each evaluation criterion, we report the mean value for the twenty replicates that we generated for each combination of graph type together with the K and n parameter values.

For MS_MixGGM, the *Pcor* method has by far the best performance on all graph types, with *perfect* accuracy (Fig. 1) and also the best sample clustering quality (Rand Index values equal to 1) and best mixing coefficient vector estimation (Hellinger distances equal to 0) (Appendix D: Figs. 3 and 4), for all values of d, K, and n. Additionally, this method is also the fastest (Appendix D: Fig. 5). Among the other methods, *Tiger_CV* and *Tiger_EBIC* have the best performance in terms of sample clustering and mixing coefficient vector estimation. Both of these methods tend to undercluster the samples (i.e., predicted number of components is greater than the ground truth value), while *Glass_CV* and *Glasso_EBIC* tend to overcluster. The *NoSparse* method has the next best

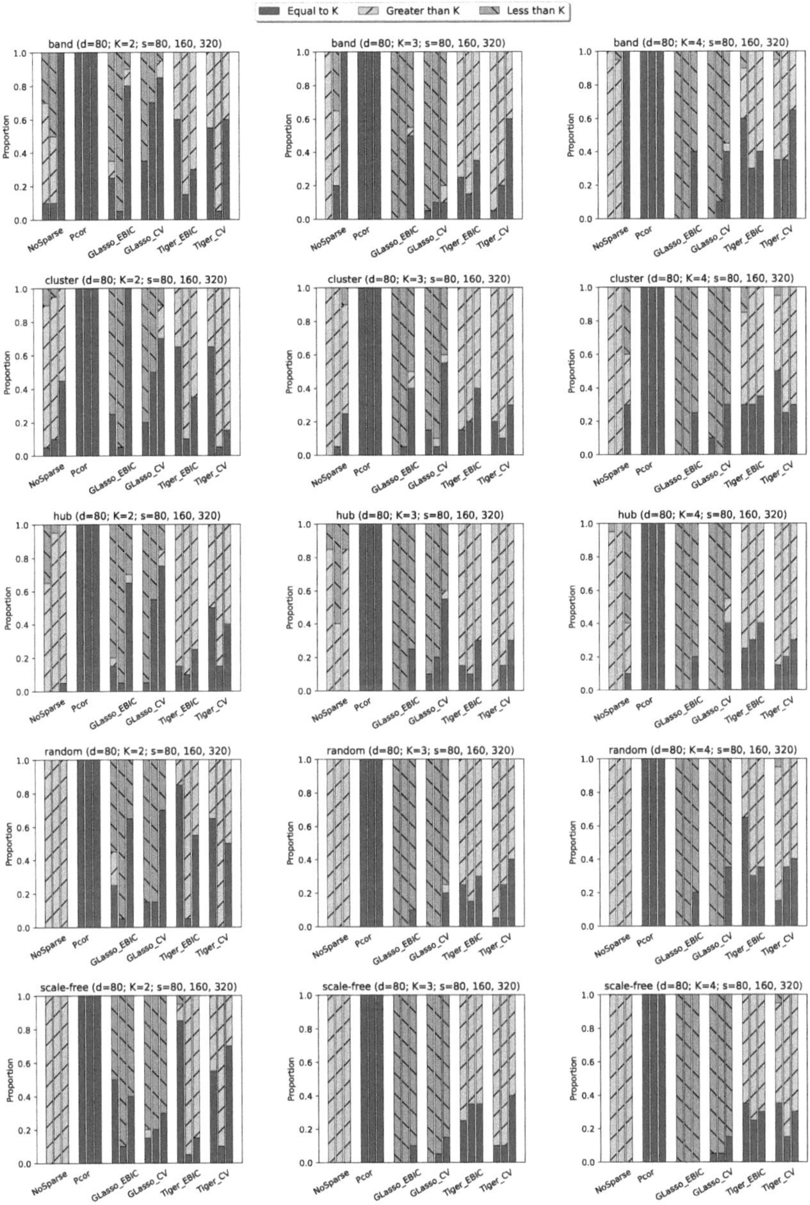

Fig. 1. Performance of the six MS_MixGGM methods (*NoSparse, Pcor, Glasso_EBIC, Glasso_CV, Tiger_EBIC,* and *Tiger_CV*) on the five graph types (*band, cluster, hub, random,* and *scale-free*). Here, the number of taxa $d = 80$, the number of components $K = 2, 3, 4$, and the number of samples per component $s = 80, 160, 320$. For each parameter combination, the proportions of predictions with value $= K, > K$, and $< K$, respectively, are shown, where K is the ground truth value for the number of components.

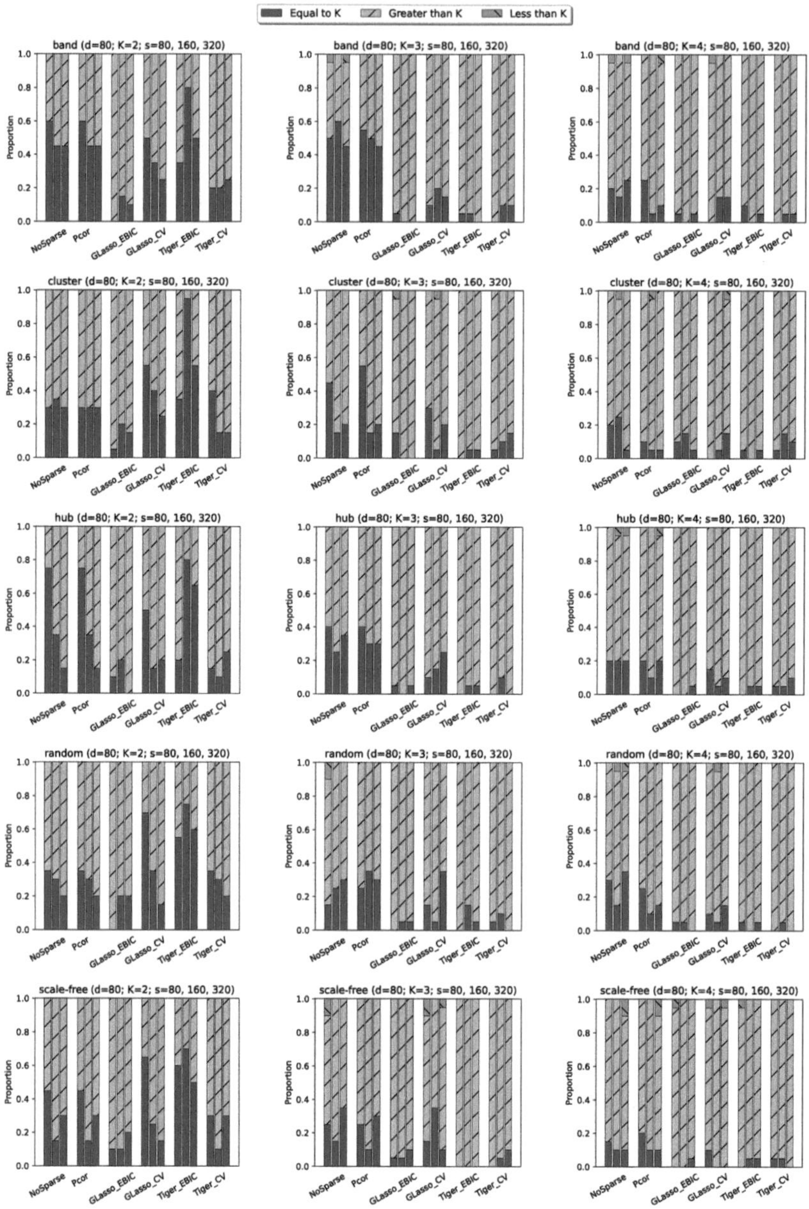

Fig. 2. Performance of the six MS_MixMPLN methods (*NoSparse, Pcor, Glasso_EBIC, Glasso_CV, Tiger_EBIC,* and *Tiger_CV*) on the five graph types (*band, cluster, hub, random,* and *scale-free*). Here, the number of taxa $d = 80$, the number of components $K = 2, 3, 4$, and the number of samples per component $s = 80, 160, 320$. For each parameter combination, the proportions of predictions with value $= K, > K$, and $<K$, respectively, are shown, where K is the ground truth value for the number of components.

performance in terms of Rand Index and Hellinger distance but with a lower accuracy compared to the *Glasso* methods. Not surprisingly, cross-validation based methods have the highest runtimes. For each method, its performance is similar across the different graph types.

For MS_MixMPLN, however, the methods show lower accuracy (Fig. 2), and none of the methods have perfect accuracy. Each method's accuracy decreases as the number of components increases. Accuracy also varies by graph type, with accuracy on the band and hub graph types being slightly higher than on cluster, random, and scale-free. The better performing methods, in terms of accuracy, are *NoSparse*, *Pcor*, and *Tiger_EBIC*. Overall, every method tends to undercluster the samples. The *NoSparse* and *Pcor* methods have the best sample clustering quality and mixing coefficient vector estimation, as assessed using the Rand Index and Hellinger distance, respectively (Appendix D: Figs. 6, 7). These two methods are also the fastest (Appendix D: Fig. 8).

We used the MS_MixGGM *Pcor* method to evaluate the CLR transformed sample-taxa matrix from the malaria study [32]. This method was run with number of components set to 4. The resulting mixing coefficient vector was $(0.72, 0, 0, 0.28)$, suggesting the presence of two underlying components. We assigned component membership to the samples based on the final α_{il} values. The samples in the resulting two clusters (of sizes 140 and 55) had very different age profiles (mean age of 10 yrs vs 0.9 yrs) of the individuals (samples) in the cohort. The analysis here supports previous findings on the presence of multiple underlying microbial networks associated with these data [25].

4 Discussion

Here we presented a variational EM framework and several algorithms for sparse precision matrix estimation for the MS_MixGGM and MS_MixMPLN computational problems. The proposed algorithms were evaluated using a large collection of simulated datasets, with sample-taxa matrices containing sample sizes that are typical of microbiome studies. For MS_MixGGM, the *Pcor* method has the best performance in every category (accuracy, sample clustering, mixing coefficient vector estimation, and runtime), while for MS_MixMPLN, both *Pcor* and *NoSparse* generally had the best performance in terms of sample clustering quality, mixing coefficient vector estimation, and runtime. All methods for MS_MPLN tend to predict number of components greater than the ground truth value (and thus, they tend to undercluster samples).

Author Disclosure Statement. The author declares they have no competing financial interests.

Funding Information. This material is based upon work supported by the National Science Foundation (NSF) under grant number DBI-2400009. Publication costs for this work were funded by NSF grant number DBI-2400009.

Appendix

A. VEM updates for MixGGM

The update formulas for the variational parameters $\alpha_i, m_{(l)}, T_{(l)}, \eta_{(l)}$, and $C_{(l)}$, along with the formulas for calculating the constituents of the ELBO function are given below. *Notation:* Tr(.) and $\psi(.)$ denote the matrix trace function and the di-gamma function respectively; $\langle F(\theta_t)\rangle$ denotes the expectation of function $F(\theta_t)$ over the factor distribution $q(\theta_t)$.

Parameter α_i for $q(S_i)$:

$$\alpha_{il} = \frac{f_{il}}{\sum_{r=1}^{K} f_{ir}}, \text{ where}$$

$$f_{il} = \exp\left\{ \log\phi_l + \frac{1}{2}\langle \log|\Omega_{(l)}|\rangle - \frac{1}{2}\mathrm{Tr}\left(\langle \Omega_{(l)}\rangle \left[Z_{:i}Z_{:i}^T - \langle\mu_{(l)}\rangle Z_{:i}^T - Z_{:i}\langle\mu_{(l)}\rangle^T + \langle\mu_{(l)}\mu_{(l)}^T\rangle\right]\right)\right\}$$

Parameters $m_{(l)}$ and $T_{(l)}$ for $q(\mu_{(l)})$:

$$T_{(l)} = \beta I + \langle\Omega_{(l)}\rangle \sum_{i=1}^{n}\langle s_{il}\rangle, \qquad m_{(l)} = T_{(l)}^{-1}\,\langle\Omega_{(l)}\rangle \sum_{i=1}^{n} Z_{:i}\langle s_{il}\rangle$$

Parameters $\eta_{(l)}$ and $C_{(l)}$ for $q(\Omega_{(l)})$:

$$\eta_{(l)} = \nu + \sum_{i=1}^{n}\langle s_{il}\rangle$$

$$C_{(l)} = V + \sum_{i=1}^{n} Z_{:i}Z_{:i}^T\langle s_{il}\rangle - \sum_{i=1}^{n} Z_{:i}\langle s_{il}\rangle\langle\mu_{(l)}\rangle^T - \langle\mu_{(l)}\rangle\sum_{i=1}^{n} Z_{:i}^T\langle s_{il}\rangle + \langle\mu_{(l)}\mu_{(l)}^T\rangle\sum_{i=1}^{n}\langle s_{il}\rangle$$

Expected Values Used in the Above Updates:

$$\langle s_{il}\rangle = \alpha_{il} \qquad \langle\mu_{(l)}\rangle = m_{(l)}$$

$$\langle\mu_{(l)}\mu_{(l)}^T\rangle = T_{(l)}^{-1} + m_{(l)}m_{(l)}^T \qquad \langle\Omega_{(l)}\rangle = \eta_{(l)}C_{(l)}^{-1}$$

$$\langle\log|\Omega_{(l)}|\rangle = d\log 2 - \log|C_{(l)}| + \sum_{j=1}^{d}\psi\left(\frac{\eta_{(l)}+1-j}{2}\right)$$

Constituents of the ELBO Function:

$$\langle\log p(Z \mid \mathcal{S}, \mathcal{M}, \mathcal{T})\rangle = \sum_{l=1}^{K}\sum_{i=1}^{n}\langle s_{il}\rangle\Big[-\frac{d}{2}\log(2\pi) + \frac{1}{2}\langle\log|\Omega_{(l)}|\rangle$$
$$-\frac{1}{2}\mathrm{Tr}\left(\langle\Omega_{(l)}\rangle\left[Z_{:i}Z_{:i}^T - \langle\mu_{(l)}\rangle Z_{:i}^T - Z_{:i}\langle\mu_{(l)}\rangle^T + \langle\mu_{(l)}\mu_{(l)}^T\rangle\right]\right)\Big]$$

$$\langle \log p(\mathcal{S}|\phi)\rangle = \sum_{l=1}^{K}\sum_{i=1}^{n}\langle s_{il}\rangle \log \phi_l$$

$$\langle \log p(\mathcal{M})\rangle = \frac{Kd}{2}\log(\frac{\beta}{2\pi}) - \frac{\beta}{2}\sum_{l=1}^{K}\mathrm{Tr}\big(\langle \mu_{(l)}\mu_{(l)}^T\rangle\big)$$

$$\langle \log p(\mathcal{T})\rangle = K\big[-\frac{\nu d}{2}\log 2 - \frac{d[d-1]}{4}\log\pi - \sum_{j=1}^{d}\log\Gamma(\frac{\nu+1-j}{2}) + \frac{\nu}{2}\log|V|\big]$$
$$+\frac{\nu-d-1}{2}\big[\sum_{l=1}^{K}\langle \log|\Omega_{(l)}|\rangle\big] - \frac{1}{2}\mathrm{Tr}\big(V\sum_{l=1}^{K}\langle \Omega_{(l)}\rangle\big)$$

$$\langle \log q(\mathcal{S})\rangle = \sum_{i=1}^{n}\sum_{l=1}^{K}\langle s_{il}\rangle \log\langle s_{il}\rangle$$

$$\langle \log q(\mathcal{M})\rangle = \sum_{l=1}^{K}\langle \log q(\mu_{(l)})\rangle = \sum_{l=1}^{K}\big[-\frac{d}{2}[1+\log 2\pi] + \frac{1}{2}\log|T_{(l)}|\big]$$

$$\langle \log q(\mathcal{T})\rangle = \sum_{l=1}^{K}\langle \log q(\Omega_{(l)})\rangle = \sum_{l=1}^{K}\Bigg[-\frac{d\eta_{(l)}}{2}\log 2 - \frac{d[d-1]}{4}\log\pi$$
$$-\sum_{j=1}^{d}\log\Gamma(\frac{\eta_{(l)}+1-j}{2})$$
$$+\frac{\eta_{(l)}}{2}\log|C_{(l)}| + \frac{\eta_{(l)}-d-1}{2}\langle \log|\Omega_{(l)}|\rangle - \frac{1}{2}\mathrm{Tr}\big(C_{(l)}\langle \Omega_{(l)}\rangle\big)\Bigg]$$

B. VEM updates for MixMPLN

The update formulas for the variational parameters $\alpha_i, m_{(l)}, T_{(l)}, \eta_{(l)}$, and $C_{(l)}$, along with the formulas for calculating the constituents of the ELBO function are given below. *Notation:* Tr(.) and $\psi(.)$ denote the matrix trace function and the di-gamma function respectively; $\langle F(\theta_t)\rangle$ denotes the expectation of function $F(\theta_t)$ over the factor distribution $q(\theta_t)$.

Parameter α_i for $q(S_i)$:

$$\alpha_{il} = \frac{f_{il}}{\sum_{r=1}^{K} f_{ir}}, \text{ where}$$

$$f_{il} = \exp\Bigg\{\log\phi_l - \frac{d}{2}\log(2\pi) + \frac{1}{2}\langle \log|\Omega_{(l)}|\rangle + \sum_{j=1}^{d}\Big[-\langle e^{\lambda_{lji}}\rangle - \log(x_{ji}!) + \langle \lambda_{lji}\rangle x_{ji}\Big]$$
$$-\frac{1}{2}\mathrm{Tr}\Big(\langle \Omega_{(l)}\rangle\Big[\langle \Lambda_{(l):i}\Lambda_{(l)\,i}^T\rangle - \langle \mu_{(l)}\rangle\langle \Lambda_{(l):i}\rangle^T - \langle \Lambda_{(l):i}\rangle\langle \mu_{(l)}\rangle^T + \langle \mu_{(l)}\mu_{(l)}^T\rangle\Big]\Big)\Bigg\}$$

Parameters $m_{(l)}$ and $T_{(l)}$ for $q(\mu_{(l)})$:

$$T_{(l)} = \beta I + \langle \Omega_{(l)} \rangle \sum_{i=1}^{n} \langle s_{il} \rangle, \qquad m_{(l)} = T_{(l)}^{-1} \langle \Omega_{(l)} \rangle \sum_{i=1}^{n} \langle \Lambda_{(l):i} \rangle \langle s_{il} \rangle$$

Parameters $\eta_{(l)}$ and $C_{(l)}$ for $q(\Omega_{(l)})$:

$$\begin{aligned} \eta_{(l)} =& \ \nu + \sum_{i=1}^{n} \langle s_{il} \rangle \\ C_{(l)} =& \ V + \sum_{i=1}^{n} \langle \Lambda_{(l):i} \Lambda_{(l):i}^{T} \rangle \langle s_{il} \rangle - \sum_{i=1}^{n} \langle \Lambda_{(l):i} \rangle \langle s_{il} \rangle \langle \mu_{(l)} \rangle^{T} \\ & - \langle \mu_{(l)} \rangle \sum_{i=1}^{n} \langle \Lambda_{(l):i} \rangle^{T} \langle s_{il} \rangle + \langle \mu_{(l)} \mu_{(l)}^{T} \rangle \sum_{i=1}^{n} \langle s_{il} \rangle \end{aligned}$$

Expected Values Used in the Above Updates:

$$\langle s_{il} \rangle = \alpha_{il} \qquad \langle \Lambda_{(l):i} \Lambda_{(l):i}^{T} \rangle = D_{(l)}^{-1} + \delta_{(l)i} \delta_{(l)i}^{T}$$

$$\langle \mu_{(l)} \rangle = m_{(l)} \qquad \langle \mu_{(l)} \mu_{(l)}^{T} \rangle = T_{(l)}^{-1} + m_{(l)} m_{(l)}^{T}$$

$$\langle \lambda_{lji} \rangle = a_{lji} \qquad \langle \Omega_{(l)} \rangle = \eta_{(l)} C_{(l)}^{-1}$$

$$\langle e^{\lambda_{lji}} \rangle = e^{a_{lji} + \frac{1}{2} b_{lji}} \qquad \langle \log |\Omega_{(l)}| \rangle = d \log 2 - \log |C_{(l)}| + \sum_{j=1}^{d} \psi\left(\frac{\eta_{(l)} + 1 - j}{2}\right)$$

$$\langle \Lambda_{(l):i} \rangle = \delta_{(l)i}$$

Constituents of the ELBO Function:

$$\begin{aligned} \langle \log p(X, \mathcal{L} \mid \mathcal{S}, \mathcal{M}, \mathcal{T}) \rangle = & \sum_{l=1}^{K} \sum_{i=1}^{n} \langle s_{il} \rangle \Big[- \frac{d}{2} \log(2\pi) + \frac{1}{2} \langle \log |\Omega_{(l)}| \rangle + \sum_{j=1}^{d} \left[- \langle e^{\lambda_{lji}} \rangle - \log(x_{ji}!) + \langle \lambda_{lji} \rangle x_{ji} \right] \\ & - \frac{1}{2} \mathrm{Tr} \left(\langle \Omega_{(l)} \rangle \left[\langle \Lambda_{(l):i} \Lambda_{(l):i}^{T} \rangle - \langle \mu_{(l)} \rangle \langle \Lambda_{(l):i} \rangle^{T} - \langle \Lambda_{(l):i} \rangle \langle \mu_{(l)} \rangle^{T} + \langle \mu_{(l)} \mu_{(l)}^{T} \rangle \right] \right) \Big] \end{aligned}$$

$$\begin{aligned} \langle \log p(\mathcal{S} | \phi) \rangle &= \sum_{l=1}^{K} \sum_{i=1}^{n} \langle s_{il} \rangle \log \phi_l \\ \langle \log p(\mathcal{M}) \rangle &= \frac{Kd}{2} \log\left(\frac{\beta}{2\pi}\right) - \frac{\beta}{2} \sum_{l=1}^{K} \mathrm{Tr}\left(\langle \mu_{(l)} \mu_{(l)}^{T} \rangle \right) \end{aligned}$$

$$\begin{aligned} \langle \log p(\mathcal{T}) \rangle =& K \Big[- \frac{\nu d}{2} \log 2 - \frac{d[d-1]}{4} \log \pi - \sum_{j=1}^{d} \log \Gamma\left(\frac{\nu + 1 - j}{2}\right) + \frac{\nu}{2} \log |V| \Big] \\ & + \frac{\nu - d - 1}{2} \Big[\sum_{l=1}^{K} \langle \log |\Omega_{(l)}| \rangle \Big] - \frac{1}{2} \mathrm{Tr}\Big(V \sum_{l=1}^{K} \langle \Omega_{(l)} \rangle \Big) \end{aligned}$$

$$\langle \log q(\mathcal{S}) \rangle = \sum_{i=1}^{n} \sum_{l=1}^{K} \langle s_{il} \rangle \log \langle s_{il} \rangle$$

$$\langle \log q(\mathcal{M}) \rangle = \sum_{l=1}^{K} \langle \log q(\mu_{(l)}) \rangle = \sum_{l=1}^{K} \Big[-\frac{d}{2}[1 + \log 2\pi] + \frac{1}{2} \log |T_{(l)}| \Big]$$

$$\langle \log q(\mathcal{T}) \rangle = \sum_{l=1}^{K} \langle \log q(\Omega_{(l)}) \rangle = \sum_{l=1}^{K} \Big[-\frac{d\eta_{(l)}}{2} \log 2 - \frac{d[d-1]}{4} \log \pi - \sum_{j=1}^{d} \log \Gamma\Big(\frac{\eta_{(l)} + 1 - j}{2}\Big) + \frac{\eta_{(l)}}{2} \log |C_{(l)}| + \frac{\eta_{(l)} - d - 1}{2} \langle \log |\Omega_{(l)}| \rangle - \frac{1}{2} \mathrm{Tr}\big(C_{(l)} \langle \Omega_{(l)} \rangle\big) \Big]$$

$$\langle \log q(\mathcal{L}) \rangle = \sum_{i=1}^{n} \sum_{l=1}^{K} \langle \log q(\Lambda_{(l):i}) \rangle = \sum_{i=1}^{n} \sum_{l=1}^{K} \Big[\frac{1}{2} \Big[\sum_{j=1}^{d} \log b_{lji} \Big] - \frac{d}{2} [1 + \log 2\pi] \Big]$$

C. Derivation of the equation for $elbo(\eta_l, C_{(l)}^{-1})$

We can ignore the constant terms in our analysis. We have

$$\int q(\Omega_{(l)}) \, \langle \log p(Z, \Theta|\phi) \rangle_{\theta_v \neq \Omega_{(l)}} \, \mathrm{d}\Omega_{(l)} - \int q(\Omega_{(l)}) \, \log q(\Omega_{(l)}) \, \mathrm{d}\Omega_{(l)}$$
$$= \int \mathbb{W}(\Omega_{(l)}|\eta_{(l)},, C_{(l)}) \, \langle \log p(Z, \Theta|\phi) \rangle_{\theta_v \neq \Omega_{(l)}} \, \mathrm{d}\Omega_{(l)} - \int \mathbb{W}(\Omega_{(l)}|\eta_{(l)},, C_{(l)}) \, \log \mathbb{W}(\Omega_{(l)}|\eta_{(l)}, C_{(l)}) \, \mathrm{d}\Omega_{(l)} \tag{9}$$

Now,

$$\begin{aligned} \langle \log p(Z, \Theta|\phi) \rangle_{\theta_v \neq \Omega_{(l)}} &= \langle \log \Big(\prod_{i=1}^{n} p(Z_{:i}|\mu_{(l)}, \Omega_{(l)})^{s_{il}} \; p(\Omega_{(l)}) \Big) \rangle_{\theta_v \neq \Omega_{(l)}} \\ &= \langle \log \Big(\prod_{i=1}^{n} \mathbb{N}_d(Z_{:i}|\mu_{(l)}, \Omega_{(l)})^{s_{il}} \; \mathbb{W}(\Omega_{(l)}|\nu, V) \Big) \rangle_{\theta_v \neq \Omega_{(l)}} \\ &= \frac{1}{2} \Big(\sum_{i=1}^{n} \langle s_{il} \rangle \Big) \log |\Omega_{(l)}| - \frac{1}{2} \mathrm{Tr}\left(\Omega_{(l)} M\right) + \frac{\nu - d - 1}{2} \log |\Omega_{(l)}| \end{aligned}$$

where

$$M = V + \sum_{i=1}^{n} Z_{:i} Z_{:i}^T \langle s_{il} \rangle - \sum_{i=1}^{n} Z_{:i} \langle s_{il} \rangle \langle \mu_{(l)} \rangle^T - \langle \mu_{(l)} \rangle \sum_{i=1}^{n} Z_{:i}^T \langle s_{il} \rangle + \langle \mu_{(l)} \mu_{(l)}^T \rangle \sum_{i=1}^{n} \langle s_{il} \rangle \tag{10}$$

Equation (9) can be then simplified using the following identities

$$\langle \Omega_{(l)} \rangle = \eta_{(l)} C_{(l)}^{-1}$$

$$\langle \log |\Omega_{(l)}| \rangle = d \log 2 + \log |C_{(l)}^{-1}| + \sum_{j=1}^{d} \psi\Big(\frac{\eta_{(l)} + 1 - j}{2}\Big)$$

$$\langle \mathrm{Tr}\left(\Omega_{(l)} M\right) \rangle = \mathrm{Tr}\left(\langle \Omega_{(l)} \rangle M\right)$$

This results in

$$elbo(\eta_{(l)}, C_{(l)}^{-1}) = \Big[\frac{\sum_{i=1}^{n}\langle s_{il}\rangle + \nu}{2}\Big] \log|C_{(l)}^{-1}| - \frac{1}{2}\mathrm{Tr}(\eta_{(l)} C_{(l)}^{-1} M) + \frac{\eta_{(l)} d}{2}$$
$$+ \Big[\frac{\sum_{i=1}^{n}\langle s_{il}\rangle + \nu - \eta_{(l)}}{2}\Big]\Big[\sum_{j=1}^{d}\psi(\frac{\eta_{(l)} + 1 - j}{2})\Big] + \sum_{j=1}^{d}\log\Gamma(\frac{\eta_{(l)} + 1 - j}{2})$$

D. Supplementary Figures

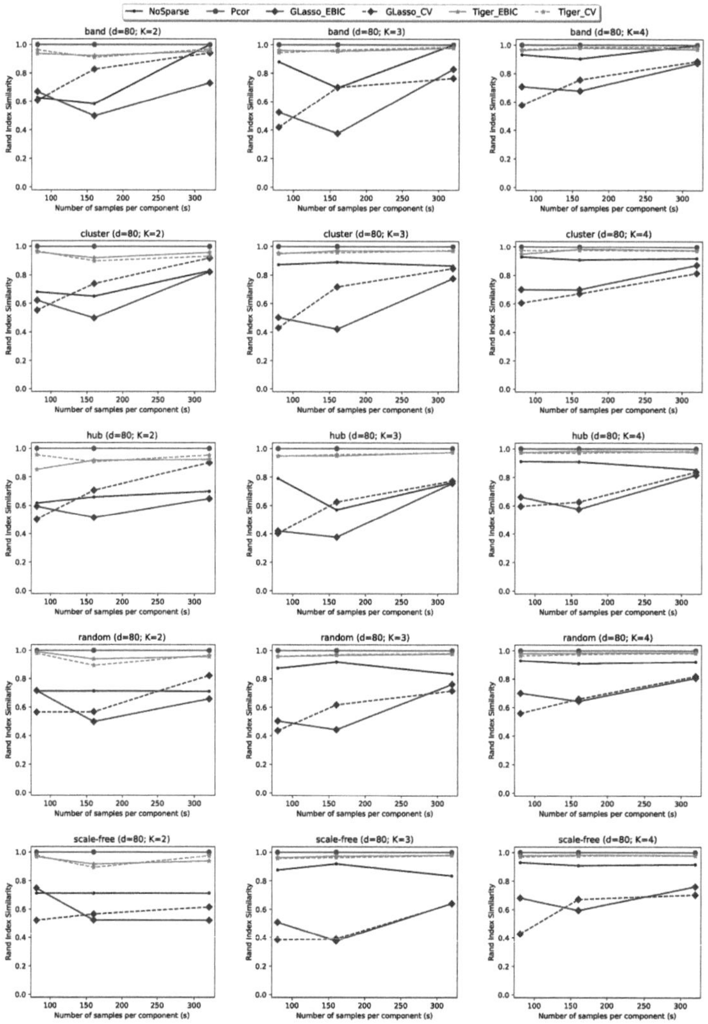

Fig. 3. Rand Index Similarity of the predicted and ground truth sample partitions for the six MS_MixGGM methods (*NoSparse, Pcor, Glasso_EBIC, Glasso_CV, Tiger_EBIC,* and *Tiger_CV*) on the five graph types(*band, cluster, hub, random,* and *scale-free*). Here, the number of taxa $d = 80$, the number of components $K = 2, 3, 4$, and the number of samples per component $s = 80, 160, 320$.

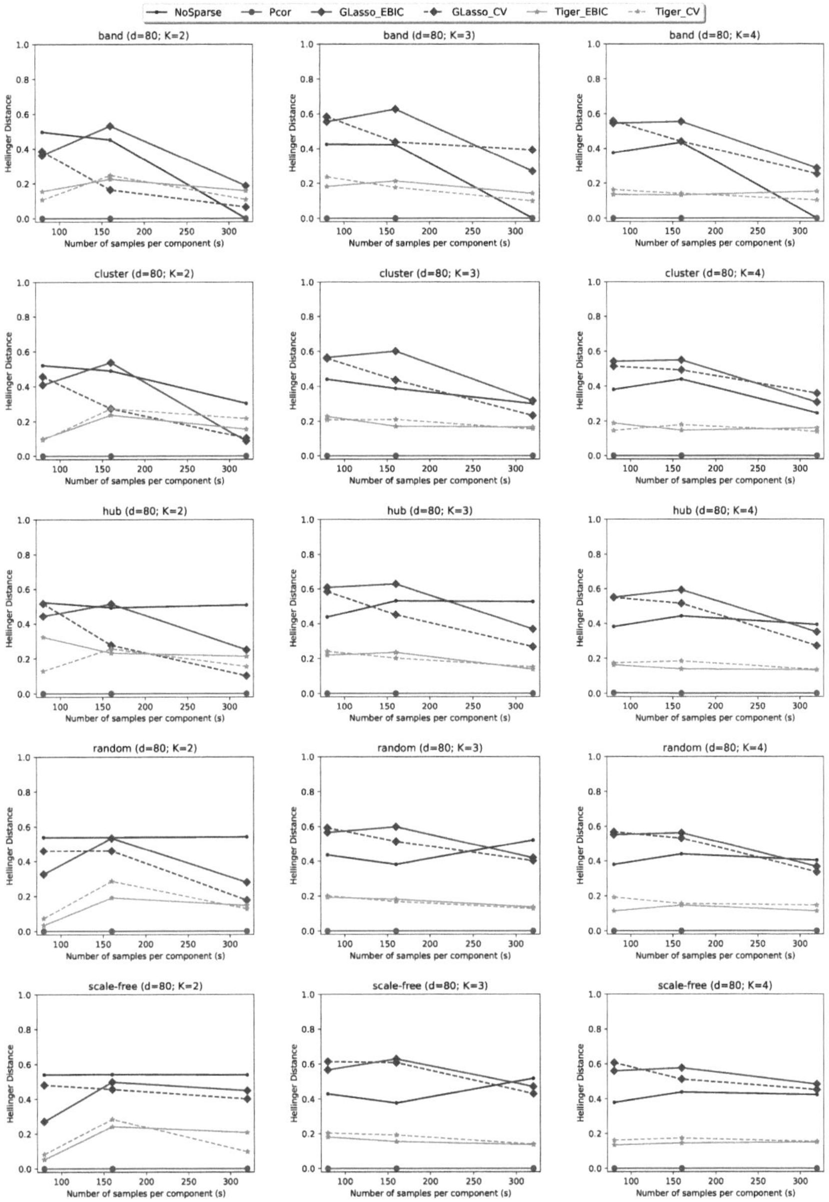

Fig. 4. Hellinger distance between the predicted and ground truth mixing coefficient vectors for the six MS_MixGGM methods (*NoSparse, Pcor, Glasso_EBIC, Glasso_CV, Tiger_EBIC,* and *Tiger_CV*) on the five graph types(*band, cluster, hub, random,* and *scale-free*). Here, the number of taxa $d = 80$, the number of components $K = 2, 3, 4$, and the number of samples per component $s = 80, 160, 320$.

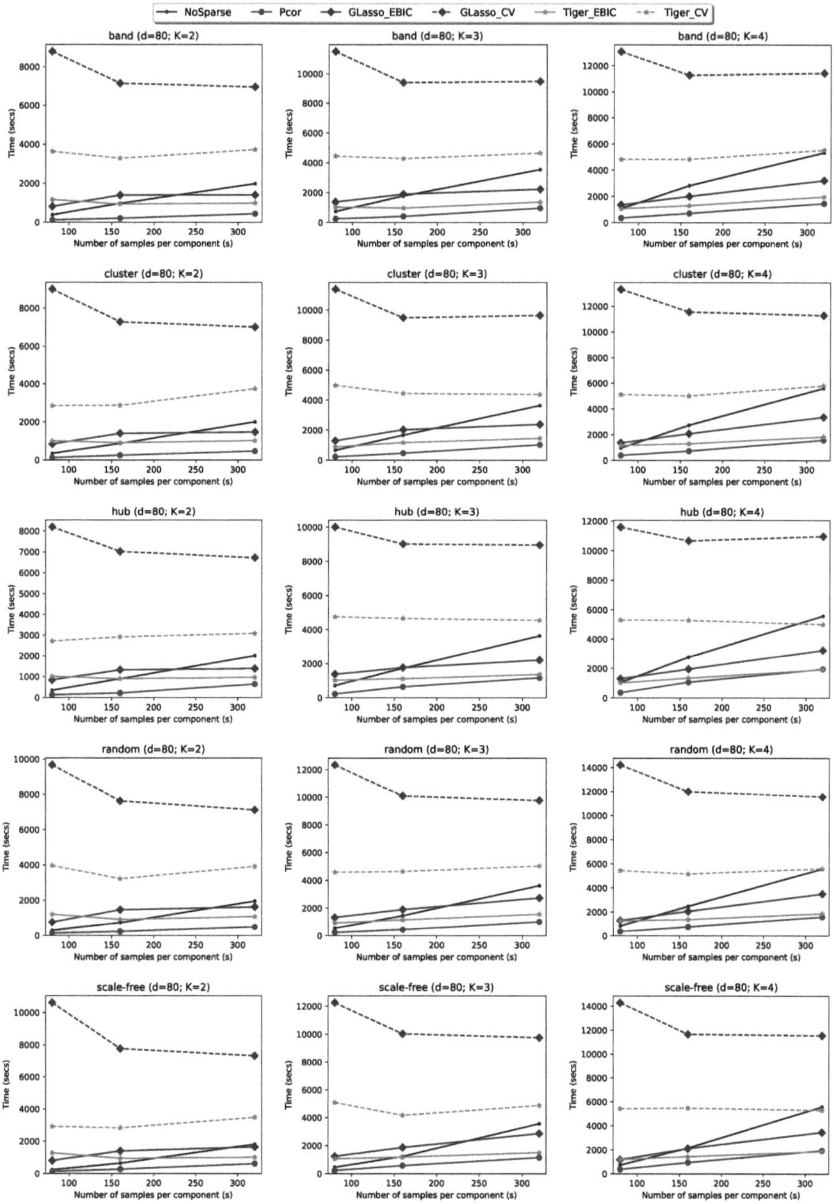

Fig. 5. Runtimes for the six MS_MixGGM methods (*NoSparse, Pcor, Glasso_EBIC, Glasso_CV, Tiger_EBIC,* and *Tiger_CV*) on the five graph types(*band, cluster, hub, random,* and *scale-free*). Here, the number of taxa $d = 80$, the number of components $K = 2, 3, 4$, and the number of samples per component $s = 80, 160, 320$.

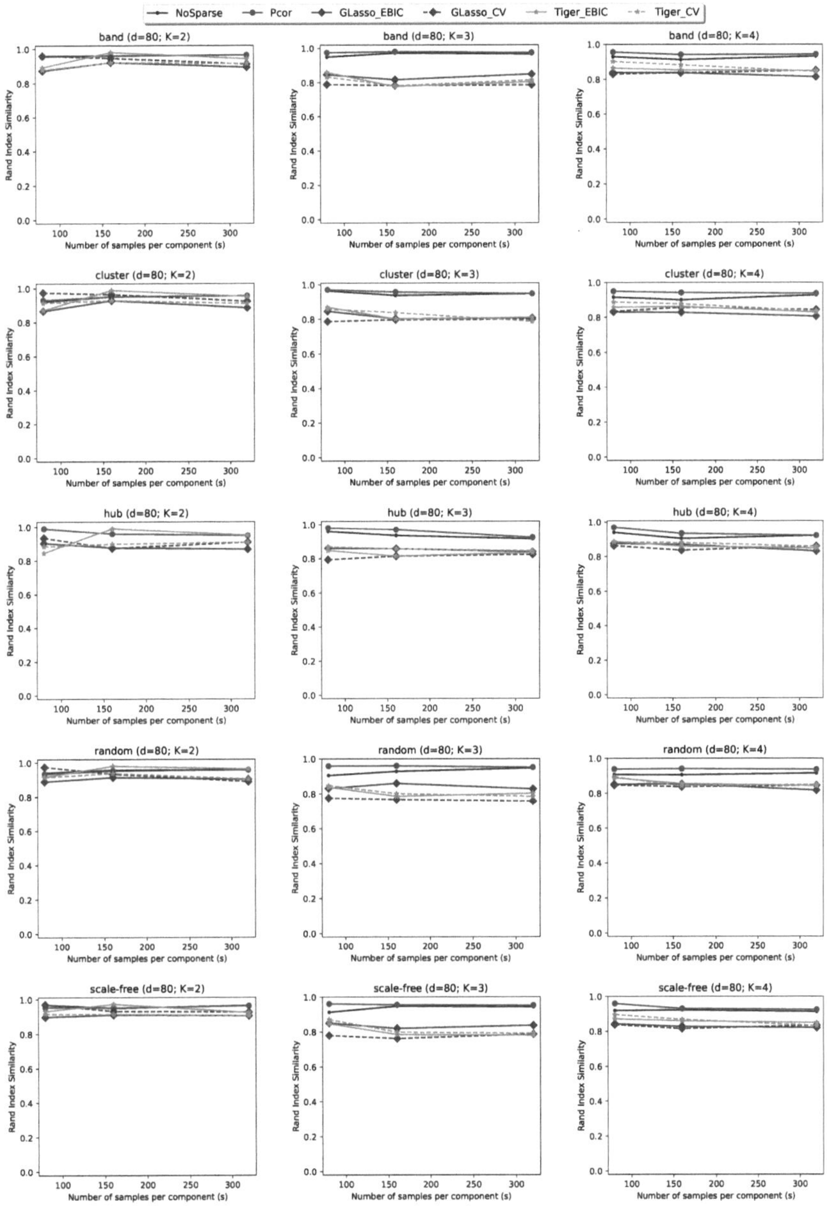

Fig. 6. Rand Index Similarity of the predicted and ground truth sample partitions for the six MS_MixMPLN methods (*NoSparse, Pcor, Glasso_EBIC, Glasso_CV, Tiger_EBIC,* and *Tiger_CV*) on the five graph types(*band, cluster, hub, random,* and *scale-free*). Here, the number of taxa $d = 80$, the number of components $K = 2, 3, 4$, and the number of samples per component $s = 80, 160, 320$.

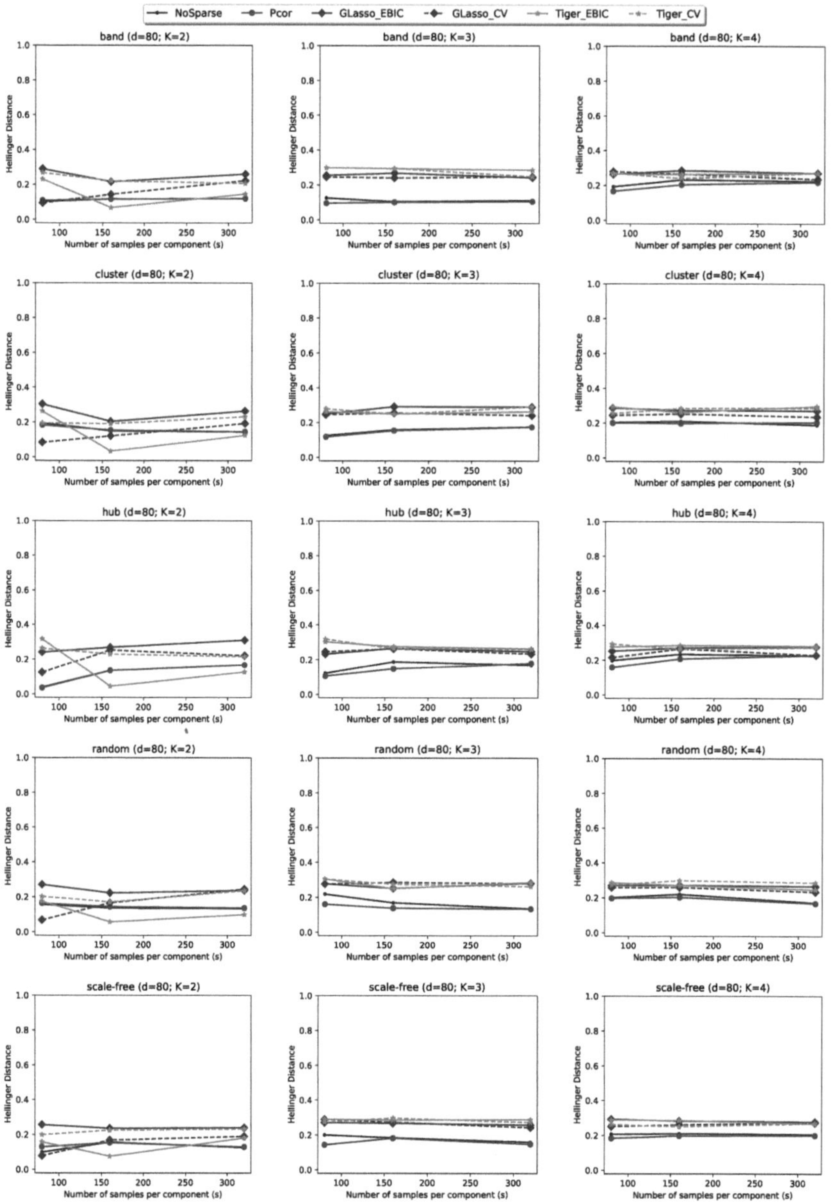

Fig. 7. Hellinger distance between the predicted and ground truth mixing coefficient vectors for the six MS_MixMPLN methods (*NoSparse, Pcor, Glasso_EBIC, Glasso_CV, Tiger_EBIC,* and *Tiger_CV*) on the five graph types(*band, cluster, hub, random,* and *scale-free*). Here, the number of taxa $d = 80$, the number of components $K = 2, 3, 4$, and the number of samples per component $s = 80, 160, 320$.

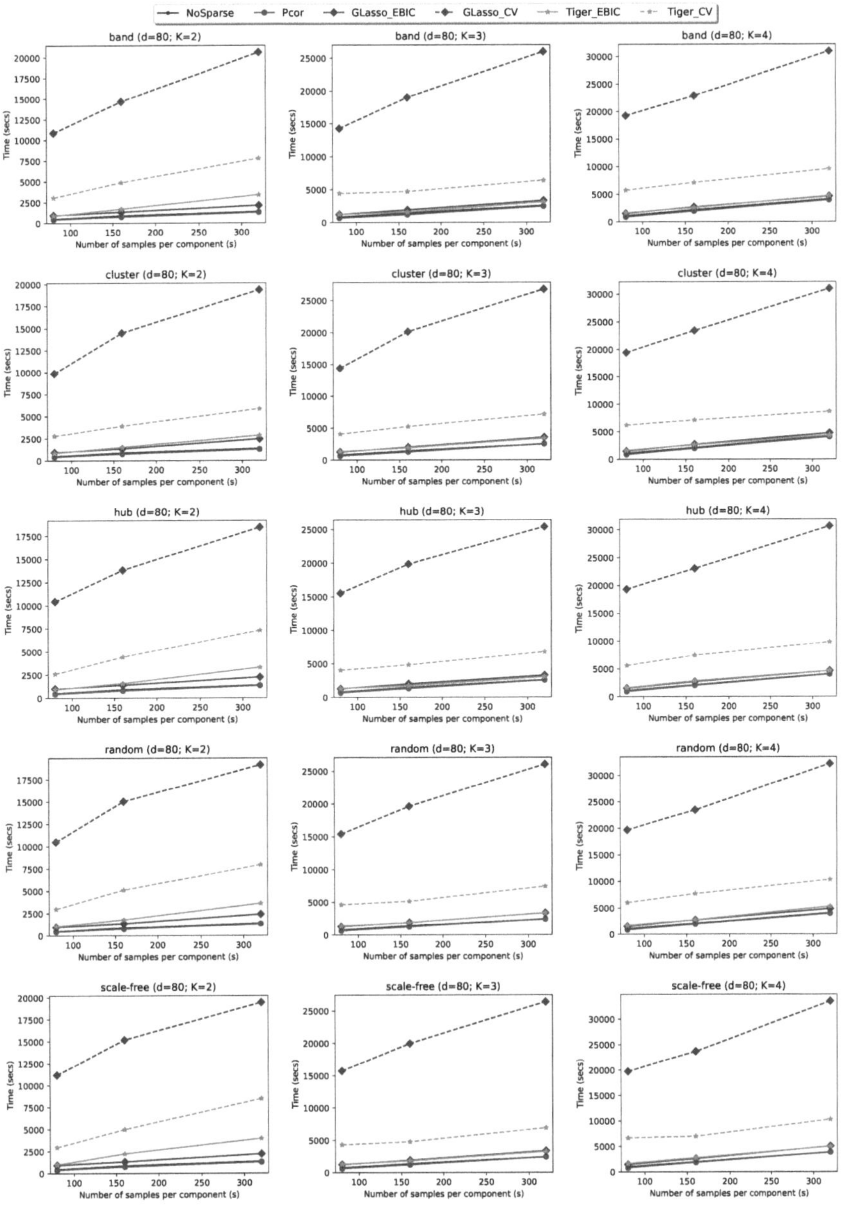

Fig. 8. Runtimes for the six MS_MixMPLN methods (*NoSparse, Pcor, Glasso_EBIC, Glasso_CV, Tiger_EBIC,* and *Tiger_CV*) on the five graph types(*band, cluster, hub, random,* and *scale-free*). Here, the number of taxa $d = 80$, the number of components $K = 2, 3, 4$, and the number of samples per component $s = 80, 160, 320$.

References

1. Aitchison, J., Ho, C.H.: The multivariate poisson-log normal distribution. Biometrika **76**(4), 643–653 (1989)
2. Bishop, C.M.: Pattern Recognition and Machine Learning (Information Science and Statistics). Springer-Verlag (2006). ISBN 0387310738
3. Biswas, S., McDonald, M., Lundberg, D.S., et al.: Learning microbial interaction networks from metagenomic count data. J. Comput. Biol. **23**(6), 526–35 (2016)
4. Chen, L., Wan, H., He, Q., et al.: Statistical methods for microbiome compositional data network inference: a survey. J. Comput. Biol. **29**(7), 704–723 (2022)
5. Corduneanu, A., Bishop, C.M.: Variational bayesian model selection for mixture distributions. In: Proceedings Eighth International Conference on Artificial Intelligence and Statistics. Morgan Kaufmann (2001)
6. Epskamp, S., Fried, E.I.: A tutorial on regularized partial correlation networks. Psychol. Methods **23**(4), 617–634 (2018)
7. Fang, H., Huang, C., Zhao, H., et al.: CCLasso: correlation inference for compositional data through lasso. Bioinformatics **31**(19), 3172–80 (2015)
8. Friedman, J., Alm, E.J.: Inferring correlation networks from genomic survey data. PLoS Comput. Biol. **8**(9), e1002687 (2012)
9. Friedman, J., Hastie, T., Tibshirani, R.: Sparse inverse covariance estimation with the graphical lasso. Biostatistics **9**(3), 432–41 (2008)
10. Hassouneh, S.A., Loftus, M., Yooseph, S.: Linking inflammatory bowel disease symptoms to changes in the gut microbiome structure and function. Front. Microbiol. **12** (2021)
11. Hibbing, M.E., Fuqua, C., Parsek, M.R., et al.: Bacterial competition: surviving and thriving in the microbial jungle. Nat. Rev. Microbiol. **8**(1), 15–25 (2010)
12. Inouye, D.I., Yang, E., Allen, G.I., et al.: A review of multivariate distributions for count data derived from the poisson distribution. WIREs Comput. Stat. **9**(3), n/a (2017)
13. Jordan, M.I.: Learning in Graphical Models. MIT Press, Cambridge, Massachusetts (1999)
14. Kurtz, Z.D., Muller, C.L., Miraldi, E.R., et al.: Sparse and compositionally robust inference of microbial ecological networks. PLoS Comput. Biol. **11**(5), e1004226 (2015)
15. Layeghifard, M., Hwang, D.M., Guttman, D.S.: Disentangling interactions in the microbiome: a network perspective. Trends Microbiol. **25**(3), 217–228 (2017)
16. Liu, H., Wang, L.: TIGER: a tuning-insensitive approach for optimally estimating Gaussian graphical models. Electron. J. Statist **11**(1), 241–294 (2017)
17. Loftus, M., Hassouneh, S.A., Yooseph, S.: Bacterial associations in the healthy human gut microbiome across populations. Sci. Rep. **11**(1), 2828 (2021)
18. Loftus, M., Hassouneh, S.A., Yooseph, S.: Bacterial community structure alterations within the colorectal cancer gut microbiome. BMC Microbiol. **21**(98) (2021)
19. MacQueen, J.: Some methods for classification and analysis of multivariate observations. In: Proceedings of the Fifth Berkeley Symposium on Mathematical Statistics and Probability, Volume 1: Statistics, Fifth Berkeley Symposium on Mathematical Statistics and Probability. University of California Press (1967)
20. Nikulin, M.S.: Hellinger Distance. EMS Press, Encyclopedia of Mathematics (2001)
21. Parisi, G.: Statistical Field Theory. Addison-Wesley, Redwood City (1988)
22. Prusty, S., Patnaik, S., Dash, S.K.: SKCV: stratified K-fold cross-validation on ML classifiers for predicting cervical cancer. Front. Nanotechnol. **4** (2022)

23. R Development Core Team: R: A Language and Environment for Statistical Computing. R Foundation for Statistical Computing, Vienna, Austria (2013)
24. Rand W.M.: Objective criteria for the evaluation of clustering methods. J. Am. Stat. Assoc. **66**(336) (1971)
25. Tavakoli, S., Yooseph, S.: Learning a mixture of microbial networks using minorization-maximization. Bioinformatics **35**(14), i23–i30 (2019)
26. Tavakoli, S., Yooseph, S.: Algorithms for inferring multiple microbial networks. In: IEEE International Conference on Bioinformatics and Biomedicine (BIBM), pp. i223–i227 (2019)
27. Tzikas, D.G., Likas, A.C., Galatsanos, N.P.: The variational approximation for bayesian inference. IEEE Signal Process. Mag. **25**(6), 131–146 (2008)
28. Venter, J.C., Remington, K., Heidelberg, J.F., et al.: Environmental genome shotgun sequencing of the sargasso sea. Science **304**(5667), 66–74 (2004)
29. Williams, D.R., Rast, P.: Back to the basics: rethinking partial correlation network methodology. Br. J. Math. Stat. Psychol. **73**(2), 187–212 (2020)
30. Wishart, J.: The generalised product moment distribution in samples from a normal multivariate population. Biometrika **20A**(1/2), 32–52 (1928)
31. Woese, C.R., Fox, G.E.: Phylogenetic structure of the prokaryotic domain: the primary kingdoms. Proc. Natl. Acad. Sci. USA **74**(11), 5088–5090 (1977)
32. Yooseph, S. Kirkness, E.F., Tran, T.M., et al.: Stool microbiota composition is associated with the prospective risk of *Plasmodium falciparum* infection. BMC Genomics **16**(631) (2015)
33. Yooseph, S., Tavakoli, S.: Variational approximation-based model selection for microbial network inference. J. Comput. Biol. **29**(7), 724–737 (2022)
34. Zhao, T., Liu, H., Roeder, K., et al.: The huge package for high-dimensional undirected graph estimation in R. J. Mach. Learn. Res. **13**, 1059–1062 (2012)
35. Zhu, Y., Cribben, I.: Sparse graphical models for functional connectivity networks: best methods and the autocorrelation issue. Brain Connect **8**(3), 139–165 (2018)

Haplotype-Based Parallel PBWT for Biobank Scale Data

Kecong Tang[1], Ahsan Sanaullah[1], Degui Zhi[2], and Shaojie Zhang[1](✉)

[1] Department of Computer Science, University of Central Florida, Orlando, FL 32816, USA
{kecong.tang,ahsan.sanaullah,shaojie.zhang}@ucf.edu

[2] McWilliams School of Biomedical Informatics, University of Texas Health Science Center at Houston, Houston, TX 77030, USA
{degui.zhi}@uth.tmc.edu

Abstract. Durbin's positional Burrows-Wheeler transform (PBWT) enables algorithms with the optimal time complexity of $O(MN)$ for reporting all vs all haplotype matches in a population panel with M haplotypes and N variant sites. However, even this efficiency may still be too slow when the number of haplotypes reaches millions. To further reduce the run time, in this paper, a parallel version of the PBWT algorithms is introduced for all versus all haplotype matching, which is called HP-PBWT (haplotype-based parallel PBWT). HP-PBWT parallelly executes the PBWT by splitting a haplotype panel into blocks of haplotypes. HP-PBWT algorithms achieve parallelization for PBWT construction, reporting all versus all L-long matches, and reporting all versus all set-maximal matches while maintaining memory efficiency. HP-PBWT has an $O((\frac{M}{T} + T)N)$ time complexity in PBWT construction, and $O((\frac{M}{T}+T+c^*)N)$ time complexity for reporting all versus all L-long matches and reporting all versus all set-maximal matches, where T is the number of threads and c^* is the maximum number of matches (of length L or maximum divergence value for L-long matches and set-maximal matches respectively) per haplotype per site. HP-PBWT achieves 4-fold speed-up in UK Biobank genotyping array data with 30 threads in the IO-included benchmarks. When applying HP-PBWT to a dataset of 8 million randomized haplotypes (random binary strings of equal length) in the IO-excluded benchmarks, it can achieve a 22-fold speed-up with 60 cores on the Amazon EC2 server. With further hardware optimization, HP-PBWT is expected to handle billions of haplotypes efficiently.

Keywords: PBWT · Parallel Computing · Haplotype Matching

1 Introduction

The positional Burrows-Wheeler transform (PBWT) [6] is an efficient data structure for finding haplotype matches and data compression. The construction of the PBWT can be done in linear time with respect to the number of haplotypes times the number of sites. The original PBWT paper also introduced

M. Alser et al. (Eds.): ICCABS 2025, LNCS 15599, pp. 130–143, 2026.
https://doi.org/10.1007/978-3-032-02489-3_10

efficient algorithms for reporting all versus all haplotype matches in a genetic panel. These algorithms have been widely applied in Identity-By-Descent (IBD) segment detection [8,11,20], genotype imputation [5,13], and haplotype phasing [4,9]. However, once the haplotype dimension of biobank panels exceeds many millions, even the PBWT may not be fast enough to satisfy the need of speed. The rapidly growing total number of genotyped individuals could reach billions in the future. This means an efficient parallel version of PBWT suitable for processing large-scale data input is in high demand.

Wertenbroek et al. [19] presented a parallelized PBWT by dividing the panel into sub-panels with ranges of sites, then applied a merging algorithm to generate the final output. Their algorithm has an $O(\frac{MN}{T} + TM \log M)$ time complexity, where M is the number of haplotypes, N is the number of sites, and T is the number of threads. However, their algorithm needs to load the whole haplotype panel into memory or read the haplotype panel multiple times to perform the dividing and merging steps. Therefore, this approach is best applicable to panels with sequencing data of a relatively small sample size [1,18].

On the other hand, the haplotype dimension of the panel is the more promising dimension to parallelize. Currently, genotyping array density datasets with millions of haplotypes [10,16,17] are widely used in both research and commercial applications. When compared to sequencing datasets, genotyping array density datasets have sparser sites, while typically having many more haplotypes, sometimes millions. Moreover, in the not-too-distant future, the number of genotyped samples could reach billions. Durbin's algorithms, even though enjoying a linear complexity to the number of haplotypes, need parallelization to scale up further.

In this paper, a haplotype-based parallel PBWT, HP-PBWT, is proposed, which is suitable for processing billions of haplotypes. The HP-PBWT is designed to break dependencies in the prefix array computation and divergence array computation in Durbin's original PBWT. Furthermore, two additional efficient algorithms are designed to report all versus all L-long matches and all versus all set-maximal matches using the parallelized PBWT. HP-PBWT is memory efficient, it maintains the efficient sweeping behavior in Durbin's original PBWT and only needs an $O(M)$ memory space. HP-PBWT has $O((\frac{M}{T} + T)N)$ run time for constructing the PBWT, and $O((\frac{M}{T} + T + c^*)N)$ run time for reporting all versus all L-long matches and reporting all versus all set-maximal matches, where T is the number of threads, and c^* is the maximum number of matches per haplotype per site and in practice $c^* \ll M$.

2 Background

The description in this work follows Durbin's original work [6]. The initial input of Durbin's PBWT is X, an M by N two dimensional binary array, which has M binary strings, x_i, that each represents a haplotype, and each haplotype has N sites. The fundamental outputs are a prefix array P_k and a divergence array D_k during the sorting process for each site $0 \leq k < N$.

The prefix array P_k stores haplotype IDs according to the colexicographic order of the haplotype prefixes of length $k+1$. P_k stores a permutation of $[0, M-1]$ such that $rev(x_{P_k[i]}[0,k])$ is lexicographically smaller than $rev(x_{P_k[i+1]}[0,k])$ for all i, where $rev(a)$ is the reverse of a string a. The term Y_k in Durbin's PBWT refers to a permutation of all sites in X at k-th location X_k, which is sorted based on the haplotype IDs in P_{k-1} such that $Y_k[i] = X_k[P_{k-1}[i]] = x_{P_{k-1}[i]}[k]$.

The divergence array D_k indicates the length of the match at k between two adjacent haplotypes in the sorted order of P_k. Therefore, if $P_k[a] = i$, $P_k[a-1] = i'$, and $D_k[i] = j$, then, $x_i[k-j+1, k] = x_{i'}[k-j+1, k]$ and $x_i[k-j] \neq x_{i'}[k-j]$. Note that the definition of the divergence array here is slightly different from Durbin's original definition. Durbin defined the divergence value as storing the starting position of the match and the values were permuted by the prefix array sorting.

A match between sequences x_a and x_b on $[i,j]$ is locally maximal if $x_a[i,j] = x_b[i,j]$ and it can't be extended in either direction. I.E. if $x_a[i-1] \neq x_b[i-1]$, and $x_a[j+1] \neq x_b[j+1]$. Given a length cut-off L, haplotypes x_a and x_b have an L-long match on $[i,j]$, if $[i,j]$ is a locally maximal match and $j-i+1 \geq L$. Durbin's Algorithm 3 outputs all L-long matches between all pairs of haplotypes. In this paper, this is referred to as outputting all versus all L-long matches.

A set-maximal match is defined on a haplotype x_a and a set of haplotypes, X. If x_a has a set-maximal match to x_b within X on $[i,j]$, then there is no larger match that contains it. I.E. $\forall x_c \in X$, $x_a[i-1,j] \neq x_c[i-1,j]$ and $x_a[i,j+1] \neq x_c[i,j+1]$. Durbin's Algorithm 4 outputs the set-maximal matches between x_d and $X \setminus \{x_d\}$ for all $x_d \in X$. In this paper, this is referred to as outputting all versus all set-maximal matches.

3 Methods

The focus of this work is to reduce the time complexity of the M dimension, the N dimension remains the same. First, a parallel prefix sum algorithm is used to compute prefix array in parallel. Second, D_k is computed by dividing Y_k into T partition blocks with T threads independently in parallel. Third, an efficient fine-grained parallel algorithm is designed to report all versus all L-long matches. At the end of this section, all versus all set-maximal matches are also reported in parallel with a similar algorithm. These algorithms have $O(\frac{M}{T} + T)$ span per site for the PBWT construction and $O(\frac{M}{T} + T + c^*)$ span per site for reporting all versus all L-long matches and reporting all versus all set-maximal matches, where T is the number of threads, and c^* is the maximum number of matches per haplotype per site and in practice $c^* \ll M$.

3.1 Parallel Prefix Array Computation

Durbin's Algorithm 1 computes P_k from P_{k-1} and Y_k, by placing $i \in P_{k-1}$ into a "zero" container if $Y_k[i] = 0$ or a "one" container if $Y_k[i] = 1$ sequentially. Then, P_k is constructed by concatenating the "zero" and "one" containers. Since the process is done sequentially, this dependency has to be removed in order to execute in parallel.

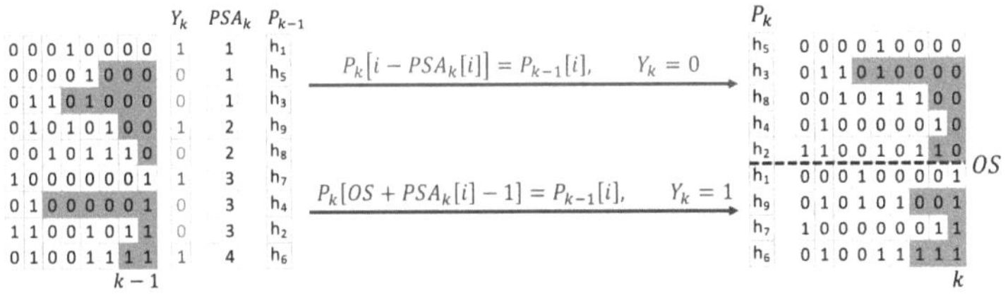

Fig. 1. Computation of new prefix array P_k from old prefix array P_{k-1} in parallel with the help of prefix sum array PSA_k. The offset is calculated by $OS = M - PSA_k[M-1]$ which is the number of zeros in Y_k.

The P_k construction of Durbin's PBWT Algorithm 1 is converted into a prefix sum problem. The first step is to run prefix sum on Y_k to create a PSA_k such that $PSA_k[i] = \sum_{j=0}^{i} Y_k[j]$. An important property of the PSA_k is, the PSA_k indicates the starting location of "1"s in P_k. The $PSA_k[M-1]$ is the number of "1"s in Y_k, and the offset $OS = M - PSA_k[M-1]$ is the number of "0"s in Y_k. The prefix array is defined as $P_k = PZ_k + PO_k$, where PZ_k is the "zero" container and PO_k is the "one" container. The PZ_k holds all haplotype indices i such that $Y_k[i] = 0$ by the order in P_{k-1}. Similarly, the PO_k holds the haplotype indices i such that $Y_k[i] = 1$. A mapping example is introduced from P_{k-1} to P_k according to PSA_k in Fig. 1. The PO_k part of P_k can be populated by $P_k[OS + PSA_k[i] - 1] = P_{k-1}[i]$. Since PSA_k only increases by one when $Y_k[i] = 1$, the value at $PSA_k[i]$ for haplotype $P_{k-1}[i]$ is the exact one-based index of $P_{k-1}[i]$ in PO_k. The PZ_K is the first part of P_k, the new location of a haplotype $P_k[i]$ is the old index i minus the number "1"s appeared before $Y_k[i]$ which is $PSA_k[i]$. Then PZ_k part of the P_k can be computed by $P_k[i - PSA_k[i]] = P_{k-1}[i]$. There is no dependency after creating the PSA_k, so P_k can be populated in parallel.

Therefore, the computation of the PBWT prefix array has been converted to a prefix sum problem. The prefix sums of Y_k are computed in parallel. First, the input is divided into T partition blocks. Then, local prefix sums within the partition blocks are calculated using one thread per partition block in parallel. Now each partition block b has its locally correct prefix sum values. To acquire the final globally correct prefix sum values, each value in partition block b has to be added a proper offset, which is the last prefix sum value of partition block $b-1$. These offsets need to be computed and added from each partition block sequentially. At the end, the corresponding offset is added to each partition block in parallel. The offset calculation is the only communication stage, and it has an $O(T)$ time complexity. This version has $O(M+T)$ work and $O(\frac{M}{T}+T)$ span per site. If T is large, then the parallel prefix sum in Section II.E of [2] (referred to as the halving merge algorithm) can be used to reduce the $O(T)$

term to $O(\log T)$ for a total work of $O(M)$ and span of $O(\frac{M}{T} + \log T)$ per site. After computing PSA_k, the haplotype IDs are mapped to the correct final prefix array location. Each haplotype takes $O(1)$ time to map and all M haplotypes are processed in parallel. Therefore the parallelized prefix array computation algorithm as described has $O(M)$ work and $O(\frac{M}{T} + T)$ span ($O(\frac{M}{T} + \log T)$ span per site if the algorithm of [2] is used), where T is the number of threads, and T also equals to the number of partition blocks. The $O(\frac{M}{T} + T)$ span parallel prefix sum algorithm is used in this work.

3.2 Parallel Divergence Array Computation

In Durbin's algorithm, D_k is computed by sweeping through P_{k-1} keeping track of the minimum matching lengths of haplotypes that have "0" and "1" in Y_k (p and q respectively in Durbin's Algorithm 2) through M haplotypes during the process of creating P_k. The algorithm checks if a haplotype still matches to its previous upper neighbor haplotype in P_{k-1} at Y_k site.

On the contrary, once a site is divided into partition blocks, and each block is assigned with one thread, to calculate all the divergence values within a single partition block, the passing-down values (p and q in Durbin's Algorithm 2) should not be acquired from the thread that assigned to the previous partition block. Otherwise, the whole process becomes sequential. Now the challenge becomes how to acquire the correct passing-down divergence values for each partition block in parallel.

Two necessary initial divergence values are calculated for each partition block by searching the previous partition block(s). To further explain, the two divergence values are: the divergence value of the first haplotype in the partition block that has "0" value in the Y_k, and the first haplotype in the partition block that has "1" value in the Y_k. Call the n-th partition block $i \in [\lceil \frac{M(n-1)}{T} \rceil, \lceil \frac{Mn}{T} \rceil)$, then, all the divergence values of the n-th partition block are $D_k[P_k[i]]$. These two divergence values can be computed by checking some upper b-th partition block(s) that $b < n$. The first divergence value is $D_k[P_k[j]]$ such that $j = Min(j \in [\lceil \frac{M(n-1)}{T} \rceil, \lceil \frac{Mn}{T} \rceil))$ and $Y_k[j] = 0$. The second divergence value is $D_k[P_k[j]]$ such that $j = Min(j \in [\lceil \frac{M(n-1)}{T} \rceil, \lceil \frac{Mn}{T} \rceil))$ and $Y_k[j] = 1$. For any of the two divergence values, $D_k[P_k[j]]$ that $Y_k[j] = v$ and $v = 0$ or 1, $D_k[P_k[j]]$ is computed by: first, find the index si that $Y_k[si] = v$, $si < j$, and $\forall i \in (si, j)$ and $Y_k[si] \neq v$; then, $D_k[P_k[j]] = Min(D_k[P_k[i]])$ that $i \in (si, j]$. For each partition block, the parallel algorithm searches the two divergence values in the previous partition block(s). After computing these two divergence values, it can simply loop through the partition block with Durbin's Algorithm 2 to compute the rest of the divergence values within this partition block in a single thread. The searching step is called upper search. This upper search may cross multiple partition blocks, and it does have a worst case $O(M)$ time complexity, but this is unlikely to happen since the selected markers in the biobank data panels usually do not have nearly singleton minor allele frequency. Optimizations are applied to prevent the long upper search from happening. The first optimization is verifying if the haplotype j is the first "1" in Y_k, by checking if $PSA_k[P_k[j]] = 1$, if so, this divergence value is set to 0, and the upper search isn't conducted.

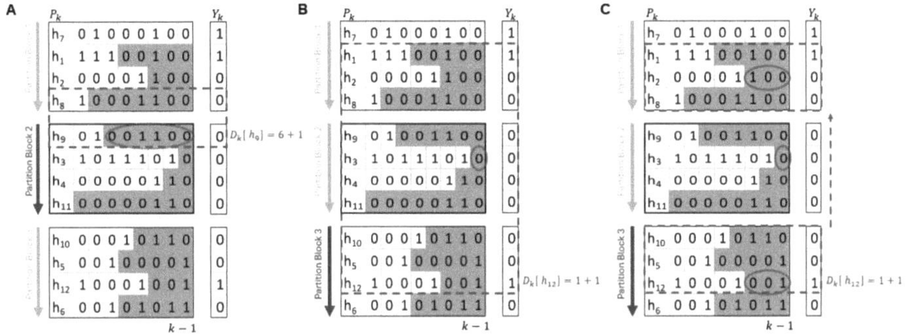

Fig. 2. Cross block upper search during divergence value computation. $D_k[h_9]$ (in A) needs a short upper search, and $D_k[h_{12}]$ (in B) needs a long upper search that cross block. Additional optimization (in C) by skipping block(s) is designed for the long upper search.

Upper search examples are shown in Fig. 2 which compute the two passing-down divergence values for $D_k[h_9]$ in partition block 2 and $D_k[h_{12}]$ in partition block 3. Computing $D_k[h_9]$ is simply $D_k[h_9] = D_{k-1}[h_9] + 1$, since the match between haplotype h_9 and h_8 continues. Haplotype h_{12} will be sorted after haplotype h_1, so $D_k[h_{12}]$ is minimum divergence value ($D_k[h_3]$) between h_{12} and h_2. It has to reach partition block 1 to find the first upper haplotype h_1 that has the same value "1" to haplotype h_{12} at k-th site.

An additional optimization is applied for the situation that there is small amount of "1" in Y_k, so the upper search for the haplotype with "1" may end up $O(M)$. Demonstrated in Fig. 2(C), assuming there could be many partition blocks that full of "0"s between h_{12} and h_1. For each partition block $B_i = [s, e]$ with an individual thread T_i, the first step is to identify whether this partition block has "1" or not. Here, a zero-block is defined as if $PSA_k[s-1] = PSA_k[e-1]$, that is to say there are no "1"s in this partition block. If the partition block B_i has "1", the thread T_i will skip all the zero-block(s) above it and find the nearest upper partition block $B_j = [s', e']$ that has "1" by checking the upper partition blocks according to the PSA_k. Then thread T_i runs from e' to s' to find the first p that $Y_k[p] = 1$, and compute the minimum divergence value $minD_0 = D_{k-1}[P_k[h]]$ that $\forall h \in (p, e']$. In the meantime, the minimum divergence values for each the skipped zero-blocks are still needed. The minimum divergence values for the zero-blocks are computed by the thread assigned to each partition block once a partition block is identified as zero-block. T_i waits to collect these minimum divergence values for the skipped zero-blocks if they are not yet calculated. Since it is unlikely that a site has a large run of "1"s in practice, in the implementation of HP-PBWT, this optimization is only applied on the "0"s.

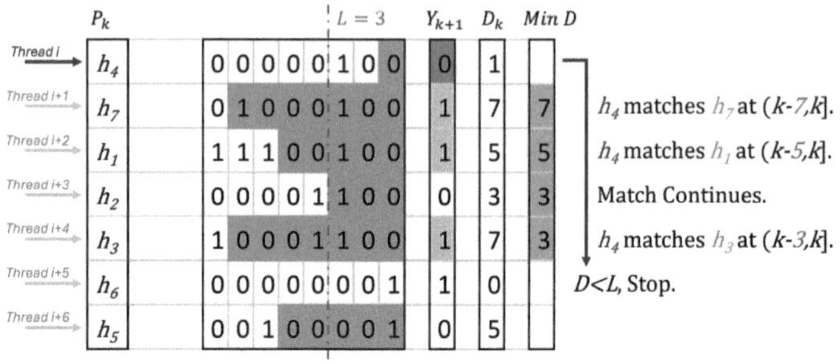

Fig. 3. Reporting all versus all L-long matches where $L = 3$ and sorted at k site in parallel. *Thread i* reports matches for haplotype h_4, it loops to haplotype h_6 since $D_k[h_6] < L$, meanwhile it compares the values of h_7, h_1, h_2 and h_3 in Y_{k+1} to h_4, then reports the match between h_4 and h_7, the match between h_4 and h_1, and the match between h_4 and h_3.

The divergence value computation in HP-PBWT has $O(M + Tr)$ work and $O(\frac{M}{T} + r)$ span per site, where r is the cost of the upper search and T is the number of threads. r is reduced to $O(\frac{M}{T} + T)$ work if the same optimization was applied for both "0"s and "1"s (note $T \ll M$). Then the final work and span of divergence value computation is $O(M + T)$ and $O(\frac{M}{T} + T)$ per site respectively.

3.3 Parallel Reporting of All Versus All L-Long Matches

Given a set of P_k, D_k, and Y_{k+1}, Durbin's Algorithm 3 outputs all matches with length $\geq L$ that end at k-th site. The algorithm has two basic steps. Step 1, acquire a precise matching block range $[s, e]$ that $\forall i \in [s, e]$ $D_k[P_k[s]] \geq L$, $D_k[P_k[e]] \geq L$, $D_k[P_k[s-1]] < L$, and $D_k[P_k[e+1]] < L$. That is to say all pairs of prefixes of haplotypes within the block have longest common suffixes of length at least length L. Step 2, output a match between every pair of haplotypes with differing values in Y_{k+1} within $[s, e]$.

In parallel execution, the simplified parallel algorithm reports matches for each haplotype in an individual thread. For each haplotype $P_k[h]$, it loops through the haplotypes $P_k[i] \in (h, M)$ until $D_k[P_k[i]] < L$, in the meantime it outputs matches if $Y_{k+1}[P_k[h]] \neq Y_{k+1}[P_k[i]]$. Figure 3 provides an example of thread i reports L-long matches for haplotype h_4 independently. In detail, this algorithm does have $O(\frac{M^2}{2})$ work, since there may be $O(M^2)$ matches at a site. If c^* is used to represent the maximum number of haplotypes that match with haplotype i length L or more ending at the current site for all i, HP-PBWT has $O(\frac{M}{T} + T + c^*)$ span per site for reporting all versus all L-long matches, where T is the number of threads (note that in practice $c^* \ll M$).

3.4 Parallel Reporting All Versus All Set-Maximal Matches

A set-maximal match is a collection of the longest matches ending at k site to a single haplotype. $[s, e]$ is a range in P_k that contains all the longest matches with haplotype i ending at k. If a match in $[s, e]$ can be extended further, then there are no set-maximal matches ending at k for haplotype i. For a haplotype $P_k[i]$, the set-maximal matches at site k are reported by: First, compute $D_{Max} = Max(D_k[P_k[i]], D_k[P_k[i+1]])$, find range $[s, e]$ such that $\forall j \in [s, e], D_k[P_k[j]] = D_{Max}$, $j \neq i$, $D_k[P_k[s]] < D_{Max}$ and $D_k[P_k[e+1]] < D_{Max}$; Second, check if any $j \in [s, e], Y_{k+1}[P_k[j]] = Y_{k+1}[P_k[i]]$, if so, haplotype $P_k[i]$ does not have set-maximal matches at site k location, and stop; Third, if it passes the second step, output a set maximal match between haplotypes $P_k[i]$ and $P_k[j]$ for all $j \in [s, e]$ with a match length of D_{Max}.

The scan up and scan down to find range $[s, e]$ are light tasks since the scan up only has to reach the location j' that $D_k[P_k[j']] \neq D_{Max}$, similarly for the scan down. At this point, it is not necessary to apply complicated parallel implementation to report all versus all set-maximal matches. Instead, the parallelization is simply done by paralleling the outer loop that loops through all the haplotypes. The total work to find all the ranges $[s, e]$ is $O(c)$, where c is the sum of the number of haplotypes that match to haplotype i with length equal to the maximum match of any haplotype and i for all i ending at the current site. So, the total work among all M haplotypes is $O(M + c)$, and the span is $O(\frac{M}{T} + T + c^*)$ per site, where c^* is the maximum number of haplotypes that match to haplotype i with length equal to the maximum match of any haplotype and i for all i ending at the current site. In practice, $c^* \ll M$.

4 Results

HP-PBWT was implemented in C# and its source code and software package are available at https://github.com/ucfcbb/HP-PBWT. To benchmark the performance of HP-PBWT, the following tests were carried out. The first test was done to ensure correctness. Then, reporting all versus all L-long matches was benchmarked using HP-PBWT, Wertenbroek et al.'s parallel PBWT, and a corrected version of Durbin's PBWT. Finally, an IO-excluded benchmark was performed for both HP-PBWT and sequential PBWT by reporting all versus all L-long matches to evaluate the scalability of the HP-PBWT-based algorithms.

4.1 Benchmark Design

Two sets of experiments were designed: IO-included and IO-excluded. The IO-included experiments test real-world scenarios with hard drive input and output. VCF files were used as input for the IO-included experiments. To further evaluate the scalability of HP-PBWT, IO-excluded benchmark of HP-PBWT was implemented and an IO-excluded sequential PBWT was implemented as well. The idea of the IO-excluded benchmark was to remove the IO influence. This is similar to the run times presented by Wertenbroek et al. in the main paper. The IO-excluded benchmark is performed by randomly generating IO-excluded panels and then running the report all versus all L-long match algorithm without outputting matches to the hard drive. Dependencies were added to HP-PBWT to make sure computations were not optimized out by the compiler.

Three measurements were used to evaluate HP-PBWT: run time, speed-up, and parallel efficiency. The run time of IO-included experiments included time for both reading the input and outputting matches. For the IO-excluded experiments, the run time was measured after generating the whole panel X in memory and without outputting matches. The speed-up was calculated by the sequential run time divided by the parallel run time. The parallel efficiency was calculated by speed-up divided by the number of cores.

The IO-included tests used chromosome 20 of the 1000 Genomes Project [1], and chromosome 20 of the UK Biobank [17]. To further test the scalability of these tools, randomly generated VCF files were also used with a fixed dimension of $N = 1000$ sites and M haplotypes that $M = 1000 \times 2^i$ where $i \in [0, 15]$. The reason that 1000×2^{15} was used as the largest input was because both Durbin's and Wertenbroek et al.'s parallel PBWT use the same VCF handling library (HTSlib [3]) that can not read datasets with $M \geq 1000 \times 2^{16}$. Thus, the largest input of these experiments was $M = 1000 \times 2^{15}$. The length cut-offs were 2000 sites for the 1000 Genomes Project chromosome 20 (the same as in Wertenbroek et al.'s benchmark), 1600 sites for the UK Biobank chromosome 20, and 30 sites for the randomly generated files.

The IO-excluded HP-PBWT was benchmarked against the IO-excluded sequential PBWT with $M = 1000 \times 2^i$ haplotypes for $i \in [0, 31]$ and a fixed dimension of $N = 100$ sites. Since the M dimension was to reach 2 billion, only 100 sites were generated for the sake of memory capacity. The length cut-off was $L = 50$. To create stable run times all the IO-included tests were executed 10 times. For the IO-excluded tests, each experiment was executed 10 times for $i \in [0, 9]$, 4 times for $i \in [10, 17]$, and 2 times for $i \in [18, 31]$.

The IO-included tests were executed on local servers with Intel^R Xeon^R CPU E5-2683 v4. The IO-excluded tests were executed in Windows Server 2022 on the Amazon EC2 servers which were powered by 3.6 GHz 3rd generation AMD EPYC 7R13 processors. To eliminate the interference of different programming languages and operating systems, the IO-excluded sequential PBWT was also programmed in C#. For HP-PBWT 10, 20, 30, 40, 50, and 60 cores were used in both IO-included and IO-excluded tests, meanwhile a 12-core setting was also used in the IO-included tests. All Wertenbroek et al.'s parallel PBWT's tests in this paper use 12 threads.

Table 1. IO-included run time (in seconds) for real datasets. "1KG" stands for 1000 Genomes Project, "UKB" refers to UK Biobank.

PBWT version	1KG Chr.20	UKB Chr.20
Durbin's	236	409
Wertenbroek et al.'s T = 12	320	652
HP-PBWT T = 10	433	227
HP-PBWT T = 12	404	199
HP-PBWT T = 20	389	130
HP-PBWT T = 30	448	104
HP-PBWT T = 40	766	133
HP-PBWT T = 50	874	131
HP-PBWT T = 60	989	139

4.2 IO-Included Benchmarks

The run times in Table 1 show Wertenbroek et al.'s parallel PBWT's best run time (from 12 threads) did not have any speed-up comparing to Durbin's version on either chromosome 20 of UK Biobank genotyping array data or chromosome 20 of 1000 Genomes Project sequencing data. With the same 12-thread setting, HP-PBWT had 2-fold speed-up on chromosome 20 of UK Biobank genotyping array data. Meanwhile HP-PBWT's best run time had about 4-fold speed-up on chromosome 20 of UK Biobank genotyping array data type with 30 threads. HP-PBWT did not have any speed-up in 1000 Genomes Project sequencing data, since the M dimension of 1000 Genomes Project is too small, it only has 5008 haplotypes, and the parallelization of HP-PBWT is not designed for this type of inputs. The speed-ups show that HP-PBWT can improve the run time of PBWT in large population biobank genotyping array data. Meanwhile, further improvements are needed for HP-PBWT to deal with sequencing data with less amount of haplotypes.

To test HP-PBWT's performance on large amount of population, randomly generated VCF files were used, since there was not any real dataset with the number of samples is greater than 1 million available to the authors. Figure 4(A) shows the run times of all tools increased when the input size M increased. The speed-up in Fig. 4(B) shows the Wertenbroek et al.'s parallel PBWT's best speed-up was about 1.6-fold, HP-PBWT had a 3-fold speed-up with 12 threads and a 7.2-fold speed-up with 60 threads on the largest input comparing to the corrected Durbin's PBWT. It shows that HP-PBWT started to gain speed-up to the corrected Durbin's PBWT at the input of $M = 64\,k$, meanwhile Wertenbroek et al.'s parallel PBWT started to gain speed-up at $M = 4\,m$ with 1.04-fold speed-up and reached its best 1.6-fold speed-up at $M = 32\,m$.

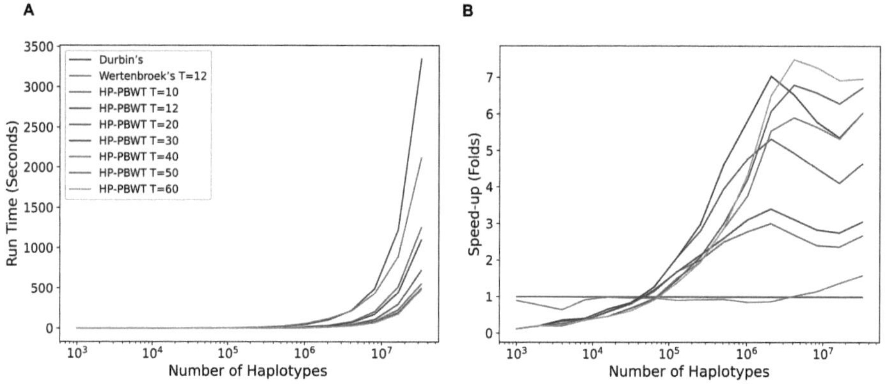

Fig. 4. IO-included run time (A) and speed-up (B) based on the random dataset across different versions of PBWT.

4.3 IO-Excluded Benchmarks

Since Durbin's PBWT does not have an IO-excluded benchmark mode and Wertenbroek et al.'s benchmark mode only reads and converts hard drive files into memory. Heavy modifications should not be applied to these versions to fit the benchmark purposes. Thus, in the IO-excluded benchmarks, the IO-excluded HP-PBWT was only benchmarked with the IO-excluded sequential PBWT.

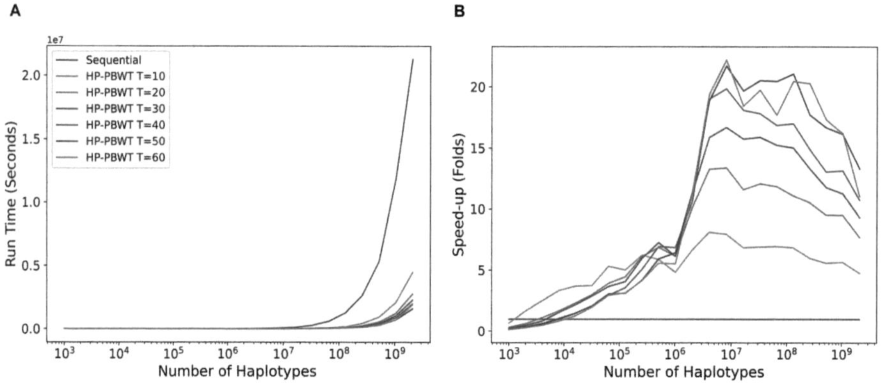

Fig. 5. IO-excluded run time (A) and speed-up (B) based on the random dataset between HP-PBWT and re-implemented IO-excluded sequential PBWT.

The run time in Fig. 5(A) and the speed-ups in Fig. 5(B) show that HP-PBWT had better performance when the size of the input increased. The best performance of HP-PBWT against the sequential PBWT was 22.2-fold at $M = 8,192,000$ with 60 threads. The benchmarks also showed the maximum number

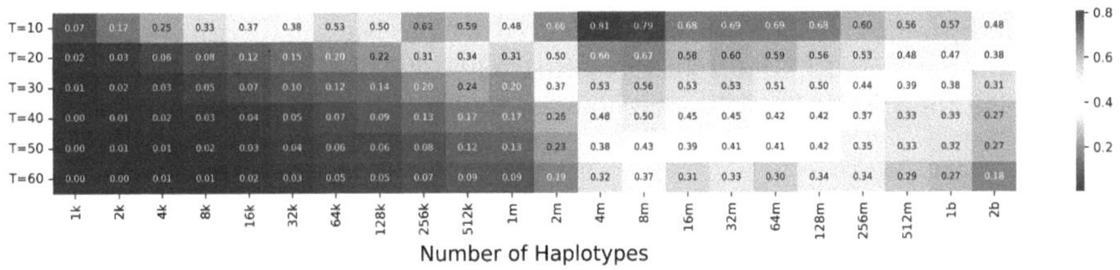

Fig. 6. Parallel efficiency from IO-excluded tests. The parallel efficiency is calculated by speed-up divided by the number of threads.

of threads (60) setting used in these experiments did not have the best run time all the time. For the largest 2 billion data input, the sequential PBWT took 5.9 h and HP-PBWT with 50 threads execution had the shortest run time, 0.4 h, a 13.3-fold speed-up. This benchmark was performed on the Amazon EC2 servers. The speed-up shown in Fig. 5(B) and parallel efficiency shown in Fig. 6 started to fall once M increased beyond 8 million. There are a couple of potential impact factors. First, the estimated $O(\frac{M}{T}+T+c^*)$ time complexity per site for reporting all versus all L-long matches is mainly dominated by the number of haplotypes M, but once M increases to a certain level c^*, the maximum number of matches per haplotype also becomes larger. On the other hand, this also reflects the parallel computational theory [7], that the more threads that are being used, the more idle worker time there is. The run time increased accordingly with the number of matches, once M reached some millions, the run time increased significantly. Second, CPU temperature also has a major impact on run time. That is why nowadays CPU and GPU cooling methods are being researched and developed constantly [12,14,15].

5 Discussion

In this paper, new algorithms were designed to break the dependencies that prevent Durbin's PBWT from being executed in parallel on the haplotype dimension M. HP-PBWT algorithms enable parallel PBWT on the panel construction of prefix arrays, divergence arrays, reporting all versus all L-long matches, and reporting all versus all set-maximal matches. HP-PBWT is both memory and IO efficient. HP-PBWT can be applied to any PBWT-based applications to leverage much larger genetic panels up to billions of haplotypes efficiently.

Currently, HP-PBWT works well on biobank scale genotyping array data, but does not have speed-up on sequencing data with small amount of population. It is likely due to the partition mechanisms in prefix array and divergence array computations that they might have too much overhead. New algorithms and optimizations can be researched for the sequencing datasets. Another possible way to speed up the whole process is to apply both haplotype-based parallel PBWT and site-based parallel PBWT algorithms simultaneously.

Acknowledgments. This work was supported by the National Institutes of Health under award numbers R01HG010086 and R01AG081398. This research has been conducted using the UK Biobank Resource under application number 24247.

References

1. 1000 Genomes Project Consortium: A global reference for human genetic variation. Nature **526**(7571), 68 (2015)
2. Blelloch, G.E.: Scans as primitive parallel operations. IEEE Trans. Comput. **38**(11), 1526–1538 (1989)
3. Bonfield, J.K., et al.: HTSlib: C library for reading/writing high-throughput sequencing data. GigaScience **10**(2), giab007 (2021)
4. Browning, B.L., Tian, X., Zhou, Y., Browning, S.R.: Fast two-stage phasing of large-scale sequence data. Am. J. Hum. Genet. **108**(10), 1880–1890 (2021)
5. Browning, B.L., Zhou, Y., Browning, S.R.: A one-penny imputed genome from next-generation reference panels. Am. J. Hum. Genet. **103**(3), 338–348 (2018)
6. Durbin, R.: Efficient haplotype matching and storage using the positional burrows-wheeler transform (PBWT). Bioinformatics **30**(9), 1266–1272 (2014)
7. Eager, D.L., Zahorjan, J., Lazowska, E.D.: Speedup versus efficiency in parallel systems. IEEE Trans. Comput. **38**(3), 408–423 (1989)
8. Freyman, W.A., et al.: Fast and robust identity-by-descent inference with the templated positional burrows–Wheeler transform. Mol. Biol. Evol. **38**(5), 2131–2151 (2021)
9. Hofmeister, R.J., Ribeiro, D.M., Rubinacci, S., Delaneau, O.: Accurate rare variant phasing of whole-genome and whole-exome sequencing data in the UK biobank. Nat. Genet. **55**(7), 1243–1249 (2023)
10. Mathieson, I., Scally, A.: What is ancestry? PLoS Genet. **16**(3), e1008624 (2020)
11. Naseri, A., Liu, X., Tang, K., Zhang, S., Zhi, D.: RaPID: ultra-fast, powerful, and accurate detection of segments identical by descent (IBD) in biobank-scale cohorts. Genome Biol. **20**(1), 1–15 (2019)
12. Ramakrishnan, B., et al.: CPU overclocking: a performance assessment of air, cold plates, and two-phase immersion cooling. IEEE Trans. Compon. Packag. Manuf. Technol. **11**(10), 1703–1715 (2021)
13. Rubinacci, S., Delaneau, O., Marchini, J.: Genotype imputation using the positional burrows wheeler transform. PLoS Genet. **16**(11), e1009049 (2020)
14. Shahi, P., et al.: Assessment of reliability enhancement in high-power CPUs and GPUs using dynamic direct-to-chip liquid cooling. J. Enhanced Heat Transfer **29**(8) (2022)
15. Siricharoenpanich, A., Wiriyasart, S., Naphon, P.: Study on the thermal dissipation performance of GPU cooling system with nanofluid as coolant. Case Stud. Therm. Eng. **25**, 100904 (2021)
16. Stoeklé, H.C., Mamzer-Bruneel, M.F., Vogt, G., Hervé, C.: 23andMe: a new two-sided data-banking market model. BMC Med. Ethics **17**, 1–11 (2016)
17. Sudlow, C., et al.: UK Biobank: an open access resource for identifying the causes of a wide range of complex diseases of middle and old age. PLoS Med. **12**(3), e1001779 (2015)
18. Taliun, D., et al.: Sequencing of 53,831 diverse genomes from the NHLBI TOPMed Program. Nature **590**(7845), 290–299 (2021)

19. Wertenbroek, R., Xenarios, I., Thoma, Y., Delaneau, O.: Exploiting parallelization in positional burrows-wheeler transform (PBWT) algorithms for efficient haplotype matching and compression. Bioinf. Adv. **3**(1), vbad021 (2023)
20. Zhou, Y., Browning, S.R., Browning, B.L.: A fast and simple method for detecting identity-by-descent segments in large-scale data. Am. J. Hum. Genet. **106**(4), 426–437 (2020)

Mammo-Bench: A Large-Scale Benchmark Dataset of Mammography Images

Gaurav Bhole(✉), S. Suba, and Nita Parekh

IIIT-Hyderabad, Gachibowli, India

{gaurav.bhole,suba.s}@research.iiit.ac.in, nita@iiit.ac.in

Abstract. Breast cancer remains a significant global health concern, and machine learning algorithms and computer-aided detection systems have shown great promise in enhancing the accuracy and efficiency of mammography image analysis. However, there is a critical need for large, benchmark datasets for training deep learning models for breast cancer detection. In this work we developed Mammo-Bench, a large-scale benchmark dataset of mammography images, by collating data from seven well-curated resources, *viz.*, DDSM, INbreast, KAU-BCMD, CMMD, CDD-CESM, DMID, and RSNA Screening Dataset. To ensure consistency across images from diverse sources while preserving clinically relevant features, a preprocessing pipeline that includes breast segmentation, pectoral muscle removal, and intelligent cropping is proposed. The dataset consists of 74,436 high-quality mammographic images from 26,500 patients across 7 countries and is one of the largest open-source mammography databases to the best of our knowledge. To show the efficacy of training on the large dataset, performance of ResNet101 architecture was evaluated on Mammo-Bench and the results compared by training independently on a few member datasets and an external dataset, VinDr-Mammo. An accuracy of 78.8% (with data augmentation of the minority classes) and 77.8% (without data augmentation) was achieved on the proposed benchmark dataset, compared to the other datasets for which accuracy varied from 25 – 69%. Noticeably, improved prediction of the minority classes is observed with the Mammo-Bench dataset. These results establish baseline performance and demonstrate Mammo-Bench's utility as a comprehensive resource for developing and evaluating mammography analysis systems.

Keywords: Mammogram Dataset · Breast Cancer Diagnosis · Computer-Aided Detection · Medical Imaging · Deep Learning

1 Introduction

Early detection of breast cancer through reliable screening methods such as mammography remains crucial for successful treatment outcomes, and hence mammogram screening programs have been established worldwide. The interpretation of mammograms,

Supplementary Information The online version contains supplementary material available at https://doi.org/10.1007/978-3-032-02489-3_11.

M. Alser et al. (Eds.): ICCABS 2025, LNCS 15599, pp. 144–156, 2026.
https://doi.org/10.1007/978-3-032-02489-3_11

however, requires careful analysis of various abnormalities such as masses (dense regions with varying shapes or patterns), calcifications (calcium deposits), architectural distortions (irregular patterns in breast tissue structure) and asymmetries (differences between corresponding regions in paired mammograms). This complexity, combined with factors like radiologist fatigue and varying expertise levels across hospitals, can lead to incorrect diagnosis. Computer Aided-Detection (CAD) systems have emerged as valuable tools to assist radiologists in identifying and classifying suspicious lesions, potentially improving detection accuracy and reducing false positives. The effectiveness of CAD systems, particularly those based on machine learning algorithms, is fundamentally dependent on the quality and size of training data. Large number of high-quality mammography images are required to capture the diverse manifestations of abnormalities and provide comprehensive annotations that include abnormality types and clinical metadata such as breast density, BI-RADS scores and molecular subtype information. Balanced representation of different case types and diverse patient demographics are crucial for developing reliable diagnostic models. Recent studies [1] have shown that such comprehensive datasets are required for developing robust deep learning models that can effectively generalize across different clinical settings and patient populations.

While several public mammography datasets exist, they often have limitations that restrict their utility for developing comprehensive CAD systems. Common challenges include small numbers of samples, inconsistencies in image quality and annotations, and significant class imbalances. Additionally, most datasets are collected from single institutions, potentially limiting their generalizability across different populations and imaging protocols. A large benchmark mammography dataset is crucial in development of machine learning methods since it would provide a "baseline" for what current algorithms can achieve on a standardized set of images and offer a reference point for setting goals and measuring progress in the research applications.

After evaluating various factors such as size, image quality, annotation completeness, and clinical metadata of mammography datasets in the public domain, we selected seven well-curated datasets: DDSM [2], INbreast [3], KAU-BCMD [4], CMMD [5], CDD-CESM [6], RSNA Screening Dataset [7] and DMID [8]. We constructed a large benchmark dataset, Mammo-Bench, that addresses the limitations of individual datasets by unifying and standardizing data from these seven sources. Before merging the images from individual resources, a rigorous pre-processing pipeline is proposed to ensure consistency and improve quality of the dataset. Annotations such as BI-RADS scores, breast density, abnormalities and asymmetries, and molecular subtypes were extracted from the respective member databases (if available), and masks for Regions of Interest (ROI) were generated by us for all the images. Mammo-Bench is a collection of 74,436 high-quality mammographic images from 26,500 patients across 7 countries. To the best of our knowledge, there is only one another resource, IRMA*, with 10,509 images, which has collated data from four resources, *viz.*, DDSM [2, 9], MIAS [10], LLNL* [11] and RWTH* [12]; however, it is not an open-source dataset.

In this work, we have constructed a large-scale unified mammography dataset by collating and standardizing data from seven well-curated resources. To achieve this, we proposed a comprehensive preprocessing pipeline that incorporates uniform image background and data format, breast segmentation and provide masks for regions of interest (RoIs), pectoral muscle removal, and intelligent cropping to ensure image consistency. Further, annotations such as BI-RADS scores, breast density, abnormality types, and molecular subtypes (extracted from individual sources) are provided to aid in breast cancer diagnosis. The detailed summary of the attributes of publicly available datasets is given in Table 1.

To show the utility of the dataset, we performed various experiments. Three-class classification tasks (Normal, Benign, Malignant) with and without data augmentation, and a hierarchical binary classification strategy was implemented to address data imbalance. Our results demonstrate that this methodology significantly outperforms multi-class classification, suggesting a more robust path forward for CAD systems.

2 Related Works

Mammography datasets are vital resources for advancing breast cancer diagnosis and CAD systems and are categorized as open-source (e.g., DDSM, INbreast, etc.), or with restricted-access (e.g., OPTIMAM*[1] [13], VinDr-Mammo* [14], etc.), depending on accessibility. Though well-curated, the publicly available mammographic datasets vary considerably in their coverage, size, annotations, quality, and accessibility, which affects their utility in diverse research applications. For example, image quality varies across datasets, for e.g., DDSM, MIAS and BancoWeb database* [15] contain older film-based digitized images, while recent datasets like CDD-CESM, LAMIS-DMDB* [16] provide high-quality digital mammograms. The size of the datasets is another critical issue, with INbreast, DMID and MIRacle dataset* [17] containing fewer images (< 500), while screening datasets like RSNA contain thousands of images. Further, annotation quality and completeness vary widely across datasets, with ROI annotations available only for few datasets and that too not for the complete set (e.g., DDSM and DMID), impeding their use in lesion detection and localization tasks. Attributes such as breast density, BI-RADS scores, abnormality type, and molecular subtype are crucial in the accurate detection of breast malignancies. However, large variation is observed in the annotation details across the mammography datasets, for example, INbreast, KAU-BCMD, CDD-CESM, RSNA and DMID provide BI-RADS scores and breast density but no information about other abnormalities, while CMMD, and DMID provide abnormality type and molecular subtype annotations. Another factor of concern is data imbalance. The screening datasets such as RSNA, DDSM, NYU Screening Dataset* [18] and Cohort of Screen-Aged Women (CSAW)* [19] are large but highly imbalanced with the number of normal images (~12×) compared to other classes. Most datasets include only mammography data while some provide multi-modal imaging data such as ultrasound, Tomosynthesis and MRI (e.g., OPTIMAM*) along with mammograms. Datasets also differ in terms of image formats, for example, RWTH dataset* (DICOM), BancoWeb* (TIFF), LLNL*

[1] *dataset with restricted access.

(image cytometry standard (ICS)) and KAU-BCMD (JPG). Majority of mammography datasets are population specific with data collected from a single or multiple hospitals from the same demography, thereby limiting their use for developing generalizable models. Therefore, an open-access database consisting of large samples with complete annotations would be valuable for researchers developing AI models in breast cancer detection using mammograms.

3 Dataset Construction

3.1 Description of the Dataset

The workflow for construction of Mammo-Bench, is given in Fig. 1. First, data was downloaded from seven mammography datasets: DDSM, INbreast, KAU-BCMD, CMMD, CDD-CESM, RSNA Screening Dataset and DMID (links provided in Supplementary Table 1). A brief description of the member databases is given below.

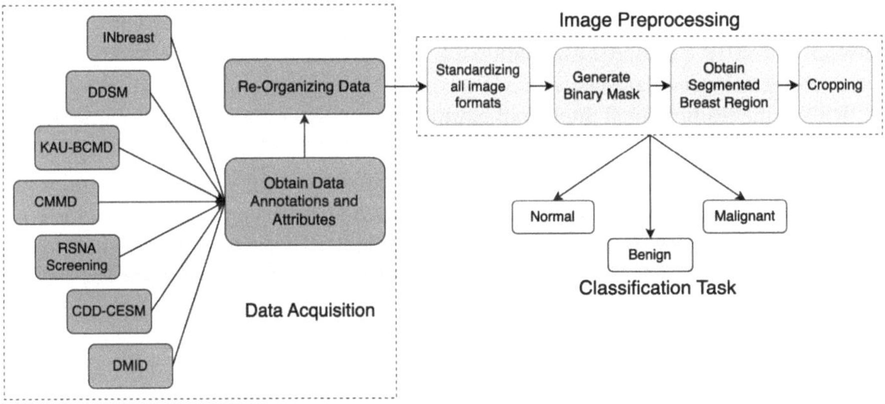

Fig. 1. Overview of the Data Flow Pipeline

DDSM, the Digital Database for Screening Mammography [3] consists of 10,400 images from 2,620 patients in GIF format collected at Massachusetts General Hospital and Wake Forest University School of Medicine between 1990–1999. It offers a balanced distribution of 2,776 normal, 3,968 benign, and 3,656 malignant cases. Metadata provided includes cancer status (benign, malignant, or normal), patient age, and ROI boundaries for 2,830 images.

INbreast dataset [2], created by the breast research group at Centro Hospitalar de São João, Portugal, includes 115 cases with 410 digital mammogram images in DICOM format. This dataset provides detailed metadata, including BI-RADS scores, breast density, pixel-level lesion contours validated by two experts, and radiology reports. However, its major limitation is dataset size and class imbalance, with 67 normal, 243 benign, 43 suspicious malignant, and 57 malignant cases.

KAU-BCMD, the King Abdulaziz University Breast Cancer Mammogram Dataset [4] consists of 1,416 cases and 2,206 images in JPG format, collected between Apr' 2019 – Mar' 2020 at the Sheikh Mohammed Hussein Al-Amoudi Center of Excellence in Breast Cancer. Metadata includes patient age, previous screenings, breast density (manually assessed by radiologists), and BI-RADS scores (determined by three radiologists' majority vote). About 172 images lack any annotation and were not included in the Mammo-Bench construction.

CMMD, the Chinese Mammography Database [5] comprises 5,202 images in DICOM format from 1,775 patients who underwent mammography examinations between July 2012 – Jan' 2016, and is categorized into two subsets: CMMD1 (1,026 cases, biopsy-confirmed benign or malignant tumors) and CMMD2 (2,956 images with molecular subtype information: Luminal A, Luminal B, HER2-positive, and Triple-negative). This dataset includes age, benign/malignant labels, and abnormality type (mass, calcification or both), with images curated by two radiologists. Notably, it lacks ROI annotations but provides valuable molecular subtype data.

CDD-CESM, the Categorized Digital Database for Low Energy and Subtracted Contrast Enhanced Spectral Mammography [6] contains 1,003 low-energy images in JPG format and corresponding subtracted images, from 326 female patients collected between Jan'2019 – Feb'2021 at National Cancer Institute, Cairo University, Egypt. It is one of the balanced datasets with 341 normal, 331 benign, and 331 malignant cases.

RSNA, the Radiological Society of North America Screening Mammography Breast Cancer Detection dataset [7] features a large collection of 54,705 DICOM images from approximately 20,000 patients, collected from BreastScreen Victoria, Australia, and Emory University, USA. Metadata includes patient age, breast density, BI-RADS score, case difficulty level, and the presence of breast implants. This screening dataset is highly imbalanced, with 24,021 normal and only 2,265 benign cases, and does not include segmentation or ROI annotations.

DMID, the Digital Mammography Dataset for Breast Cancer Diagnosis Research [8], from Samved Hospital, India, contains 510 images in both DICOM and TIFF formats. It includes metadata on case types (normal, benign, malignant), BI-RADS scores, breast density, and abnormality type (calcification, mass and both). Each abnormality region is annotated with center coordinates, radius, and ROI mask. This dataset also provides balanced representation of normal and abnormal cases.

Table 1. Summary of annotations and attributes of the seven publicly available datasets considered for the construction of Mammo-Bench

Dataset Features	DDSM	INbreast	KAU-BCMD	CMMD	CDD-CESM	RSNA	DMID	Mammo-Bench
Origin	USA	Portugal	Saudi Arabia	China	Egypt	USA/Australia	India	Diverse
Year	2001	2012	2021	2021	2022	2022	2023	2025
No. of Cases	2,620	115	1,416	1,775	326	20,000	NA	**26,500**
No. of Images	10,400	410	2,206	5,202	1,003	54,705	510	**74,436**
Img. Format	JPG	DICOM	JPG	DICOM	JPG	DICOM	DICOM	JPG
View & Lat.	✓	✓	✓	✓	✓	✓	✓	✓
N/B/M labels	✓	✓	✓	✓	✓	✓	✓	✓
BI-RADS	✗	✓	✓	✗	✓	✓	✓	✓
Breast Density	✓	✓	✓	✗	✓	✓	✓	✓
Abnormality	✗	✗	✗	✓	✗	✗	✗	✓
Mol. Subtype	✗	✗	✗	✓	✗	✗	✗	✓
ROI Mask	✓	✓	✗	✗	✗	✗	✓	✓
Age	✓	✗	✓	✓	✓	✓	✗	✓
Asymmetry	✗	✗	✗	✗	✗	✗	✓	✓

3.2 Data Acquisition

After preliminary filtering of images with no available annotation (172 from KAU-BCMD, 1 from RSNA), 74,436 images from ~26,500 cases were considered for the construction of Mammo-Bench. Typically, four images per subject corresponding to two views for each breast are present: Mediolateral Oblique (MLO) and Craniocaudal (CC). The mapping between BI-RADS score and disease status labels is performed as given in Table 2 (according to [2, 4, 6, 8]) and all the cases with BI-RADS score 4a, 4b and 4c, have been relabeled to score 4. A summary of the annotations provided in the dataset is given in Table 2 and is briefly described below.

Class Labels: Defines the disease status of the cases (images) as Normal, Benign, Suspicious Malignant, or Malignant.

Breast Density: Breast density gives information about the different types of breast tissue, such as fat, fibrous, and glandular tissues. It is categorized as per the ACR standards into four categories A to D, as defined in Table 2.

BI-RADS Score: It is a standardized rating system used by radiologists to describe breast imaging test results and ranges from 0 to 6.

Abnormality: An abnormality is anything unusual found in breast tissue such as mass lumps or tiny calcium deposits.

Molecular Subtype: Based on the expression of the receptors, Progesterone (PR), Estrogen (ER), and human epidermal growth factor receptor 2 (HER2), breast cancer is grouped into four molecular subtypes, *viz.*, Luminal A, Luminal B, HER2-positive, and Triple-negative, following the St. Gallen International Expert Consensus [20].

Table 2. Description of various attributes of the images in the Mammo-Bench dataset

Labels	No. of Images	Class	Images in the Class
Normal/Benign/ Malignant	46,017	Normal (N)	29,264
		Benign (B)	8,334
		Suspicious Malignant (SM)	235
		Malignant (M)	8,184
Density	43,911	ACR A (Fatty)	5,372
		ACR B (Fatty + Scattered Areas of Fibroglandular Density)	18,299
		ACR C (Heterogeneously Dense)	16,418
		ACR D (Extremely Dense)	3,822
BI-RADS Score	30,383	0 (Additional Diagnosis Required)	8,250
		1 (Normal Findings)	18,325
		2 (Benign)	2,670
		3 (Probably Benign)	455
		4 (Suspicious Malignant)	358
		5 (>95% chance Malignant)	313
		6 (Biopsy Proven Malignant)	12
Abnormality	5,712	Mass	3,344
		Calcification	747
		Both	1,411
Molecular Subtype	2,956	Luminal A	600
		Luminal B	1,482
		HER2-enriched	532
		Triple Negative	342

After the BI-RADS score – status label mapping, the number of images with Normal, Benign, Suspicious Malignant and Malignant class labels are 46,017 images (originally only 30,383 images had BI-RADS score). The number of images with breast density information is 43,911 of which 38,117 images also have Normal (26,662), Benign (6,597), Malignant (4,061) and Suspicious Malignant (233) labels. Breast density annotation is valuable for studying relationships between breast density and cancer risk. Assessment of breast density is an important factor to the radiologists, as it is also used as a clinical indicator in deciding BI-RADS scores. It also helps in making informed decisions about additional screenings as dense breasts may occlude the presence of masses and calcified lesions in the mammography images making early detection of breast cancer challenging. Additionally, 5,712 images are annotated for specific abnormality types (mass, calcification, both), enabling detailed radiological assessment tasks. Molecular subtype information is available for only 2956 images from a single member database, CMMD. It helps mapping the morphological differences between the subtypes thereby providing a promising step towards precision medicine.

3.3 Preprocessing

To improve the quality of mammography images collected from diverse datasets and ensure uniformity, following preprocessing steps were performed (Fig. 2).

Data Format: Initially, all images were converted to JPG format from their original formats, and images with white backgrounds were standardized to black.

Breast Segmentation: The OpenBreast toolkit [21] was used to isolate the breast region through a multi-step segmentation process, as depicted in the second column of Fig. 2. This process begins with background detection using a threshold-based method on the intensity histogram, followed by chest wall detection using a Hough-based line detector that represents edge pixels in parametric space. The process also ensures retention of the nipple region, identified as the farthest contour point from the detected chest wall.

Pectoral Muscle Removal: The pectoral muscle, commonly observed in MLO views and other irrelevant anatomical structures, which may confound in image analysis, are removed using the OpenBreast toolkit.

Mask Application: The segmentation process results in a binary mask for each image (third column in Fig. 2). These masks effectively isolate the breast region while excluding irrelevant anatomical structures, and adapt to different breast shapes and orientations, both for CC views (Fig. 2 (a)) and MLO views (Fig. 2 (b)).

Image Cropping: An adaptive cropping algorithm based on contour detection [22] which removes extraneous background, retaining only the breast region is implemented (rightmost column in Fig. 2). This algorithm applies a binary threshold determined through histogram analysis and empirical testing to identify the largest contour as the breast region, and crops the image along the computed bounding rectangle with appropriate padding to preserve the entire breast tissue.

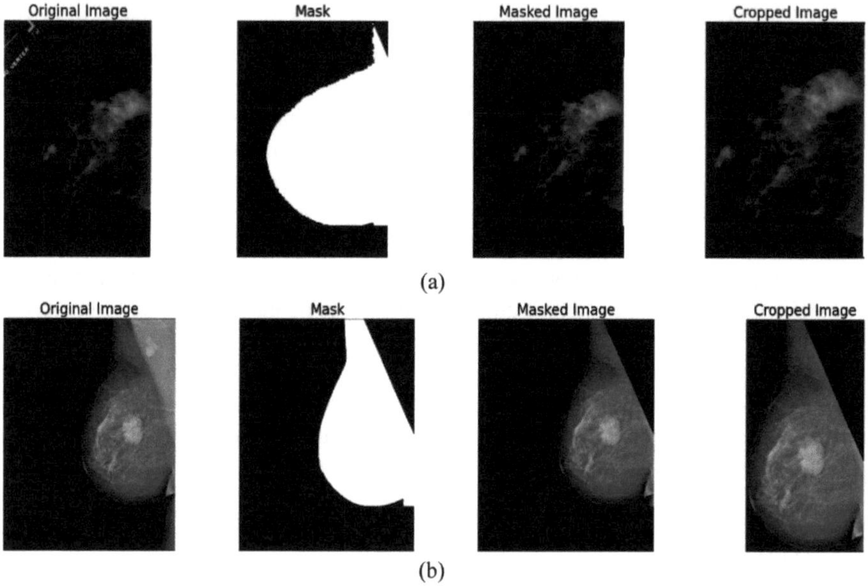

Fig. 2. Illustration of the preprocessing pipeline: From left to right: Original mammogram, binary mask, breast region segmentation, and final cropped image for (a) Left-CC view (DDSM) and (b) Right-MLO view (INbreast). The effective removal of pectoral muscle in the MLO view and preservation of essential breast tissue in both views may be noted.

3.4 Organization of the Dataset

The data is organized into four main folders: 'Original Dataset' - contains the raw images, 'Masks' - generated breast region segmentation masks, 'Preprocessed Dataset' - processed images after applying the workflow (given in Fig. 1), and 'CSV Files' - provide the clinical metadata. Each directory of images is organized into subfolders based on individual datasets. Within these subfolders, each image is assigned a unique identifier in the format 'dataset_imgID.jpg'. A CSV file containing associated annotations from the member databases and paths to map images between Mammo-Bench and the individual datasets is provided. This is accompanied by helper codes in python for easy accessibility of the data. To support classification tasks with varied annotation labels, this file has been split into five separate CSV files, each tailored to specific classification tasks such as BI-RADS scores (0–6), breast density (A-D), abnormality)mass, calcification, both), and molecular subtype (Luminal A and B, Her2-enriched, Triple Negative). This organization of Mammo-Bench allows assessment of the impact of these attributes in classifying Normal, Benign, and Malignant cases.

4 Performance Evaluation

To assess the utility of the proposed dataset, Mammo-Bench, we performed the classification of the images (Normal, Benign, Malignant) using ResNet101 architecture pre-trained on the ImageNet database. Comparative performance evaluation was conducted by training ResNet101 on Mammo-Bench and three member datasets, *viz.*, INbreast, CDD-CESM, DMID and one external dataset, VinDr-Mammo. First, to assess the impact of data imbalance, three experiments were performed using Mammo-Bench: three-class classification with and without data augmentation and hierarchical binary classification. Two data augmentation strategies were implemented on the minority class, namely, random rotation up to 10 degrees, and color jittering with brightness and contrast adjustments. Both transformations included resizing images to 224×224 and normalization using standard ImageNet statistics. For all experiments, the data was split into 80:20 for train-test sets, and performance was assessed using the metrics, precision, recall, F1-score, accuracy and Mathew's correlation coefficient (MCC). In all cases the models were trained for 50 epochs, and the best weights were used for testing. The loss function used was categorical/binary cross-entropy and the optimizers were Adam and SGD.

4.1 Three-Class Classification

In Table 3, performance of the ResNet101 model on Mammo-Bench dataset, with and without data augmentation, is given. As is seen from Table 3, there is only a marginal improvement in accuracy from 0.778 to 0.789 and MCC value from 0.59 to 0.61, on data augmentation. No difference in the precision (~0.86) and recall (~0.92) values of the normal class is observed, while marginal improvement in recall (0.74 to 0.77) for the malignant class and in precision (0.50 to 0.54) for the benign class is observed. This is probably due to the significant number of images in the two minority classes and ability of the deep architecture of ResNet101 in handling data imbalances to certain extent. For comparison, ResNet101 was independently trained on four datasets, *viz.*, INbreast, CDD-CESM, DMID, and VinDr-Mammo and the averaged value of the performance metrics is given in Table 4. It may be noted that there is a significant drop in the performance of ResNet101 on these datasets, with accuracies ranging from 0.25 (DMID) to 0.69 (VinDr-Mammo). Though performance on VinDr-Mammo is reasonably good, it may be noted from Supplementary Table 1, that the recall of Benign and Malignant classes is ~0, i.e., all the images are predicted as normal. These results clearly indicate the importance of a large benchmark dataset with good representation of all the classes for robust model building.

4.2 Hierarchical Binary Classification

Next, we performed a two-step binary classification to handle class imbalance. First the model was trained to classify images into normal and abnormal (benign and malignant) cases. Merging benign and malignant classes helped in balancing the classes to a certain extent (Table 3). Next, another model was trained on the correctly predicted abnormal images from the previous step, to classify the images as benign or malignant. It may be noted that the model achieved an accuracy of 0.89 to distinguish between the normal and abnormal cases, and 0.73 to distinguish between Benign and Malignant classes. There is a significant improvement in the recall and precision values of the Benign and Malignant classes compared to 3-class classification results. Notably, performance of the Benign class improved significantly without affecting the performance of other two classes. This again shows the utility of Mammo-Bench in building strategic models with superior performances.

Table 3. Performance comparison of ResNet101 model for three-class classification without augmentation, with augmentation of minority classes and hierarchical binary classification using Mammo-Bench

Approach	Class	Precision	Recall	F1-Score	Accuracy	MCC
Three-Class Classification	Normal	0.865	0.928	0.895	0.778	0.595
	Benign	0.502	0.369	0.425		
	Malignant	0.708	0.743	0.725		
Three-Class Classification with Augmentation of Minority Classes	Normal	0.869	0.929	0.898	0.788	0.614
	Benign	0.546	0.382	0.45		
	Malignant	**0.709**	0.777	0.741		
Hierarchical Binary Classification	Normal	**0.876**	**0.954**	**0.913**	0.891	0.774
	Abnormal	0.92	0.798	0.854		
	Benign	**0.78**	**0.67**	**0.72**	0.736	0.479
	Malignant	0.7	**0.81**	**0.75**		

Table 4. Performance comparison of ResNet101 on various mammography datasets

Dataset	Total Images	Precision	Recall	F1-Score	Accuracy	MCC
INbreast	367	0.462	0.607	0.498	0.607	−0.004
CDD-CESM	1,003	0.504	0.5	0.491	0.5	0.256
VinDr-Mammo	19,238	0.677	0.698	0.592	0.698	0.105
DMID	390	0.55	0.25	0.234	0.25	0.155
Mammo-Bench	34,721	**0.760**	**0.778**	**0.766**	**0.778**	**0.595**

5 Discussion

FAIRness of the Dataset: Mammo-Bench adheres to the FAIR (Findable, Accessible, Interoperable, and Reusable) principles [23]. In terms of Findability, the dataset is described with rich metadata, including detailed information on patient demographics, abnormality types, breast density categories, and BI-RADS classifications. Its structure, with unique image IDs and uniform file structure and paths enhance its findability and organization. Accessibility is ensured through documentation of the dataset's composition and preprocessing methods, while Interoperability is enhanced by the standardized preprocessing pipeline applied across source datasets. The preprocessing ensures consistent image format and quality. The documentation (including description of data sources, preprocessing methods and annotations) ensures reusability and compatibility across different analysis platforms and machine learning frameworks.

The primary advantage of this dataset lies in its extensive collection of images, which can be utilized for developing and training deep learning models across various detection and classification tasks. It is free of any unlabeled or missing files, ensuring its reliability and value for a wide range of analyses. The dataset's standardized and user-friendly format further enhances its uniqueness as a comprehensive resource for mammography image analysis. A limitation of this dataset is that class imbalance persists even after combining multiple datasets. This requires the need for data balancing strategies during model training.

6 Conclusion

In this paper we present a large-scale, unified mammogram benchmark dataset consisting of 74,436 images. The comprehensive preprocessing pipeline proposed in this work has helped in standardizing the images from diverse sources, and providing regions of interest (ROIs) for the abnormalities in the images and removing irrelevant features. The usability of the proposed dataset is demonstrated by training ResNet101 on it and four other smaller datasets. The organization of the dataset allows for various classification tasks and assesses the impact of breast density and abnormality information on prediction. This makes Mammo-Bench a valuable resource for the development and evaluation of CAD systems. While Mammo-Bench represents a significant step forward in mammography datasets, certain limitations persist, particularly regarding class imbalance and demographic imbalances across different geographical regions (Supplementary Fig. 1). This may introduce potential biases in the model development and future work may attempt to address these issues. Additionally, a promising direction for future research involves the integration of multimodal data, such as combining mammography images with ultrasound, MRI, or genomic data, to enable personalized and precise screening and propose novel diagnostic strategies.

Author Contributions:. G. B. collated and analyzed the data. G.B. was responsible for training the models and writing the code for data handling. S.S. and N.P. supervised the training of the models, and all the technical details. G.B. wrote the initial draft. S.S. and N.P. supported the study, edited, reviewed and worked on the revised manuscript. All authors subsequently critically edited the report. All authors have read and approved the final report.

Dataset Availability. Link to the dataset: Mammo-Bench (Login required for access).

Code Availability: The preprocessing code: https://github.com/Gaurav2543/Mammo-Bench.

Disclaimer: The mammographic images used were sourced from public datasets with appropriate permissions and usage compliance. This dataset is intended for research purposes only and should not be used for direct clinical diagnosis.

In the updated version of Mammo-Bench, the RSNA dataset is removed. However it can be downloaded from the source and the proposed preprocessing pipeline can be applied to it to get it in the same format.

Competing Interests. The authors declare no competing interests.

References

1. Carriero, A., et al.: Deep learning in breast cancer imaging: state of the art and recent advancements in early 2024. Diagnostics **14**, 848 (2024)

2. Heath, K.B., et al.: The digital database for screening mammography. In: Proceedings of the fifth international workshop on digital mammography. In: Yaffe, M.J. (ed.) 212–218, Medical Physics Publishing (2001)
3. Moreira, I.C., et al.: INbreast. Acad. Radiol. **19**, 236–248 (2012)
4. Alsolami, A.S., et al.: King Abdulaziz university breast cancer mammogram dataset (KAU-BCMD). Data **6** 111 (2021)
5. Cui, C. et al.: The Chinese Mammography Database (CMMD): An online mammography database with biopsy confirmed types for machine diagnosis of breast (2021). https://www.cancerimagingarchive.net/collection/cmmd/
6. Khaled, R., et al.: Categorized contrast enhanced mammography dataset for diagnostic and artificial intelligence research. Sci. Data **9**, 122 (2022)
7. Carr, C., et al.: RSNA Screening Mammography Breast Cancer Detection (2022). https://kaggle.com/competitions/rsna-breast-cancer-detection
8. Oza, P., et al.: Digital mammography dataset for breast cancer diagnosis research (DMID) with breast mass segmentation analysis. Biomed. Eng. Lett. **14**, 317–330 (2024)
9. Heath, M., et al.: Current status of the digital database for screening mammography. In: Karssemeijer, N., Thijssen, M., Hendriks, J., Van Erning, L. (eds.) Digital Mammography, pp. 457–460. Springer, Dordrecht (1998)
10. Suckling, J., et al.: Mammographic Image Analysis Society (MIAS) database v1.21 (2015). https://www.repository.cam.ac.uk/handle/1810/250394
11. Ashby, A.E., et al.: UCSF/LLNL high resolution digital mammogram library. In: Proceedings of 17th International Conference of the Engineering in Medicine and Biology Society, Montreal, Que., Canada, pp. 539–540. IEEE (1995)
12. Oliveira, J.E.E., et al.: Toward a standard reference database for computer-aided mammography. In: Presented at the Medical Imaging, San Diego, CA March 6 (2008)
13. Halling-Brown, M.D., et al.: OPTIMAM mammography image database: a large-scale resource of mammography images and clinical data. Radiol. Artif. Intell. **3**, e200103 (2021)
14. Nguyen, H.T., et al.: VinDr-Mammo: a large-scale benchmark dataset for computer-aided diagnosis in full-field digital mammography. Sci. Data **10**, 277 (2023)
15. Matheus, B.R.N., Schiabel, H.: Online Mammographic images database for development and comparison of CAD schemes. J. Digit. Imaging **24**, 500–506 (2011)
16. Imane, O., et al.: LAMIS-DMDB: a new full field digital mammography database for breast cancer AI-CAD researches. Biomed. Signal Process. Control **90**, 105823 (2024)
17. Antoniou, Z.C., et al.: A web-accessible mammographic image database dedicated to combined training and evaluation of radiologists and machines. In: 2009 9th International Conference on Information Technology and Applications in Biomedicine, Larnaka, Cyprus, pp. 1–4. IEEE (2009)
18. Wu, N., et al.: The NYU Breast Cancer Screening Dataset v1.0
19. Sorkhei, M., et al.: CSAW-M: An ordinal classification dataset for benchmarking mammographic masking of cancer (2021)
20. Coates, A.S., et al.: Tailoring therapies - improving the management of early breast cancer: St Gallen international expert consensus on the primary therapy of early breast cancer 2015. Ann. Oncol. **26**, 1533–1546 (2015)
21. Pertuz, S., et al.: Open framework for mammography-based breast cancer risk assessment. In: IEEE EMBS International Conference on Biomedical & Health Informatics (BHI), Chicago, IL, USA, pp. 1–4. IEEE (2019)
22. Pei, C., et al.: Segmentation of the breast region in mammograms using marker-controlled watershed transform. In: The 2nd International Conference on Information Science and Engineering, Hangzhou, China, pp. 2371–2374. IEEE (2010)
23. Wilkinson, M.D., et al.: The FAIR Guiding Principles for scientific data management and stewardship. Sci. Data **3**, 160018 (2016)

MetaEdit: Computational Identification of RNA Editing in Microbiomes

Arpit Mehta[1(✉)], Vitalii Stebliankin[1], Kalai Mathee[2,3], and Giri Narasimhan[1]

[1] Bioinformatics Research Group (BioRG), Florida International University, Miami, FL 33199, USA
{ameht014,vsteb002,giri}@fiu.edu

[2] Biomedical Sciences Institute, Florida International University, Miami, FL 33199, USA

[3] Lifetime Omics Inc., Miami, FL, USA

Abstract. RNA editing is a pivotal post-transcriptional mechanism that plays a critical role in the regulation of some genes by altering their mRNA sequences, thereby influencing the resulting protein sequence, structure, and the functional and cellular responses. While extensively studied in eukaryotes, its significance and prevalence in prokaryotic microbiomes remain underexplored. Given the crucial role of microbiomes in various biological processes and their potential impact on human health and disease, understanding RNA editing within these communities could reveal new insights into microbial gene regulation and adaptation. The lack of studies to detect RNA editing in microbiomes motivates the need for developing bioinformatic strategies to bridge this research gap. This study introduces MetaEdit, a computational tool designed to detect RNA editing in bacterial microbiomes. We apply MetaEdit to metatranscriptomic and metagenomic datasets to identify and characterize RNA editing events in the human gut microbiome. Our results demonstrate the presence of RNA editing in *Escherichia coli* and provide a foundation for future investigations into the functional implications of RNA editing in microbiomes. Our findings are supported by previously reported research but need validation with laboratory experiments. The developed pipeline is generic and can be applied to find RNA editing in any sequencing datasets containing both metagenomic and metatranscriptomic data.

Availability: Pipeline is available from https://biorg.cs.fiu.edu/metaedit/.

Supplementary information: None.

Keywords: Bioinformatics · RNA-Editing · Microbiome

1 Introduction

RNA editing is a type of post-transcriptional modification that alters the mRNAs (or another type of RNAs called transfer RNAs) caused by the insertion, deletion

M. Alser et al. (Eds.): ICCABS 2025, LNCS 15599, pp. 157–170, 2026.
https://doi.org/10.1007/978-3-032-02489-3_12

or substitution of single nucleotides [10]. It occurs in a wide range of organisms. In conjunction with alternative splicing, they increase the diversity in proteomes [16]. In humans, RNA editing affects both coding and non-coding transcripts by the deamination of adenosine to inosine (A-to-I) and cytosine to uridine (C-to-U). A-to-I editing occurs when members of adenosine deaminase (ADAR) family of enzymes work on double-stranded RNA (dsRNA) to perform A-to-I conversion [10]. Recent transcriptomic studies have identified a number of "recoding" sites at which A-to-I editing results in non-synonymous changes in protein-coding sequences, thus creating modified proteins [7].

RNA Editing in Eukaryotes. RNA editing in eukaryotes has been studied and is known to play a role in human health and disease [26]. From regulating neuronal dynamics [15] to playing a fundamental role in innate and adaptive immunity [18], RNA editing affects a lot of biological processes. A recent study [20] demonstrated that a panel of RNA editing-based blood biomarkers could accurately discriminate major depressive disorder (MDD) from depressive bipolar disorder (BD) with high sensitivity and specificity, emphasizing the potential of RNA editing for reducing misdiagnosis and optimizing treatments for specific diseases. Identifying RNA editing sites in eukaryotes is achieved by variant calling on RNA-seq data with sufficient depth. Tools analyzing sequencing data to detect RNA editing have been developed [6]. Well-characterized genomic references for human, mouse and other model eukaryotes ensure that editing calls are accurate and consistent. Databases containing RNA editing sites for these organisms are available as well [6].

RNA Editing in Bacteria. For decades, RNA editing in bacteria was reported only at a single location on a tRNA, mediated by the enzyme tRNA-specific adenosine deaminase (TadA) [28]. RNA editing sites in bacterial mRNA were unknown. Recently, Bar-Yaacov and colleagues demonstrated that non-synonymous A-to-I editing can also occur in bacterial (*Escherichia coli*) mRNA [27]. The *hokB* gene encoding a toxin that confers antibiotic tolerance to the bacteria, undergoes A-to-I editing mediated by TadA, and was the first editing event discovered in bacterial mRNA. Furthermore, this editing increased with cell density and caused an increase in toxicity of the culture. They detected other editing sites in an *E. coli* strain. A-to-I RNA editing site was also reported in the *fliC* gene of *Xanthomonas oryzae pv. oryzicola*, which increases the bacterium's pathogenicity and tolerance to oxidative stress, thereby making it more virulent [21]. Thus, RNA editing in bacteria creates novel proteins, imparting increased antibiotic resistance, stress tolerance, and pathogenicity. We conjecture that more RNA editing remains to be discovered, not just in individual bacterial strains, but also in bacterial communities (microbiomes), which can impact more complex biological processes in their hosts.

RNA Editing in Microbiomes. Microbiomes are complex naturally-occurring microbial communities, often containing thousands of species of bacteria and other microorganisms. The interactions between the microbes in the microbiome adds a layer of intricacy to the modeling and understanding of these communities. Host-microbiome interactions contribute additional layers of complexity to this fascinating ecosystem. The interactions between microbes and with the host are facilitated by the genes that are expressed, the proteins that are produced, the metabolites that are released, the mutations they carry, and the other biological processes they trigger. Recent multiomics studies have revealed that host-microbiome interactions play a crucial role in the health of the host and the host environment. Given that RNA editing changes the mRNA products, it can play a significant role in microbiomes by impacting translated proteins, alternative splicing, miRNA regulation, protein-protein and protein-drug interactions, and protein diversity. However, **RNA editing has never been documented in microbiomes**. The extent of RNA editing in microbiomes and their potential role in disease is unknown. Also unknown is whether microbial interactions within a microbiome can impact the process of RNA editing. Even within controlled laboratory settings with known monoclonal strains of bacteria, only two studies have successfully established the occurrence of RNA editing in bacterial systems. This motivates the study of RNA editing in microbiomes along with the effects of RNA editing on the metatranscriptome, metaproteome, metabolome, host-microbe interactions, and on disease.

Identification of RNA editing in microbiome can be like looking for the proverbial "needle in a haystack". The human gut microbiome consists of thousands of bacterial species (estimated at 15,000 to 36,000) [32], with an estimated number of bacterial cells ranging from 10^{13} to 10^{14}, roughly as much as the number of human cells (3.0×10^{13}) [25]. The undetected unknowns [34] are microbes which are so low in abundance that there is little or no genetic information available on them. In a recent extensive study to reveal the human microbiome [9], around 77% of the sequence did not match with any genomic reference available in public databases, making the process of finding the source bacteria of the sequencing read a tedious and computationally intensive process. Compared to the roughly 25,000 human genes, the gut microbiome has at least a 100 times more genes in its repertoire [33]. Some genes are also common between the host and the microbiome, making it an arduous task to annotate the aligned reads. *Strain level variability* add another difficulty level in the process, because variability in low-abundance strains of a bacterial species can be falsely read as an RNA-editing site. It is therefore clear that without adequate care, RNA editing predictions can lead to erroneous inferences. Thus, the detection of RNA editing in microbiomes requires stringent criteria to avoid false positives. *Contamination* from the host genome, ribosomal reads, and shared host-microbiome regions are additional confounding factors. Prior filtering of these reads is missing in current tools. PCR biases and sequencing errors affecting quality of the reads are additional challenges to consider. Finally, sufficiently strong evidence for editing events are needed to support claims for novel site discovery.

To bridge the research gap in identifying RNA editing within complex microbial communities, we developed MetaEdit, a novel computational pipeline designed to detect RNA editing in microbiomes. Unlike previous approaches, which have primarily focused on single bacterial strains or eukaryotic systems, MetaEdit is tailored to work with metagenomic and metatranscriptomic data, enabling it to capture RNA-editing events across diverse bacterial communities within microbiomes. By integrating information from both DNA and RNA sequencing, METAEDIT can distinguish true RNA-editing sites from DNA mutations and strain variability, addressing the challenges posed by contamination, low-abundance species, and sequencing artifacts. Using this tool on iHMP microbiome data, we successfully identified numerous potential RNA-editing sites and observed associations between specific editing events and disease states. These findings demonstrate the utility of METAEDIT in expanding our understanding of RNA editing in microbiomes and underscore its potential for uncovering novel regulatory mechanisms and biomarkers for microbiome-related diseases.

2 Methods

In this study, we developed an open source bioinformatics tool called METAEDIT. This tool works under the premise that evidence for an RNA editing site requires integration of two distinct omics data sets. First, the edited base must be detected in the metatranscriptomic reads, and must be mixed with the unedited base since the editing cannot be happening in every sampled cell. Second, the edited base must NOT be detected in the metagenomic reads. If the edited base is detected, then we cannot differentiate between a mutation present in a mutant strain from an RNA editing site. By identifying "consistent" discrepancies between the sequenced reads from genomic DNA and the sequenced reads from the mRNA transcripts, the algorithm can pinpoint edits occurring at the RNA level only. Additional care is needed in filtering out potential false positives arising from DNA mutations or sequencing errors. METAEDITâĂŹs approach is particularly novel in microbiome research, as it enables the detection of RNA editing in complex microbial communities, offering a more precise and scalable method for exploring gene regulation and microbial adaptation in diverse environmental contexts.

The METAEDIT tool is designed to reliably identify RNA-editing sites in a bacterium of interest, using a cohort of metagenomics and metatranscriptomics samples (see Algorithm 1). It takes as input metatranscriptomics and metagenomics raw reads from a collection of samples (`RNA.fq` and `DNA.fq`), along with the reference genome and annotations of the target bacteria (`Ref.fa` and `Ref.gtf`), the reference genome of the host (`Host.Ref.fa`), and user-defined parameters for coverage threshold (`cov.threshold`) and RNA-edit threshold (`edit.threshold`). The algorithm consists of six steps.

1. **Quality Control and Adaptor Removal**: Sequencing reads from each DNA and RNA sample undergoes quality control and adaptor trimming to ensure that only high-quality reads remain.
2. **Alignment to Reference**: Filtered DNA and RNA reads are aligned to the bacterial reference genome.

3. **Base Distribution Calculation**: For each sample, the distribution of bases (A, G, C, and T) across genomic positions is generated.
4. **Coverage Calculation**: The coverage at each genomic position is calculated to assess sequencing depth and ensure robust variant calling.
5. **RNA-Editing Site Detection and Aggregation across Samples**: RNA-editing sites are predicted by the CallEditing function. The existence of the same edit across multiple samples is used to strengthen the prediction.
6. **Annotation**: Each candidate RNA-editing site is annotated with information about its genomic location and the corresponding amino acid change in the encoded protein.

Steps 1–4 above are executed independently for each sample, while Step 5 aggregates results across the study to ensure reliable identification of RNA-editing sites. The CallEditing function combines coverage information from all samples, comparing RNA and DNA reads at each position to identify consistent discrepancies indicative of RNA editing. This aggregation is crucial for distinguishing true RNA-editing events from false positives due to DNA mutations or strain variation, as it confirms that RNA-editing sites show consistent editing across the cohort and lack any DNA mutations at the corresponding positions. Each step is detailed below

Algorithm 1. RNA editing data mining and detection

```
Function MetaEdit()
Input: RNA.fq, DNA.fq, Ref.fa, Ref.gtf, Host.Ref.fa, cov.threshold,
  edit.threshold
while no more samples do
    ▷ Step 1: Read quality checks for RNA and DNA samples
    filtered.RNA.fq = ReadQC(RNA.fq, Host.Ref.fa)
    filtered.DNA.fq = ReadQC(DNA.fq, Host.Ref.fa)
    ▷Step 2: Align DNA and RNA reads to bacterial reference genome
    RNA.bam = AlignRef(filtered.RNA.fq, Ref.fa)
    DNA.bam = AlignRef(filtered.DNA.fq, Ref.fa)
    ▷ Step 3: Generate distribution of A,G,C,T bases at each genomic position
    pileup.RNA = Pileup(RNA.bam)
    pileup.DNA = Pileup(DNA.bam)
    ▷ Step 4: Calculate coverage at each position using pileup file
    cov.RNA = CovTable(pileup.RNA)
    cov.DNA = CovTable(pileup.DNA)
end while
▷ Step 5: Identify edit sites from generated coverage tables based on criteria in
  cov.threshold
RNA.edit = CallEditing(cov.RNA, cov.DNA, cov.threshold, edit.threshold)
▷ Step 6: Identify Amino acid changes in protein resulting from editing
Result.annotate.tab = Annotate(RNA.edit, Ref.gtf)
Output files RNA.edit, Result.annotate.tab
```

Quality Control and Contamination Removal: The raw FASTQ reads from the metagenomic and metatranscriptomic sequencing data were trimmed using Trim Galore [31] to remove any reads with low Phred Score (<20). Trimmed reads that were <70% of original length were completely removed. Sequencing adapters were removed using CutAdapt [17] to clean up the raw reads. Low-quality bases at the end of the reads were removed as well. The reads were then aligned to the assembly of the human reference genome HG38 and the human reference transcriptome to remove all host contamination. Metatranscriptomic reads that mapped to small and large subunit ribosomal RNA sequences using the SILVA (rRNA) database were also removed. The prefiltered data from iHMP2 may have already performed all these steps as part of the original processing, as documented in the pipeline [23]. However, to ensure consistency, we apply quality control procedures even to the filtered samples.

Alignments, Base Distribution, and Coverage. The QC-filtered reads The filtered reads were then aligned to selected bacterial reference genomes using `bowtie2 v.2.4.1` with default settings [4]. The aligner outputs an alignment file (BAM) with details of reads aligned with information on positions on the reference genomes. `Samtools v 1.9` [14] was used to create a pileup file to calculate per base coverage for the reference genome. The pileup file was then converted to a coverage table using utilities from aimap [21] that created a base count file containing the name of the reference genome, genomic position, the base in the reference genome at that position, total counts, total read counts per nucleotide (`A,G,C,T`), and forward and reverse aligned read counts per nucleotide.

Identifying `A`-to-`I` Edit Sites: Using MetaEdit, we identified genomic positions that exhibited more than one base in the metatranscriptomic reads, and then ensured that these candidate positions do not exhibit any variant bases in the metagenomic reads. These sites were then checked for sample depth and read quality in both metagenomic and metatranscriptomic reads. Sites were filtered out if the percentage of variant reads (from metatranscriptomic data) was not well above the published estimates of sequencing error rate for the technology/instrument used. Sites that survived the filtering mentioned above were termed RNA-editing candidate sites. For this preliminary study, we only focused on `A`-to-`I` editing sites. In order to use an entire cohort of samples, strong editing sites are those where the variants are *present* in metatranscriptomic reads of multiple samples, but are *absent* in the metagenomic reads from all samples. An *EditScore* was then calculated for each candidate site in each sample by dividing the edited base count with total base count. *EditScores* are computed for the same site across all samples. *EditScores* above the sequencing error rate of 0.03 qualify a location as potential editing site. An RNA editing site is detected if the following conditions are met: (a) the edits are present in the MTX reads from a minimum of MinSampleThreshold samples (arbitrarily chosen to be 3 for our experiments), (b) the coverage is above the sequencing error threshold, and (c) we can confirm the absence of the same mutation in the MGX reads from any sample, including ones in which no editing is spotted. A differentially edited site would be present in a minimum of MinSampleThreshold samples (arbitrarily chosen to be 3 for our experiments) of one group and absent in the other group. The details of RNA-editing calling process can be found in Algorithm 2. A MinSampleThreshold value of greater than 1 is used to guard against poor sample quality or data quality from a single subject.

Annotations: MetaEdit annotates all identified RNA editing sites using microbial reference gene annotation files in GFF3 format [30], associating each site with the relevant genes and proteins involved in the editing. Information on whether the mutation is synonymous or non-synonymous, along with predicted functional changes based on the sequence alteration, is cataloged.

Algorithm 2. The CallEditing Process

```
Function CallEditing(cov.RNA, cov.DNA, cov.thresh, edit.thresh)
for each refBase, pos in cov.RNA do
    if refBase==T then
        DNA.cond = (cov.DNA[pos][T] > cov.thresh) && (max(cov.DNA[pos][C]) < cov.thresh)
        RNA.cond = (cov.RNA[pos][T] + cov.RNA[pos][C] > cov.thresh)
        if DNA.cond and RNA.cond then
            editScore = cov.RNA[pos][C] / (cov.RNA[pos][T]+cov.RNA[pos][C])
        else
            editScore = 0   ▷ Either low coverage in DNA/RNA, or a mutation present in DNA
        end if
    end if
    if refBase == A then
        DNA.cond = (cov.DNA[pos][A] > cov.thresh) && (max(cov.DNA[pos][G]) < cov.thresh)
        RNA.cond = (cov.RNA[pos][A] + cov.RNA[pos][G] > cov.thresh)
        if DNA.cond and RNA.cond then
            editScore = cov.RNA[pos][G]/(cov.RNA[pos][A]+cov.RNA[pos][G])
        else
            editScore = 0   ▷ Either low coverage in DNA/RNA, or mutation in DNA is detected
        end if
    end if
    if editScore < edit.thresh then
        editScore = 0   ▷ Low RNA-editing signal
    end if
end for
ReportEditScore(editScore, pos, cov.RNA, cov.DNA)
```

3 Results

In this work, we developed MetaEdit, a metagenomics and metatranscriptomics-based computational pipeline designed to identify RNA-editing sites within microbiomes. Applying MetaEdit to gut microbiome data from the iHMP study, we successfully identified RNA-editing sites in *Escherichia coli*, including previously reported loci. These findings confirm the reliability of MetaEdit and offer new insights into the extent of RNA editing in microbiomes. Additionally, we observed associations between specific RNA-editing events and disease states, suggesting potential biomarker applications for conditions such as Inflammatory Bowel Disease (IBD).

Dataset: We used data from the iHMP [8] study, which contained gut microbiome sequence data from 132 participants (105 IBD subjects and 27 non-IBD subjects). We only included samples for which both metagenomics and metatranscriptomics samples were available, resulting in a selection of 834 distinct instances from 109 participants. After fixing a racial imbalance in the study, we focused on 748 samples from 96 individuals.

MetaEdit *Results:* For our preliminary study, we selected *E. coli* as a reference bacterial genome, as it was known to have RNA-editing [27]. Our goal was to confirm the occurrence of editing events in *E. coli* even in the context of microbiomes. We applied MetaEdit to 748 paired metagenomic (MGX) and metatranscriptomic (MTX) data sets from the iHMP project. The minimum coverage threshold was set at 30, and the minimum RNA edit score at 0.03. This analysis uncovered 8,419 potential RNA-editing sites. Among these, 331 sites appeared in at least three samples with coverage exceeding 30 reads. Further analysis revealed that RNA-editing at 43 of these sites resulted in amino acid changes in the encoded proteins.

Table 1 lists the top *E. coli* RNA editing sites, ranked by RNA read coverage across the cohort. The columns of the table are described as follows. The "Type" column categorizes the sequences based on their biological roles, including coding DNA sequences (CDS), transfer RNAs (tRNA), and transfer-messenger RNAs (tmRNA). "Position" column specifies the nucleotide position within the bacterial reference genome. "Base" column describes the type of RNA-editing observed (either `A`-to-`G`or `T`-to-`C`). The "Gene name" column identifies the gene impacted by RNA-editing, and "AA change" describes the resulting amino acid alteration. The column "edit level" shows the average RNA-edit level across all participants. The column "RNA cov" specifies the total RNA coverage at the given position. The rest of the columns shows the aggregate number of reads across the cohort for bases A, C, G, and T in RNA and DNA samples, respectively. The table is organized as follows: **A)** the top 10 RNA-editing instances in the CDS biotype; **B)** the top 10 instances in tRNA and tmRNA; and **C)** additional known RNA-editing sites that do not fall within the top 10 hits. The top section of the table (labeled A) presents the top 10 edit sites in coding DNA sequence (CDS) genes, while the section labeled B highlights the top 10 sites in tRNA and transfer-messenger RNAs (tmRNA). In addition, part C indicates additional known RNA-editing sites that do not fall within the top 10 hits. We identified six RNA-editing sites that have been previously reported in the genes *Hok/gef, argV, clpB, ppiD, nudE,* and *ilvC*, as documented by Bar-Yaacov et al. [27]. Of these, *Hok/gef, argV,* and *clpB* met our criteria for coverage threshold filters. Our confidence in reporting predicted RNA-editing sites is supported by the total absence of even a single nucleotide polymorphisms (SNPs) in the corresponding DNA samples. Specifically, for T-to-C editing events, a substantial presence of both C and T reads was observed in RNA samples (Table 1, columns 10 and 12), almost zero amount of T base was detected in DNA samples (Table 1, column 16). Similarly, for A-to-G editing events, RNA samples showed a significant count of both A and G reads (Table 1, columns 9 and 11), with DNA samples exhibiting only the A base (Table 1, column 13). In the next subsections, we examine a subset of RNA-editing sites found in *E. coli* with potential biological significance.

RNA Editing in the hok/gef Gene: The top RNA-editing site was *hok/gef* gene at position 3,607,424 in *E. coli* (Table 1, row 1). This gene was previously reported to have an RNA edit site, however the experiment was conducted on clonal bacterial cultures with overexpression of *tadA* to induce editing [27]. Hence, this is the first time the site was confirmed to be active in the context of a natural microbiome. The hok/gef family of a toxin-antitoxin system has been well studied. This editing changes a tyrosine (Tyr) encoded by the TAC codon into a cysteine (Cys) encoded by the TGC codon. This is consistent with Bar-Yaacov's discovery of tadA-specific edit site.

The tRNA-ARG Site: One of the RNA-editing sites identified was located at position 3,542,570 within the tRNA gene for arginine in *E. coli* (Table 1, last row). Although the edit event at this site does not alter the amino acid in the protein product, it was included in the table because it had been previously reported by Bar-Yaacov et al. [27]. Therefore, it serves as further validation of our pipeline. Prior to the work of Bar-Yaacov et al., editing in tRNA genes was only reported in eukaryotes. It is known to be mediated by the enzyme tRNA-specific adenosine deaminase (tadA) [28].

RNA-Editing Sites as Potential Biomarkers: The two previously mentioned edit sites in *E. coli* at locations 3,607,424 and 3,542,570, were investigated as potential biomarkers for Inflammatory Bowel Disease (IBD). Given that different patients may have varying numbers of time points, comparing RNA edits at the sample level could introduce a bias. Therefore, edit levels were aggregated and averaged across participants. The non-parametric MannâĂŞWhitney U-test was used to compute statistical significance. The distribution of these average edit levels at the two sites for the subjects from the iHMP study is shown in Fig. 1. We observed that the edit levels were significantly higher in patients with Crohn's Disease (CD) and Ulcerative Colitis (UC) compared to controls for site 3,542,570 of *argV* gene, suggesting that it could serve as a potential biomarker for IBD (Fig. 1, panel B). Conversely, the distribution of edit levels among CD, UC, and control groups did not show significant differences for site 3,607,424, indicating it is not a likely biomarker candidate (Fig. 1, panel A). Additionally, a bimodal distribution of edit levels was noted in CD and UC patients for each plot, potentially indicating that edit events were observed predominantly in patients with

Table 1. Top RNA Edit sites predicted for *E. coli*. The column names are described in the text.

	Type	Position	Base	Gene	AA Change	Edit Score	RNA Cov	RNA A	RNA C	RNA G	RNA T	DNA A	DNA C	DNA G	DNA T
A	CDS	3607424	T-C	***Hok/gef***	Y46C	0.1801	10,804	1	1,945	1	8,857	0	0	0	410
	CDS	3044947	T-C	*ompC*	N309D	0.6504	1,625	0	1,055	3	567	0	7	0	301
	CDS	4191210	T-C	*tufB*	N52D	0.0565	1,578	3	89	0	1,486	0	3	0	535
	CDS	2464921	A-G	*gapA*	K138E	0.3354	1,465	973	1	491	0	265	0	0	0
	CDS	3044940	T-C	ompC	D311G	0.1025	1,271	0	130	3	1,138	0	1	0	340
	CDS	2464904	A-G	*gapA*	K132R	0.3243	1,175	794	0	381	0	267	0	1	0
	CDS	3436520	T-C	*grcA*	K88E	0.2383	1,130	0	269	1	860	0	1	1	375
	CDS	1305585	T-C	*mokC*	E16G	0.5121	1,076	0	549	4	523	0	2	0	101
	CDS	2194213	T-C	*mokC*	E16G	0.5138	982	0	502	5	475	0	0	0	126
	CDS	3045379	T-C	*ompC*	N165D	0.8095	763	0	616	2	145	0	12	0	231
B	tmRNA	3475768	A-G	*ssrA*	-	0.4460	228,199	126,298	185	101,661	55	358	0	1	0
	tmRNA	3476014	A-G	*ssrA*	-	0.2579	29,781	21,969	171	7,633	8	369	0	9	0
	tRNA	4060905	T-C	*metY*	-	0.1548	28,198	12	4,356	52	23,778	0	0	0	725
	tRNA	2712254	A-G	*serU*	-	0.0819	17,547	16,100	4	1,437	6	234	0	0	0
	tRNA	865013	T-C	*lysT*	-	0.1394	13,350	4	1,860	3	11,483	0	0	0	450
	tRNA	3669924	A-G	*metZ*	-	0.1257	13,342	11,650	11	1,675	6	510	0	1	0
	tRNA	3670177	A-G	*metV*	-	0.0906	11,813	10,696	42	1,066	9	368	1	2	0
	tRNA	3542693	A-G	*serV*	-	0.0714	11,389	10,348	245	796	0	392	0	0	0
	tRNA	3670034	A-G	*metW*	-	0.1714	10,312	8,529	15	1,764	4	581	0	0	0
	tmRNA	3475976	A-G	*ssrA*	-	0.0881	9,368	8,521	16	823	8	758	0	1	0
C	tRNA	3542570	T-C	***argV***	-	0.7176	8,327	2	5,971	4	2,350	0	0	0	560
	CDS	3452003	T-C	***clpB***	Y808C	0.0918	686	0	63	0	623	0	0	0	383
	CDS	529829	A-G	***ppiD***	E621G	0.0444	45	43	0	2	0	638	0	3	0
	CDS	4238024	A-G	***nudE***	W150R	0.0882	34	31	0	3	0	357	0	0	0
	CDS	4752350	T-C	***ilvC***	I270T	0.0323	34	31	0	1	0	30	0	0	0

dysbiosis. This observation supports a hypothesis that RNA editing plays a role in responding to environmental stress [27] and possibly in gut disorders such as IBD.

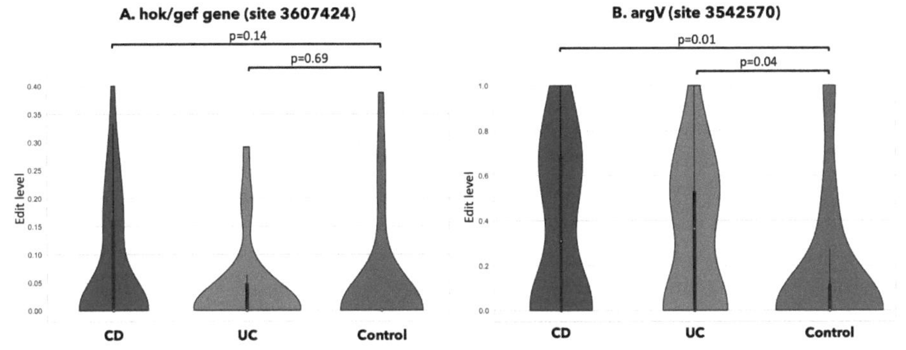

Fig. 1. Violin plots showing differential RNA editing levels in two genes, hok/gef (A) and argV (B), across three cohorts: Crohn's Disease (CD), Ulcerative Colitis (UC), and Control. Statistical comparisons between cohorts were conducted using the Mann–Whitney U-test, with p-values indicated for significant differences.

4 Discussion

Applying MetaEdit to the iHMP dataset enabled us to identify numerous RNA-editing sites in Escherichia coli, including previously reported editing sites in genes such as *hok/gef* and *argV*. These findings not only confirm the functionality of MetaEdit but also validate the occurrence of RNA editing in a microbiome setting, which had previously been reported only in isolated bacterial strains under laboratory conditions. The observation of differential editing in specific genes between Inflammatory Bowel Disease (IBD) and control groups further underscores the potential of RNA-editing events as biomarkers for disease. For instance, the higher edit levels observed in IBD patients at certain sites suggest a potential link between RNA editing and gut inflammation or dysbiosis, aligning with hypotheses that RNA editing could play a role in microbial responses to environmental stressors within the host gut environment.

While MetaEdit represents a significant advancement, there are still areas for future improvement and expansion in this study. While MetaEdit addresses strain variation by aggregating coverage across samples in MGX and MTX data, it still relies on reference genomes for alignment and analysis. This dependence on reference genomes introduces potential bias, as any significant divergence between a microbial strain in the sample and the reference genome could affect the accuracy of RNA-edit identification. Future improvements could include incorporating de novo assembly approaches or leveraging pan-genome references to reduce bias and enhance METAEDITś applicability to diverse and novel strains within microbiomes. Additionally, although METAEDIT provides strong in silico evidence for RNA editing, experimental validation through laboratory techniques is necessary to conclusively confirm these edits.

An evaluation of the *sensitivity* and *specificity* of MetaEdit is essential for its broader adoption in microbiome data analysis. Lack of ground truth of the location and occurrence of RNA editing makes it challenging to estimate sensitivity of METAEDIT. Since we are not aware of prior work detecting RNA edit sites in microbiome sequence data, sensitivity can be assessed by benchmarking MetaEdit in synthetic data sets where edit sites are spiked randomly at varying strengths. Specificity could be evaluated by testing MetaEdit on negative control datasets, such as non-editing bacterial strains or synthetic data without known RNA edits, to ensure minimal false positives. However, we are unaware of the existence of such microbiome data sets. Again, specificity can be assessed by benchmarking MetaEdit in synthetic microbiome data sets with a large number of strains and with many hot spots with relatively high mutation rates. These strategies could help validate the tool's accuracy and enhance its reliability for studying RNA editing in diverse microbial communities.

In conclusion, METAEDIT is a valuable addition to the toolbox for microbiome research, offering a reliable method for RNA-editing detection in complex microbial communities. The findings from this study open new avenues for understanding the functional roles of RNA editing in microbiomes and its potential implications for host health, especially in disease contexts like IBD. Therefore, METAEDIT could contribute to the identification of novel biomarkers and therapeutic targets related to RNA editing in microbiome-associated diseases.

Disclosure of Interests. The authors have no competing interests to declare that are relevant to the content of this article.

References

1. Aguiar-Pulido, V., Huang, W., Suarez-Ulloa, V., Cickovski, T., Mathee, K., Narasimhan, G.: Metagenomics, metatranscriptomics, and metabolomics approaches for microbiome analysis: supplementary issue: bioinformatics methods and applications for big metagenomics data. In: Evolutionary Bioinformatics, vol. 12, pp. EBO–S36436. SAGE Publications Sage UK: London, England (2016)
2. Bellacosa, A., Moss, G.: RNA repair: damage control. Curr. Biol. **13**(12), 1142–1151 (2003)
3. Berg, G., et al.: Microbiome definition re-visited: old concepts and new challenges. Microbiome **8**(1), 1–22 (2020)
4. Langmead, B., Salzberg, S.L.: Fast gapped-read alignment with Bowtie 2. Nat. Methods **9**(4), 357–359 (2012)
5. Crick, F.: Central dogma of molecular biology. Nature **561**(3), 437–447 (1970)
6. Diroma, M.A., Ciaccia, L., Pesole, G., Picardi, E.: Bioinformatics resources for RNA editing. In: RNA Editing, pp. 177–191. Springer (2021)
7. Grice, L.F., Degnan, B.M.: The origin of the ADAR gene family and animal RNA editing. BMC Evol. Biol. **17**(4), 27–34 (2015)

8. HMP Integrative: The integrative HMP (iHMP) Research network consortium: the integrative human microbiome project: dynamic analysis of microbiome-host omics profiles during periods of human health and disease. Cell Host & Microbe **16**(3), 276–289 (2014)
9. Pasolli, E., et al.: Extensive unexplored human microbiome diversity revealed by over 150,000 genomes from metagenomes spanning age, geography, and lifestyle. Cell **176**(3), 649–662 (2019)
10. Gott, M.J., Emeson, R.B.: Functions and mechanism of RNA editing. Ann. Rev. Genet. **5**(2), 121–132 (2000)
11. Knight, R., Vrbanac, B.C., Aksenov, A., Dorrestein, P.C.: Best practices for analysing microbiomes. Nat. Rev. Microbiol. **401**(6755), 788–798 (2018)
12. Kiss, T.: Small nucleolar RNA-guided post-transcriptional modification of cellular RNAs. EMBO **20**(14), 3134–3142 (2001)
13. Khandekar, S.S., Daines, R.A., Lonsdale, J.T.: Bacterial β-ketoacyl-acyl carrier protein synthases as targets for antibacterial agents. Curr. Protein Pept. Sci. **4**(1), 21–29 (2003)
14. Li, H., et al.: 1000 genome project data processing subgroup: the sequence alignment/map format and SAMtools. Bioinformatics **25**(16), 2078–2079 (2009)
15. Blow, M., Futreal, P.A., Wooster, R., Stratton, M.R.: A survey of RNA editing in human brain. Genomics Res. **14**(12), 2021–2028 (2004)
16. Mallela, A., Nishikura, K.: A-to-I editing of protein coding and noncoding RNAs. Nat. Rev. Mol. Cell Biol. **17**, 83–96 (2017)
17. Martin, M.: Cutadapt removes adapter sequences from high-throughput sequencing reads. EMBnet. J. **17**(1)
18. Madani, N., Kabat, D.: An endogenous inhibitor of human immunodeficiency virus in human lymphocytes is overcome by the viral Vif protein. J. Virol. **72**(12), 10251–10255 (1998)
19. Marchesi, J.R., Ravel, J.: The vocabulary of microbiome research: a proposal. BMC Microbiome **9**(8), 1735–1780 (2005)
20. Nicolas, S.: AI algorithm combined with RNA editing-based blood biomarkers to discriminate bipolar from major depressive disorders in an external validation multicentric cohort. J. Affect. Disord. **356**(1), 385–393 (2024)
21. Nie, W., et al.: A-to-I RNA editing in bacteria increases pathogenicity and tolerance to oxidative stress. PLoS Pathog. **16**(8), e1008740 (2020)
22. Pearl, J.: The seven tools of causal inference with reflections on machine learning. Commun. ACM **62**(3), 54–60 (2018)
23. Lloyd-Price, J., et al.: Multi-omics of the gut microbial ecosystem in inflammatory bowel diseases. Nature **569**(7758), 655–662 (2019)
24. Ruiz-Perez, D., et al.: Dynamic bayesian networks for integrating multi-omics time series microbiome data. Msystems **6**(2), e01105–e01120 (2021)
25. Sender, R., Fuchs, S., Milo, R.: Revised estimates for the number of human and bacteria cells in the body. PLoS Biol. **14**(8), e1002533 (2016)
26. Christofi, T., Zaravinos, A.: RNA editing in the forefront of epitranscriptomics and human health. J. Transl. Med. **17**(1), 1–15 (2019)
27. Bar-Yaacov, D., et al.: RNA editing in bacteria recodes multiple proteins and regulates an evolutionarily conserved toxin-antitoxin system. Genome Res. **27**(10), 1696–1703 (2017)
28. Wolf, J., Gerber, A.P., Keller, W.: tadA, an essential tRNA-specific adenosine deaminase from Escherichia coli. EMBO J. **21**(14), 3841–3851 (2002)
29. Li, H., Durbin, R.: Fast and accurate short read alignment with Burrows-Wheeler transform. Bioinformatics **25**(14), 1754–1760 (2009)

30. Stein, L.: Generic Feature Format Version 3 (GFF3) (2020). https://github.com/The-Sequence-Ontology/Specifications/blob/master/gff3.md
31. Krueger, F.: Trim Galore a wrapper tool around Cutadapt and FastQC to consistently apply quality and adapter trimming to FastQ files (2012). https://www.bioinformatics.babraham.ac.uk/projects/trim_galore/
32. Frank, D.N., St. Amand, A.L., Feldman, R.A., Boedeker, E.C., Harpaz, N., Pace, N.R.: Molecular-phylogenetic characterization of microbial community imbalances in human inflammatory bowel diseases. Proc. Natl. Acad. Sci. **104**(34), 13780–13785 (2007)
33. Gill, S.R., et al.: Metagenomic analysis of the human distal gut microbiome. Science **312**(5778), 1355–1359 (2006)
34. Thomas, A.M., Segata, N.: Multiple levels of the unknown in microbiome research. BMC Biol. **17**(1), 1–4 (2019)

Drug-Centric Prior Improves Drug Response Modeling in Partially Overlapping Pharmacogenomic Screens

Dharani Thirumalaisamy[1,2], Sunil K. Joshi[5], Stephen E. Kurtz[1,3], Tania Q. Vu[1,2], Jeffrey W. Tyner[1,4], Mehmet Gönen[6], and Olga Nikolova[1,2,3](✉)

[1] Oregon Health and Science University, Portland, OR, USA
{thirumal,kurtzs,vuta,tynerj,nikolova}@ohsu.edu
[2] Department of Biomedical Engineering, Portland, OR, USA
[3] Division of Oncological Sciences, Portland, OR, USA
[4] Cell, Developmental and Cancer Biology, Portland, OR, USA
[5] Department of Medicine, Stanford University, Stanford, CA, USA
[6] College of Engineering, Koç University, Istanbul, Turkey

Abstract. With the accumulation of large-scale genomic data such as whole-genome RNA sequencing, copy number, and mutation profiles for tens of thousands of samples, associated with screening thousands of small molecules and other perturbagens, arises the question of how to best leverage partially overlapping datasets generated at different facilities. As research groups across the world continue to generate drug screens of variable size and quality, the need for approaches that can learn from such partially overlapping experiments and improve the signal to noise ratio emerges with increasing importance. We present an application of a Bayesian group factor analysis model, where we employ a drug-centric prior to transfer information about drugs screened in the same samples in multiple datasets. We show that joint models leveraging partially overlapping pharmacogenomic datasets from the Broad and Sanger institutes can overall improve drug signature identification.

Keywords: Bayesian modeling · Mitogen-activated protein kinase (MAPK) pathway · Pharmacogenomics · Drug response · Group factor analysis

1 Introduction

The ability to predict how individual patients will respond to different therapies is critical for effective patient treatment. Evidence points to response being determined in part by patient-specific genomic alterations and gene expression changes [3,13]. Heterogeneous diseases like cancer pose additional challenges for tailoring treatments to each patient that benefit from large scale profiling both in the space of molecular features and the available and effective therapies. High-throughput efforts have been undergoing for over a decade to characterize the

M. Alser et al. (Eds.): ICCABS 2025, LNCS 15599, pp. 171–182, 2026.
https://doi.org/10.1007/978-3-032-02489-3_13

genomic landscape and pharmacogenomic dependencies in the most common forms of human cancer by way of analyzing thousands of cancer cell lines led by multiple institutions around the world. These include The NCI-60 Human Tumor Cell Lines Screen (NCI60), The Cancer Cell Lines Encyclopedia (CCLE) [1], The Cancer Therapeutics Response Portal (CTRP), The Genomics of Drug Sensitivity in Cancer (GDSC, also known as Cancer Genome Project or CGP) [7], and The Genentech Cell Line Screening Initiative (gCSI). With cell lines being immortalized, biological material from genomically identical cell lines is effectively available without constraints, allowing for continuous screening with new therapeutic compounds as they continue to emerge. This has led to the accumulation of partially overlapping resources, both in the space of cell lines that various laboratories have screened, and in the compounds that have been investigated. This opens the possibility of integrating smaller-scale studies with existing resources, and potential to rescue signal where it otherwise may be underpowered for statistical learning.

The Challenging Landscape of High-Throughput Pharmacogenomics: High-throughput pharmacogenomics, where thousands of samples are molecularly characterized and further subject to drug screening are inherently noisy, posing challenges for deriving robust and generalizable predictions of genetic links with drug efficacy. Analysis of the consistency of biological findings across some of the above mentioned large-scale studies quantified some of these challenges [4,8]. In particular, despite differences in the biological assays and experimental protocols used, gene expression measurements for the same cell lines have been found to have high consistency across different sequencing efforts (median Spearman correlation of 0.85 on 471 pan-cancer cell lines screened in both CCLE and GDSC) [8]. Similarly, analysis of 64 commonly tested for mutation genes across 471 cell lines revealed overall consistent calls (Cohen's kappa coefficient of $k = 0.65$). In contrast, drug response was not as consistent as well as the downstream conclusions about associations between genomic drivers and drug response. For the 15 drugs screened in common between CCLE and GDSC datasets, the reported Spearman correlations ranged from $0.61 - 0.03$ with an average of 0.3. An example of a subset of this data is shown in Fig. 1 for 5 drugs screened in both CCLE and GDSC. The 2D scatter plots illustrate spacial trends pointing to values of drug response which are indistinguishable in one or the other cohort. These can lead to deviations in downstream conclusions.

Contributions: We previously showed that integrating genomic data via a gene-wise prior can improve drug response prediction and gene signature identification [12]. We employed a multitask Bayesian group factor analysis model (GBGFA) with a feature-wise prior that favors genes that carry information in multiple measurement platforms. Here, we explore the utility of this feature-wise prior in not only integrating molecular modalities, but also, for the explicit modeling of drugs that have been screened in multiple datasets. Because our model is multitask, we posit that jointly with the drug-wise prior, our framework can help elucidate true biological signal even in conflicting cases and lead to more generalizable downstream associations.

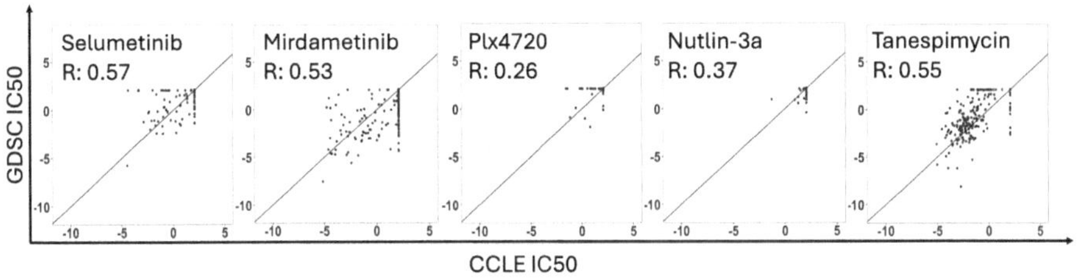

Fig. 1. Drug response to 5 drugs screened in both CCLE (X-axis) and GDSC (Y-axis) in 184 common cell lines as measured by half-maximal inhibitory concentration ($IC50$). Spearman correlation (R) is also reported.

Table 1. Curated information for drugs of interest.

Drug	Target	Targeted pathway	Drug family	Biomarkers	References
Selumetinib	MAP2K1/2	MAPK/ERK signaling pathway	S/T kinase	MAP2K1, BRAF, KRAS, HRAS RIT1, GNA11, GNAQ, NRAS	[15,16]
Mirdametinib	MAP2K1	MAPK/ERK signaling pathway	S/T kinase	BRAF, HRAS, DDX43, NF1 GNA11, KRAS, NRAS	[2,5]
Plx4720	BRAF	MAPK/ERK signaling pathway	S/T kinase	BRAF, GNAQ, KRAS, NRAS	[9,16]
Nutlin-3a	MDM2	Transcription	Other	BRAF, MDM2, TP53	[10,14]
Tanespimycin	HSP90	PI3K-Akt signaling pathway	Heat shock protein	NRAS, ALK, PIK3CA PTEN, NQO1	[6,11]

2 Materials and Methods

To evaluate the proposed application, we examined six publicly available datasets, which include both the molecular characterization and drug response measurements, jointly for the same subset of observations. We arrived at employing pan-cancer molecular data and drug response measurements from CCLE [1] and GDSC [7]. These data were independently acquired by two institutes using different experimental protocols and technologies to detect cell viability in response to drug perturbations.

Pharmacogenomic Data: We consider three modalities, namely gene expression, mutation, and drug response. Gene expression was measured using RNA-sequencing (RNA-seq), mutation data was measured by sequencing, and drug response was summarized as half-maximal inhibitory concentration (IC_{50}), which indicates the drug concentration needed to achieve 50% cell death. For each cohort, drugs were ordered in the number of cell lines in which they were screened. A total of five drugs were found to have been screened in at least 220 cell lines, simultaneously in CCLE and GDSC. These were: selumetinib (AZD6244), tanespimycin (17AAG), nutlin-3a, mirdametinib (PD-0325901), and plx4720 (precursor of vemurafenib). Missing values in gene expression and mutation matrices were filtered by gene to produce complete data, yielding the following data dimensions: CCLE expression for 19,221 genes, which we denote with e; CCLE mutation for 1,667 genes denoted h; GDSC expression for 19,077 genes denoted se; GDSC mutation for 310 genes denoted sh; with a total of 184 cell lines. We denote drug responses in CCLE and GDSC as $ic50$ and $sic50$, respectively. For the five drugs of interest, we surveyed information about their respective targets, targeted pathways, and drug families from previously published studies and curated additional biomarkers from civicDB (see Table 1).

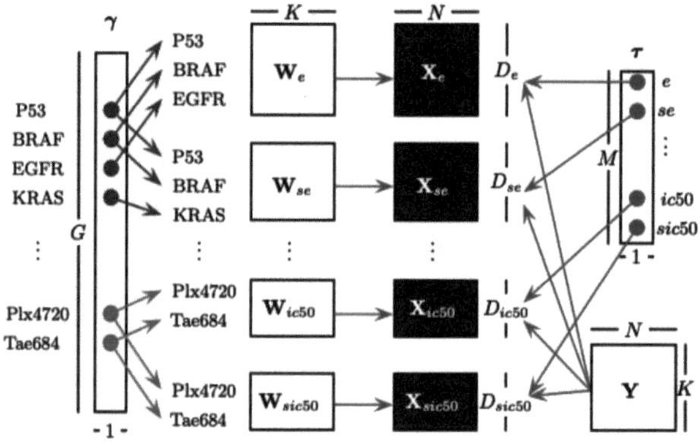

Fig. 2. Schematic representation of the proposed application, illustrating vector and matrix representation of variables and priors with their respective dimensions. Partially overlapping genomic and drug features in γ are shown in purple and orange, respectively. In this example, variables and priors indexed by e and $ic50$ refer to CCLE gene expression and drug response, respectively. Variables and priors indexed by se and $sic50$ refer to GDSC gene expression and drug response, respectively.

Notation: We briefly review the notation in the previously published gene-wise prior Bayesian group factor analysis (GBGFA) model as it pertains to our application. We consider M different sets of measurements (also called views) $\mathcal{X}_1, \mathcal{X}_2, \ldots, \mathcal{X}_M$. Broadly, views can represent measurements by different biological assays (modalities), or measurements of the same modality acquired in different experiments. If m indexes the views, each view $\mathcal{X}_m$ has D_m features and N observations. For each view $\mathcal{X}_m$, we assume an independently and identically distributed sample $\mathbf{X}_m = \{x_{m,n,d_m} \in \mathcal{X} | m \in [1 \ldots M], n \in [1 \ldots N], d_m \in [1 \ldots D_m]\}$, where m, n and d_m enumerate the views, observations, and features, respectively. Views can have partially overlapping sets of features (Fig. 2). For example, omic modalities such as copy number and gene expression could profile a shared subset of genes. Similarly, two or more biological experiments could measure the cell viability post drug perturbation for a subset of shared drugs. We utilize this property here by employing the formerly defined prior γ. In previous work, the use of γ was focused on genes, simultaneously assayed across multiple genomic platforms. However, this prior is a general, feature-centric prior, and here we examine its utility to effectively transfer information for any subset of drugs that were simultaneously tested across multiple experiments. Thus, $\gamma \in \mathbb{R}^{G \times 1}$, where G is the set of unique features profiled across any of the M views, which can include not only genes but also drugs. The GBGFA is a generative model that uses γ and view-specific projection matrices $\mathbf{W}_m \in \mathbb{R}^{D_m \times K}$ to learn a shared latent space representation of all views $\mathbf{Y} \in \mathbb{R}^{K \times N}$ with K latent components.

3 Results

We evaluate the utility of using a drug-centric prior to transfer information between noisy pharmacogenomic datasets. First, we perform predictive performance evaluation. Second, we assess consensus biomarker rankings. Third, we investigate the prior effect on multitasking. Lastly, we explore a latent subspace focused on a subgroup of drugs with a joint targeted pathway. Models were initialized with values of 1e−14 for $\alpha_\gamma, \beta_\gamma, \alpha_\tau$ and β_τ and K=24.

Comparative Analysis of Predictive Performance: We assess the effect of the drug-wise prior on drug prediction when integrating data from independent high-throughput screens with conflicting signal. We consider the joint cohort molecular feature space $\{e, h, se, sh\}$. We train a joint model inclusive of both cohorts' drug response data $\{ic50, sic50\}$. We wish to analyze the correctness of drug response prediction in each cohort's respective drug response matrix without the direct use of the complementary drug response matrix during prediction. Thus, after training jointly, in the prediction step, we remove all components associated with one cohort's drug response, and predict the other using only the shared prior γ and latent space $\mathbf{Y}$. We compare this performance to baseline models trained on single cohort drug response matrix. To assess the predictive accuracy, we perform 5-fold cross validation, comparing predicted values for one partition after training on the remaining four. To ensure that our results are independent of how the data was partitioned, we randomly split the data into partitions 10 different times and summarize the overall results. For each drug, we report the proportion of variance explained (POV) which uses normalized mean squared root error (NRMSE): POV = 1 − NRMSE. The results of these experiments are summarized in Fig. 3.

Feature Selection for Drug Signatures Identification: We assess the utility of the drug-centric prior in recapitulating consensus biomarkers of drug

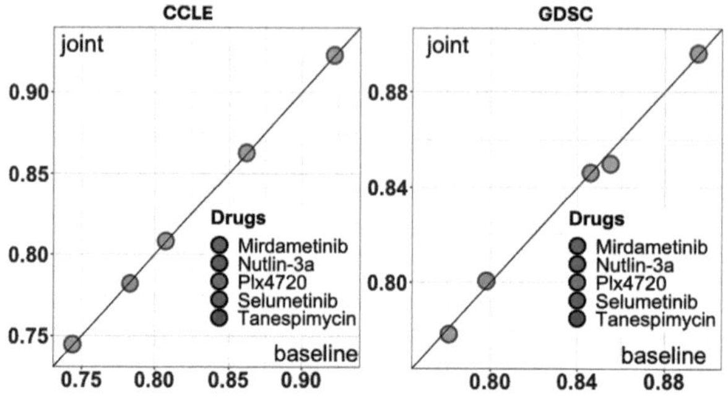

Fig. 3. Drug response prediction results. Comparison of the joint (Y-axis) to baseline (X-axis) models in CCLE (left) and GDSC (right) are summarized, reporting the POV for each drug in 5-fold cross validation experiment with 10 random data splits.

response. As before, we compare *joint* models trained on both cohorts drug response $\{ic50, sic50\}$ to *baseline* models trained using a single cohort drug response $\{ic50\}$ or $\{sic50\}$. In addition, we evaluate how integration of various molecular features affects the recapitulation of biomarkers. We consider models trained on the complete combinatorial space of the four molecular feature modalities $\{e, h, se, sh\}$: e, h, se, sh, e_h, e_se, e_sh, se_h, se_sh, h_sh, e_h_se, e_h_sh, se_h_sh, se_e_sh, $e\_h\_se\_sh$ for a total of 15 unique combinations. For each drug i, we first calculate gene scores $\boldsymbol{s}_{i,m} \in \mathbb{R}^{D_m \times 1}$ in every view m as the product of the view-weights and drug response vector $\boldsymbol{s}_{i,m} = \mathbf{W_m} \boldsymbol{w}_{i,dr}$, where dr denotes a generic drug response matrix representing either $ic50$ or $sic50$. We next calculate the cumulative density function (CDF) using the kernel density as input to function CDF of the `R` package `Spatstat`. A final gene score is computed by optimizing values on both tails of view-specific gene CDFs: $\boldsymbol{s}_i^* = \arg\min_m\{\min\{\mathrm{CDF}_{\boldsymbol{s}_{i,m}}, 1 - \mathrm{CDF}_{\boldsymbol{s}_{i,m}}\}\}$. Gene scores for consensus biomarkers outlined in Table 1 are compared across tested models. The results of these experiments are summarized in Fig. 4.

Analysis of Drug Prior and Multitasking: Our model is a multitask group factor analysis model, where the prediction of all drugs is cast and solved as one problem simultaneously, and the model can recapitulate additional information from drugs with similar response profiles to yield more robust predictions. Here, we evaluate the effect of the drug prior on the multitask property. Specifically, we analyze the drug weight matrices $\mathbf{W_{ic50}}$ and $\mathbf{W_{sic50}}$ to assess if drugs with similar mechanisms of action cluster closer together. We perform this analysis within each cohort. As before, we evaluate a joint model trained on both cohorts molecular features and drug response $\{e, h, se, sh, ic50, sic50\}$ and compare it to each cohort's respective baseline model inclusive of a single drug response dataset, i.e., $\{e, h, se, sh, ic50\}$ and $\{e, h, se, sh, sic50\}$. We compare the results against two baselines: *(i)* clustering of the original drug response matrix and *(ii)* models trained on single cohort molecular features and drug response data.

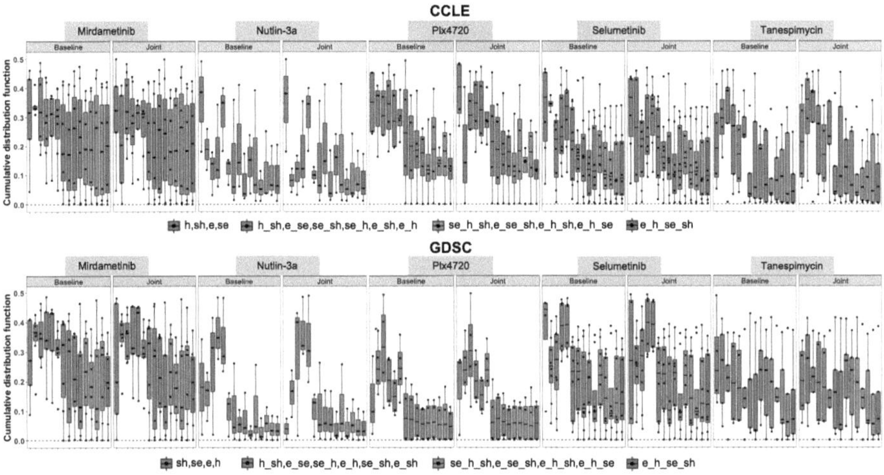

Fig. 4. Consensus biomarker analysis results. Joint and baseline models for all combinations of the molecular modalities e, h, se, sh (X-axis) and drug biomarker CDF (Y-axis), compared in CCLE (top) and GDSC (bottom), are reported. Each bar represents a model, colored by the number of input modalities, and ordered as in the legend.

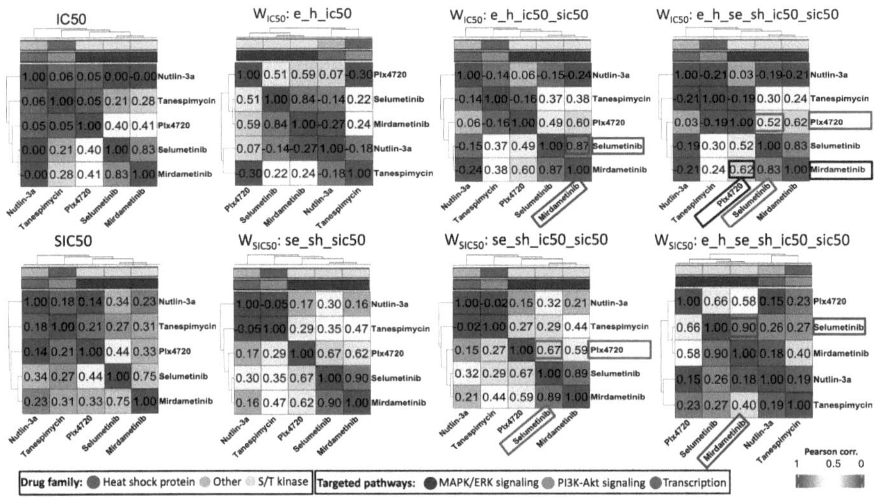

Fig. 5. Factor loading analysis results. Unsupervised hierarchical clustering results for the original drug response matrix (column 1), factor loadings for the single cohort drug response model (column 2), factor loadings for the joint cohort drug response model (column 3), and full joint model inclusive of all molecular features and drug response data (column 4) are reported for CCLE (top row) and GDSC (bottom row). Pearson correlation was used as the distance metric. Ground truth in terms of drug families and targeted pathways are summarized at the top of each heatmap as column annotations.

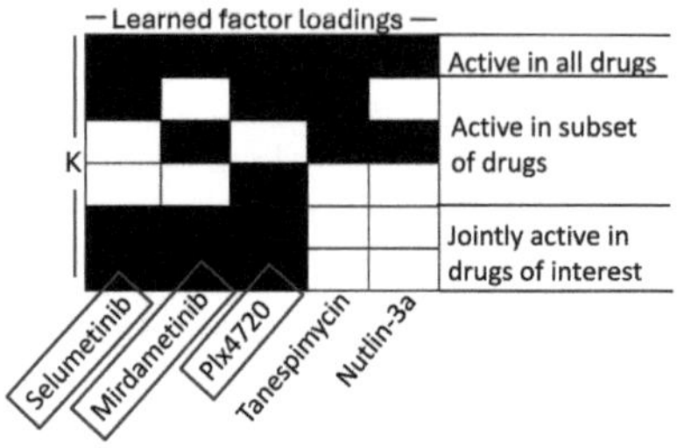

Fig. 6. Schematic representation of a targeted latent subspace. An example factor loadings matrix $\mathbf{W}_i^T$ is shown, where latent components (factors) are shown on the Y-axis and features (drugs) on the X-axis. Three S/T kinases which target the same signaling pathway are highlighted in red. Active and inactive latent components are shown respectively in black and white. After sorting, patterns are revealed indicating subsets of factors active in different drugs.

We perform hierarchical clustering on the learned factor loading matrices and analyze the targets, targeted pathways, and drug family information curated from current literature summarized in Table 1. The results of these experiments are summarized in Fig. 5.

Targeted Pathway Analysis of an Interpretable Latent Subspace: Our model learns interpretable latent space representations. Using this property, we assess the utility of the drug-centric prior in recapitulating the targeted pathway for a subset of drugs with a similar mechanism of action. We compare the joint model inclusive of both cohorts molecular features and drug response $\{e, h, se, sh, ic50, sic50\}$ against its respective baselines $\{e, h, se, sh, ic50\}$ in CCLE and $\{e, h, se, sh, sic50\}$ in GDSC. The observed data $\mathbf{X}_i$ is factorized as a product of the latent variable $\mathbf{Y}$ and factor loadings $\mathbf{W}_i$ (see Fig. 2). Consider the schematic representation of $\mathbf{W}_{ic50}^T$ shown in Fig. 6. The factor loadings are group-wise sparse, so that each factor $\boldsymbol{f}_{m,k}$ is active (dark shading) only in some subset of drugs (or all of them). The factors active in just one of the drugs model the structured noise, variation independent of all other drugs, whereas the rest model the dependencies. The nature of each of the factors is learned automatically. In our previous analysis, we use the complete factor loading matrices to calculate gene scores $\boldsymbol{s}_{i,m}$ and hierarchical clustering. Here, we focus on three drugs: selumetinib, mirdamatinib, and plx4720, all of which are S/T kinases and target the MAPK/ERK signaling pathway (see Table 1 for reference). We first identify the latent components in which the three drugs of interest are jointly active (illustrated as the bottom two rows in Fig. 6). We do so by first calculating the 25th and 75th percentiles of the learned factor loadings. Factors in the top and bottom percentiles were deemed *active* and set to $\boldsymbol{f}_{m,k} = 1$, and all in between were deemed *inactive* and set to $\boldsymbol{f}_{m,k} = 0$. We then use the subspace defined by the three drugs of interest and factors active jointly only within those drugs, to calculate gene scores $\boldsymbol{s}'_{i,m}$ for each drug i and view m

and their respective CDFs, as previously described in Sect. 3. Final gene scores were calculated using $s_i^{'*} = argmin_{\forall m} CDF_{s'_{i,m}}$ and selected the top 100 genes to perform pathway enrichment analysis. We used `R` package `enrichR` with the curated `C2` genes of the `MSigDB` and `KEGG` database. We perform this analysis for each drug response factor loadings matrix $\mathbf{W}_{ic50}$ and $\mathbf{W}_{sic50}$. We compare the results from the joint model latent subspace analysis to those of the baseline models.

4 Discussion

Predictive Performance Analysis Indicates No Loss in Predictive Accuracy: We analyze the predictive performance of our joint model inclusive of drug response matrices from two different cohort. Spearman correlation comparing the two cohorts in five drugs is illustrated in Fig. 1. Spatial patterns can be seen for all drugs, revealing each cohort capturing different parts of the drugs' active range in subsets of cell lines (see vertical and horizontal lines forming from one or the other cohort considering multiple cell lines to have indistinguishable drug response). Such inconsistencies are particularly challenging to resolve. Notably, for two of the drugs plx4720 and nutlin-3a the correlation is especially low (0.26 and 0.37, respectively). We compare our joint model to two baselines as described in Sect. 3. We use 5-fold cross validation and 10 random data splits. For each drug, we report the POV and summarize the results in Fig. 3. The comparison revealed slight improvement in the prediction of one of the five drugs in both cohorts. While this result in itself is not a significant improvement, we emphasize that biologically relevant improvements presented in subsequent experiments are achieved without sacrificing prediction accuracy.

Drug Prior Improves Consensus Biomarkers Identification: The identification of genes that drive resistance or are predictive of sensitivity for a given drug is a fundamental question in computational biology. The application of such signatures can help stratify patients to treatments and nominate novel targets for drug therapies. In the absence of solid *ground truth* to probe this question, we examine consensus biomarkers of drug response which were expertly curated and summarized in Table 1. We set out to evaluate if the drug-centric prior improves the recapitulation of these biomarkers. In addition, we evaluate how the inclusion of various modalities affects biomarker discovery. We perform this analysis comparing our joint model and baselines as described in Sect. 3 across 15 different sets of molecular modalities. We note that the number of biomarkers ranges from 3–8 among the drugs. The results are summarized in Fig. 4. For each drug, CDF (Y-axis) is shown for consensus biomarkers across 15 sets of modality combinations (X-axis) across baseline and joint models. An overall trend of decreasing CDF can be observed upon visual inspection of the results in Fig. 4 for all drugs, indicating improved biomarker ranking. We consider the models grouped by the number of molecular modalities they were trained on, i.e. 1-modality (e, h, se, sh), 2-modalities $(e\_h, e\_sh$, etc.), 3 and 4 modalities, respectively. Our results revealed that when compared within these groups, the joint models did better than baseline in both CCLE (4 out 5) and GDSC (3 out of 5) for the single modality group, and in GDSC for the 3-modality group (4 out of 5). These results were not statistically significant. However, when analyzing the effect of integrating different modalities, we observed a steady improvement as more data was integrated. Namely, we compared 1-modality group to 2-modalities group, 2-modalities group to 3-modalities group, and etc. We observed that the improvements were statistically significant when transitioning from 1 to 2, 1 to 3, and 1 to 4 modality groups (t-test p-values, sig. level of < 0.05). The statistical significance of the effect decreased as we compared transitions between 2, 3 and 4 modalities. These results indicate that the feature-wise prior is beneficial in both the drug- and gene- feature spaces and effectively leverages both types of information.

Drug Prior Informs Analysis of Drugs with Similar Mechanisms of Action: Multitasking is a useful property of our model, which takes advantage of patterns in drug response profiles in a multi-drug panel and learns more robust and generalizable representations. Here, we investigate if the drug-centric prior offers additional advantage by assessing hierarchical clustering of the learned factor loading matrices as compared to the original drug response matrix (see experimental details in Sect. 3). The results of these experiments are summarized in Fig. 5, where the top row illustrates CCLE and bottom GDSC models. We focus our attention on the three S/T kinases that target MAPK/ERK signaling: selumetinib, mirdametinib, and plx4720. We evaluate two joint models: joint-drug model inclusive of molecular features from one cohort and both drug response datasets (Fig. 5 column 3), and full-joint: inclusive of joint molecular features and drug responses (Fig. 5 column 4). We compare them to the original drug response matrix (Fig. 5 column 1) as well as a single cohort trained model (Fig. 5 column 2). In CCLE, for the drug pair selumetinib–mirdametinib, we observe the highest correlation in the joint-drug model (0.87). Similarly, for the drug pair selumetinib–plx4720, we observe the highest correlation for the full-joint model (0.52), an improvement from the original $ic50$ matrix of 0.40. Finally, for the drug pair mirdametinib–plx4720, we observe the highest correlations for the full-joint (0.62) and joint-drug (0.60) as compared to the original $ic50$ matrix of 0.41. In summary, in all three pair-wise comparisons, the drug prior offers advantage over the single cohort baseline model (Fig. 5 column 2 top) and the original $ic50$ matrix. In GDSC, for the drug pair selumetinib–mirdametinib, the highest correlation is a tie between the full-joint and single cohort models (0.90), improvement from 0.75 in the original $sic50$ matrix. Similarly, for the drug pair selumetinib–plx4720, we observe a tie joint-drug and single cohort models (0.67), an improvement from the original $sic50$ matrix of 0.44. Finally, for the drug pair mirdametinib–plx4720, we observe the highest correlations for the single cohort model (0.62) over joint-drug (0.59), both $\sim 2X$ improvements over the original $sic50$ matrix correlation of 0.33. In summary, we do not observe an advantage in GDSC. The results in both CCLE and GDSC are a function of integrating of data from both cohorts (both molecular features and drug response), and each cohort informs the other to a different extent. We observe that the integration of GDSC data with CCLE informs the model well leading to advantages described above, while the CCLE data does not seem to additionally inform GDSC predictions.

Drug Prior and Interpretable Latent Subspace Analysis Improve the Identification of the Drug Targeted Pathway: The latent space representations that our model learns can be further interpreted to investigate subsets of drugs with similar mechanisms of action, effectively amplifying relevant signal. We explore the utility of the drug-centric prior in conjunction with this property to recapitulate the targeted pathway of a subset of S/T kinases in our dataset. We consider signatures learned from the joint subspace in $\mathbf{W}_{ic50}$ (CCLE joint subspace) and $\mathbf{W}_{sic50}$ (GDSC joint subspace) and their ability to recapitulate the MAPK/ERK signaling pathway targeted jointly by all three drugs by way of pathway enrichment analysis. We compare these results to their baseline counterparts as previously described in Sect. 3. In CCLE, our results indicate that while for mirdametinib we recover the targeted pathway with a less significant p-value as compared to the baseline model (0.037 vs 0.02), for selumetinib, we observe an improvement from 0.023 to 4.9e−4 and for plx4720 – from 0.028 to 0.022. Similarly, in GDSC, we recover the targeted pathway with higher statistical significance for all three drugs, improving from 0.022 to 0.011, from 0.033 to 0.002, and from 0.032 to 9.9e−4, for selumetinib, mirdametinib, and plx4720, respectively. Overall, our results reveal improved statistical significance in both CCLE and GDSC.

Conclusion: High-throughput pharmacogenomic data are inherently noisy due to differences in experimental assays, protocols, and procedures. Producing consistent measurements of drug response using cell viability assays has been shown to be particularly challenging. Here, we explore the utility of integrating multiple drug screens via a drug-centric prior in addition to multiple molecular modalities. Our results indicate that information transfer via drug-wise prior, in addition to multiple modality integration, has overall positive effect in three tested scenarios, at no sacrifice in drug response prediction accuracy. We demonstrate a proof-of-concept that a drug-centric prior in conjunction with multitask properties in a Bayesian group factor analysis model improves the robustness of predictive drug signatures. Improved drug response signature can lead to novel therapeutic targets, more tailored combination therapies, and improve patients to treatment stratification, especially for patients who lack known disease biomarkers.

Acknowledgments. This work was partially funded by NIH NCI K22CA258799.

Disclosure of Interests. No competing interest is declared.

References

1. Barretina, J., et al.: The cancer cell line encyclopedia enables predictive modelling of anticancer drug sensitivity. Nature **483**(7391), 603–607 (2012)
2. Berman, H.M., et al.: The protein data bank. Nucleic Acids Res. **28**(1), 235–242 (2000)
3. Collisson, E.A., et al.: Subtypes of pancreatic ductal adenocarcinoma and their differing responses to therapy. Nat. Med. **17**(4), 500–503 (2011)
4. Cancer Cell Line Encyclopedia Consortium Broad Institute: Pharmacogenomic agreement between two cancer cell line data sets. Nature **528**(7580), 84–87 (2015)
5. Deng, Y., et al.: Mek inhibitor mirdametinib suppresses map kinase pathway activity and inhibits tumor growth in atypical teratoid/rhabdoid tumors. Cancer Res. **84**(6_Supplement), 6498–6498 (2024)
6. Erlichman, C.: Tanespimycin: the opportunities and challenges of targeting heat shock protein 90. Expert Opin. Investig. Drugs **18**(6), 861–868 (2009)
7. Garnett, M.J., et al.: Systematic identification of genomic markers of drug sensitivity in cancer cells. Nature **483**(7391), 570–575 (2012)
8. Haibe-Kains, B., et al.: Inconsistency in large pharmacogenomic studies. Nature **504**(7480), 389–393 (2013)
9. Kaplan, F., Shao, Y., Mayberry, M., Aplin, A.: Hyperactivation of mek-erk1/2 signaling and resistance to apoptosis induced by the oncogenic b-raf inhibitor, plx4720, in mutant n-ras melanoma cells. Oncogene **30**(3), 366–371 (2011)
10. Kojima, K., Konopleva, M., McQueen, T., O'Brien, S., Plunkett, W., Andreeff, M.: Mdm2 inhibitor nutlin-3a induces p53-mediated apoptosis by transcription-dependent and transcription-independent mechanisms and may overcome ATM-mediated resistance to fludarabine in chronic lymphocytic leukemia. Blood **108**(3), 993–1000 (2006)
11. Modi, S., et al.: Hsp90 inhibition is effective in breast cancer: a phase ii trial of tanespimycin (17-AAG) plus trastuzumab in patients with HER2-positive metastatic breast cancer progressing on trastuzumab. Clin. Cancer Res. **17**(15), 5132–5139 (2011)
12. Nikolova, O., Moser, R., Kemp, C., Gönen, M., Margolin, A.A.: Modeling gene-wise dependencies improves the identification of drug response biomarkers in cancer studies. Bioinformatics **33**(9), 1362–1369 (2017)
13. Roden, D.M., George, A.L., Jr.: The genetic basis of variability in drug responses. Nat. Rev. Drug Discovery **1**(1), 37–44 (2002)
14. Secchiero, P., Bosco, R., Celeghini, C., Zauli, G.: Recent advances in the therapeutic perspectives of nutlin-3. Curr. Pharm. Des. **17**(6), 569–577 (2011)
15. Yeh, T.C., et al.: Biological characterization of ARRY-142886 (AZD6244), a potent, highly selective mitogen-activated protein kinase kinase 1/2 inhibitor. Clin. Cancer Res. **13**(5), 1576–1583 (2007)
16. Zhou, Y., et al.: TTD: therapeutic target database describing target druggability information. Nucleic Acids Res. **52**(D1), D1465–D1477 (2024)

Improving Inter-helical Residue Contact Prediction in α-Helical Transmembrane Proteins Using Structural Neighborhood Crowdedness Information

Aman Sawhney(✉) and Li Liao(✉)

Department of Computer and Information Sciences, University of Delaware, Smith Hall, 18 Amstel Avenue, Newark, DE 19716, USA
{asawhney,liliao}@udel.edu
https://www.cis.udel.edu/

Abstract. Residue contact maps are a useful compressed representation that can be used as constraints for structural modeling, but can also help identify inter-helical binding sites and are hence effective on their own. In this work, we hypothesize that crowdedness around a target residue pair influences whether it is a contact point. We developed two measures of crowdedness in a residue's 3D neighborhood: bin counts - defined in terms of relative residue distance; and residue contact number for inter-helical TM proteins - the number of residues in a specified relative distance. Since unsupervised language models such as MSA transformer, trained on millions of sequences, are very accurate but also complementary to our approach, we combined MSA transformer score with our proposed features to assess the impact of crowdedness on residue contact prediction. We found that crowdedness measures can in fact increase the upper bound performance by at least 7.65% average precision in cross validation experiments and by at least 11.59% average precision in held out experiments. Further, we developed a method to "transfer" this information when ground truth crowdedness measures are unavailable. Our approach outperformed MSA transformer by at least 1.15% average precision in cross validation experiments and 1.85% average precision in held-out experiments.

Keywords: Contact map prediction · Protein structure modeling · Alpha helix · Transmembrane proteins

1 Introduction

Membrane proteins are encoded by about 20 to 30 % of genes in all genomes [9]. Transmembrane (TM) proteins are essential to a variety of cell processes such as catalysis, signal transduction and, transporting molecules & ions through the cell membrane [10]. Not surprisingly, membrane proteins are the targets of three-fifths of all clinically approved drugs [22]. Comprehension of the 3-D structure of TM protein is essential to facilitating the development of drugs [7]. However, removal of membrane proteins from their native environment can alter their integrity and prevent crystallization which

M. Alser et al. (Eds.): ICCABS 2025, LNCS 15599, pp. 183–194, 2026.
https://doi.org/10.1007/978-3-032-02489-3_14

is needed for X-ray crystallography [7]. This has led to a sequence-structure gap and the number of solved TM protein structures remains disproportionately low.

A residue contact map is a 2-D representation of geometric constraints that can be used in the absence of 3-D structures. Folding engines like Rosetta [17] input binary contact maps and turn them into folded proteins [1]. Additionally, when observed in different functional states, TM helices have been found to tilt and bend [16] in which case a residue contact map can be useful for detecting inter-helical binding sites and be useful by itself.

A range of features derived from physio-chemical attributes, sequence data and co-evolutionary information [20] have been used to estimate residue contacts. Evolutionary coupling approaches estimate residue pair contact propensities using multiple sequence alignments [19]. Raptor X [21] and TrRosetta [5] used Residual Networks for residue contact prediction with great success. Recent methods such as AlphaFold2 [6], with the use of transformers and deep multiple sequence alignments can estimate the 3-D coordinates of residues to a high degree of accuracy.

In this work, we explored the idea that crowdedness around a target residue pair influences whether or not it is a contact point. We proposed bin counts, defined in terms of relative distance to a residue, as a measure of crowdedness in a residue's neighborhood. For a pair of residues, we concatenated the bin counts for each residue in the pair. Further, we adapted the residue contact number, defining it for inter-helical residues, as another measure of crowdedness. Unsupervised protein language models such as MSA transformer [15] and ProtBert [2] utilize the attention mechanism to train on millions of sequences and can capture residue pair relationships very well. Hence, we combined our crowdedness measures with these to assess if in fact discriminative information is added. We found that residue crowdedness measures are indeed predictive and can elevate the performance by at least 7.65% and up to 14.4% in terms of average precision in our cross validation experiments and by at least 11.59% and up to 12.87% in terms of average precision in our held out experiments. Since realistically ground truth crowdedness measures will not be available during testing, we adopted a k nearest neighbor mean approach to substitute them using the training set. Using this approach to "transfer" crowdedness information, the classification performance increased by at least 1.15% in terms of average precision in our cross validation experiments and by at least 1.85% in terms of average precision in our held out experiments.

2 Materials and Methods

2.1 Dataset

In this study, we utilized the recognized DeepHelicon dataset [20]. 5606 α-helical TM protein chains with a resolution better than 3.5Å were filtered from the PDBTM database [8]. To ensure that the protein chains were structurally dissimilar, a limiting TM score of 0.4 and a 23% sequence identity threshold were imposed. This resulted in a dataset composed of 222 protein chains, with a varying count of TM helices (2-17). The dataset consists of three sub-datasets a) TRAIN - which can be used as the training set and has 165 sequences. For clarity, we refer to it as S_L dataset; b) TEST - which can be used as a held-out set and has 57 sequences. For clarity, we refer to it as S_{M1}

dataset; and c) PREVIOUS - which can be used as a held-out set and has 44 sequences. For clarity, we refer to it as S_{M2} dataset.

For each protein chain, protein sequence, annotations identifying the positions which are located within TM regions and residue pairs that are in contact, and the 3-D structure (PDB format) - with coordinates for every residue's heavy atoms, are present in the dataset.

If the heavy atoms of a residue pair are within a certain distance of one another, then the residue pair are regarded to be in contact. In the DeepHelicon dataset [20], a contact point is defined as two residues which are separated by a minimum of 5 residues in sequence and for which the minimum distance between any pair of their heavy atoms measures less than $5.5A^o$ [20].

Following our previous work [18], sequences with no inter-helical contact positions and for which there was a mismatch between positions used by DeepHelicon and ones annotated as TM positions were removed. Further, in this work we use MSA transformer [15] to generate contact scores and ESM transformer [12] to generate residue embeddings, both limit the sequence length to 1024 residues, hence sequences longer than 1024 residues were removed as well. Finally, due to resource constraints, for a few sequences an effective multiple sequence alignment (input to MSA transformer) could not be generated. Hence, these were removed as well. This results in 154, 36 & 53 sequences in S_L, S_{M2} and S_{M1} datasets respectively. A summary of these changes can be found in Table 1.

$$CR = \frac{\#contact\ points}{\#residue\ pair\ positions} \tag{1}$$

Table 1. Dataset statistics - protein chain count and contact ratio (CR) for S_L, S_{M1} and S_{M2} datasets.

Dataset	#Sequences	#Filtered Sequences	$CR \times 100$
S_L	165	154	2.13
S_{M1}	57	53	2.08
S_{M2}	44	36	1.99

2.2 Features

MSA Transformer Score (MSA). MSA transformer is an unsupervised protein language model which takes as input a multiple sequence alignment. It is trained with a variant of the masked language modeling objective across several protein families on around 26 million multiple sequence alignments. The attention weights in the MSA transformer are predictive of contact between residue pair positions, a linearly weighted combination of these attention weights - MSA transformer score (single number) can be outputted from the model and used as contact probability [15]. For each chain in our dataset, multiple sequence alignments were generated using three iterations of HHblits searches against the UniProt20 (version 2016) database. From each multiple sequence alignment, a maximum of 128 sequences were selected using a diversity maximizing greedy strategy - starting with a reference and adding the sequence with the highest average hamming distance to the current set of sequences [15]. Then we used ESM-MSA-1b to generate MSA transformer scores for all relevant residue pair positions in our dataset i.e. positions that are in helical regions, on different helices and sequence separated by at least 5 positions.

Bin Counts as Features (BC). In 3D space, a residue pair (i, j) may have several residues around it. It follows that the relative distance of these surrounding residues to (i, j), as well as their number, may have an impact on if (i, j) is a contact point.

For each relevant residue pair in our dataset (inter-helical and sequence separated by 5 positions), we computed the least distance between any of their heavy atoms. Any residue that forms a relevant pair is considered to be a relevant residue. We used this information to compute the number of residues that are within a relative distance range to each relevant residue. We call these distance ranges - "Bins"; and we call the number of residues within those relative distance ranges - "Bin counts". The bins used in this work are: Bin 1 - [0 - 2.5Å], Bin 2 - [2.5 - 3.5Å], Bin 3 - [2.5 - 4.5Å], Bin 4 - [4.5 - 5.5Å], Bin 5 - [5.5 - 6.5Å], Bin 6 - [6.5 - 7.5Å] and Bin 7 - [7.5 - 10Å]. This is visually depicted in Fig. 2a. The bin count vector for a residue R_x is a vector of length 7 as depicted via an example in Fig. 2b.

To investigate if such features can help discriminate contact from non-contact, we computed bin counts for all relevant residues in the S_{M_1} dataset using the procedure described above. For a residue pair, we point-wise added the bin counts to create a feature vector of length 7. Then we plotted density normalized histograms for counts in each of the 7 bins separately for contact and non-contacts. We only depict neighbor counts ≤ 10 in these histograms for a clear visualization. Figure 1 depicts one such histogram - Bin 5 - pictorially and in a tabular format. From Fig. 1 it is evident that there is a difference in the count distribution when residue pairs form a contact point vs when they do not. We then hypothesized that bin counts used as features have discriminative power and can aid in the classification of a residue pair as a contact point. Hence, we constructed bin counts as features - bin counts for each residue in a residue pair were concatenated to form a feature vector of length 14.

Contact Numbers as Features (CN). Contact number for a residue i can be defined as the number of contacts within a specific distance to it [4]. In our case, we update this definition to only count residues that are sequence separated by 5 positions and positioned on different helices. Corresponding to the bin counts we count the number of residues at a relative distance of $\leq$ 10Å. For a residue pair, the contact numbers of each residue in the pair are concatenated to form a feature vector of length 2.

MSA Transformer Score and Bin Counts (MSA + BC). Language models such as MSA transformer [15] and ProtBert [2] utilize attention mechanism and are trained on millions of sequences. As a result, these models can capture residue pair relationships

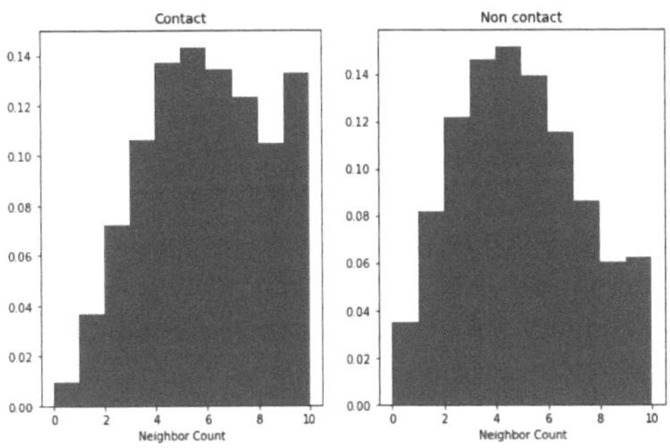

(a) Normalized histogram of residue pair bin counts distribution - contacts vs non-contacts

Neighbor count	Contact normed %	Non-contact normed %
0	0.0097	0.035
1	0.0365	0.0819
2	0.0723	0.1217
3	0.1064	0.1463
4	0.137	0.1514
5	0.1429	0.1396
6	0.1345	0.1153
7	0.123	0.0863
8	0.1049	0.0603
9	0.1328	0.0621

(b) Normalized histogram values for residue pair bin counts distribution - contacts vs non-contacts

Fig. 1. Residue pair bin counts distribution for Bin 5, contact points and non-contact points separated.

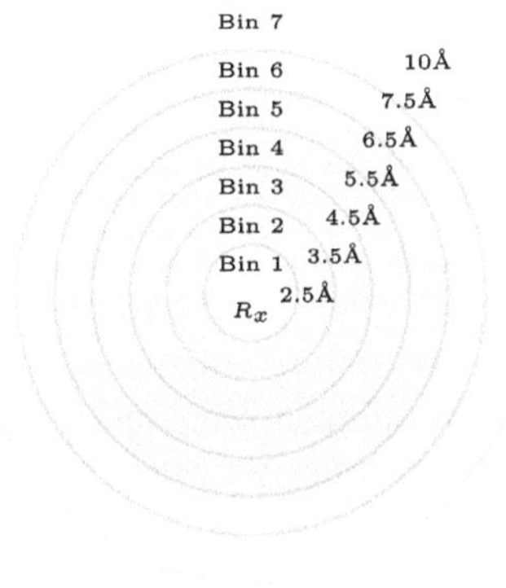

(a) Bins are defined in terms of relative distance to a residue R_x

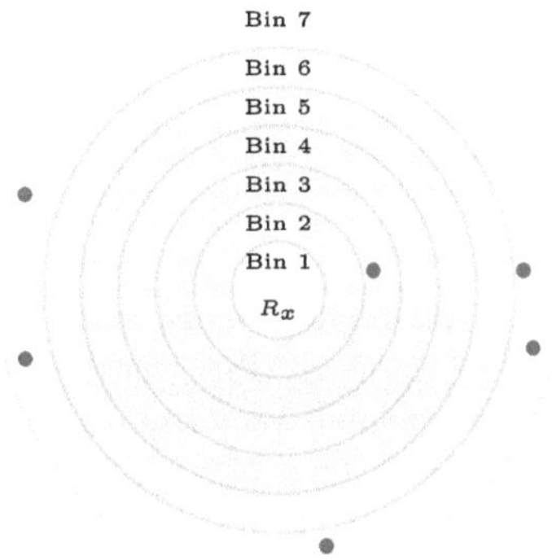

Bin counts = [0, 0, 1, 0, 0, 0, 5]
Contact number = 6

(b) Bin counts for a residue R_x (residue 91 in sequence 1jb0L depicted here)

Fig. 2. Crowdedness around a residue measured as bin counts - count of residues at a certain distance from the target residue.

very well. Since this information is complementary to the information provided by bin counts, we use MSA transformer score along with bin counts as well.

During training, we constructed a feature vector of length 15 by concatenating MSA transformer score and bin counts as features (BC). This is depicted in Fig. 3a. While testing, we would not have access to BC which are extracted from experimentally determined structures. ESM transformer is a single sequence language model that can be used to extract residue level embeddings for any sequence [11,12]. In this work, we used the pre-trained model esm2_t33_650M_UR50D to extract residue level embeddings (1280 length) for every relevant residue in our dataset. Then during testing, for a residue pair (i, j), we used cosine similarity between the ESM embeddings to find k residues from the training set that are most similar to residue i and k residues that are most similar to residue j. Then we computed the point-wise mean for the k residues corresponding to both residue i and residue j and used these mean vectors as substitutes for their bin counts which are unknown to us. As a baseline, we also computed the point-wise mean of k randomly sampled neighbors for each residue in the residue pair and used as a substitute for the unknown bin counts.

$$\underbrace{\boxed{MSA_{i,j}\ ,B_1^i,\ \cdots\ ,B_7^i\ ,B_1^j\ \cdots\ ,B_7^j}}_{1 \times 15} = F_{(i,j)}$$

(a) MSA + BC - bin counts for Residue i and residue j were concatenated with MSA transformer score

$$\underbrace{\boxed{MSA_{i,j}\ ,CN_i,\ ,CN_j}}_{1 \times 3} = F_{(i,j)}$$

(b) MSA + CN - contact numbers for Residue i and residue j were concatenated with MSA transformer score

Fig. 3. Feature vectors.

MSA Transformer Score and Contact Numbers (MSA + CN). Similar to Sect. 2.2, here too we used MSA transformer score along with CN. During training, we constructed a feature vector of 13 by concatenating MSA transformer score and contact numbers as features (CN). This is depicted in Fig. 3b. While testing, we would not have access to ground truth contact numbers, hence we used a procedure similar to the one described in the previous section. For a residue pair (i, j), we used cosine similarity between the ESM transformer embeddings to find k residues from the training set that are most similar to residue i and k residues that are most similar to residue j. We computed the point-wise mean of each of these k residues corresponding to i and j and used the resulting vector as a substitute for their contact numbers. As a baseline, we also computed the point-wise mean of randomly sampled k neighbors for each residue in the residue pair and used that as a substitute for unknown contact numbers.

2.3 Classification

We approached the prediction of whether an inter-helical TM residue pair position is a contact point as a binary classification task using supervised learning. Only residue pair positions that are sequence separated by a minimum of 5 residues and reside on separate helices were considered.

The MSA transformer scores (Sect. 2.2) were used to calculate the baseline performance. When using bin counts as features, a feature vector of length 14 was constructed (as described in Sect. 2.2); when using Contact numbers as features, a feature vector of length 2 was constructed (as described in Sect. 2.2); when MSA transformer score and bin counts were used together a feature vector of length 15 was formed (as described in Sect. 2.2); and, when MSA transformer score and contact numbers were used together a feature vector of length 3 was constructed (as described in Sect. 2.2).

Every feature set was normalized, with each feature scaled to a $[-1, 1]$ range prior to classification. A neural network classifier with 5 hidden layers and leaky Relu activation function was used with binary cross entropy as the loss criterion. We trained in batches of 256 samples for a total of 100 epochs for cross validation experiments and 200 epochs for experiments on held-out datasets. The network was trained using Adam optimizer with a learning rate of 0.0001. The weights of the network were initialized using Xavier uniform distribution and gradients were clipped to the range $[-1, 1]$ to prevent exploding and vanishing gradients. We used the PyTorch package for our implementation [13]. We used the same architecture in all cases with the input layer size adjusted to match the feature set size.

We assessed our performance on each dataset - S_L (154 sequences), S_{M1} (53 sequences) and S_{M2} (36 sequences) - using cross validation (5 folds). In each fold, 20% of the sequences were randomly selected and set aside for validation, while the remaining 80% were used for training. Further, a model was retrained on the entire S_L dataset and its performance evaluated on held out - S_{M1} and S_{M2} datasets.

In each experiment, we used BC and CN generated using experimentally determined structures. These provide upper bound performance for when these features are used separately or combined together with MSA transformer score. When using MSA + BC and MSA + CN feature sets, the bin counts and contact numbers are generated using experimentally determined structures during training, while during testing the mean value of k training residues - randomly picked or selected using cosine similarity of ESM transformer embeddings (as described in Sect. 2.2) - were used as a substitute for each residue in the residue pair.

Performance Metrics. Average Precision was used to evaluate the classification performance in all experiments. The precision-recall curve is summarized by computing a weighted mean of precision scores at different thresholds. The weight corresponding to a threshold's precision value is set as the increase in recall from the previous threshold [14],

$$Average\ Precision = \sum_n (R_n - R_{n-1})P_n \tag{2}$$

where precision at the nth threshold is denoted by P_n and recall by R_n.

This metric allows us to evaluate the classifier's performance in a threshold-free manner - a comprehensive evaluation that isn't dependent on selecting a specific threshold. Going through the ranked list of test examples, the variation of precision with recall can be illustrated as a curve. Average precision represents the area under the precision-recall curve. In cases with skewed data [3] - where the proportion of one class is much larger - average precision may provide a more accurate picture than ROC score which can overestimate the performance. We report the mean score across all sequences.

3 Results

To test out our hypothesis that crowdedness measures - incorporating bin counts (BC) and contact number (CN) with MSA transformer score - can help improve the residue contact prediction over using MSA transformer individually, we designed cross-validation experiments. We trained a neural network classifier on residue pairs (contact pairs as positive and non-contact pairs as negative) represented by different types of features - including BC, CN, MSA + BC, MSA + CN - and evaluated the trained classifier's performance on both the cross validation test set and a held out set. The classification performance was evaluated using Average Precision. For both, MSA + BC and MSA + CN, we computed substitute features during testing using cosine similarity to select k most similar residues from the training set corresponding to both residues in a residue pair. For the MSA + BC case, we also computed substitute features during testing by randomly selecting k residues from the training set corresponding to both residues in a residue pair, as a baseline.

The effect of varying the number of similar residues used to compute substitute features for the cases - when selection is at random with MSA + BC, or using cosine similarity with MSA + BC and with MSA + CN - for cross validation experiments are depicted in Fig. 4a, 4b and 4c for S_L, S_{M_1} & S_{M_2} respectively. For all datasets, random selection eventually converges close to the MSA transformer's performance and does not outperform it. For all datasets, MSA + BC curve outperforms MSA + CN by a very slight margin. For S_L dataset, peak performance is obtained when MSA + BC features are computed using 200 neighbors; for S_{M_1} dataset the best performance is obtained with 50 neighbors; and for S_{M_2} the best performance is obtained with 100 neighbors.

The performance when the number of similar residues (computed using cosine similarity) are varied for MSA + BC and MSA + CN features on held out datasets is depicted in Fig. 5a and 5b for S_{M_1} & S_{M_2} respectively. For both held out datasets - S_{M_2} and S_{M_1} - MSA + CN is slightly outperformed by MSA + BC. The best performance for S_{M_1} dataset is obtained when MSA + BC features are computed using 200 neighbors and for S_{M_2} dataset when using MSA + BC with 300 neighbors.

The classification performance for cross validation experiments are reported in Table 2a and for held out datasets in Table 2b. Individually, both BC and CN are predictive with at least 14.5% average precision in all experiments but with a performance much lower than the MSA transformer score which scores at least 61.8% average precision in all experiments.

Randomly selecting residues from the training set negatively impacted classification performance, with even the best performing case under performing MSA transformer score in each cross validation experiment. This implies that residues used to compute substitute BC or CN need to be chosen in a relevant way in order for the substitution to enable a gain in performance. In cross validation experiments, MSA + BC (cosine) outperforms MSA transformer score by 1.41%, 1.84% & 1.15% in terms of average precision for S_{M_1}, S_{M_2} and S_L datasets respectively. In the held out experiments, MSA + BC (cosine) outperforms the MSA transformer score by 1.91% & 2.26% in terms of average precision for S_{M_1} and S_{M_2} respectively. This implies that in both experiments we were able to positively "transfer" knowledge from ground truth crowdedness measures.

For cross validation experiments, the upper bound performance of MSA + BC outperforms the MSA transformer score by 10.25%, 7.87% & 14.4% in terms of Average precision for S_{M_1}, S_{M_2} and S_L datasets respectively. In the held out experiments, the upper bound performance of MSA + BC outperforms by 12.87% & 11.59% in terms of Average precision for S_{M_1} & S_{M_2} datasets respectively. This implies that there is a significant potential for further improvement during testing; we aim to pursue this in the future. The upper bound performance of MSA + CN is much lower than that of MSA + BC, which makes intuitive sense as bin counts are able to capture neighborhood crowdedness in a much more nuanced way.

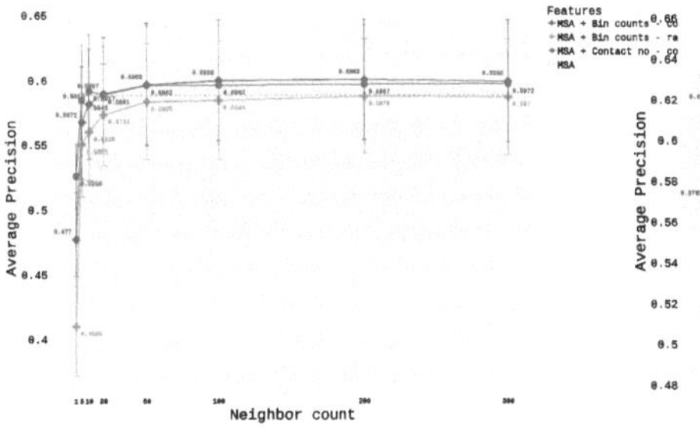

(a) S_L dataset

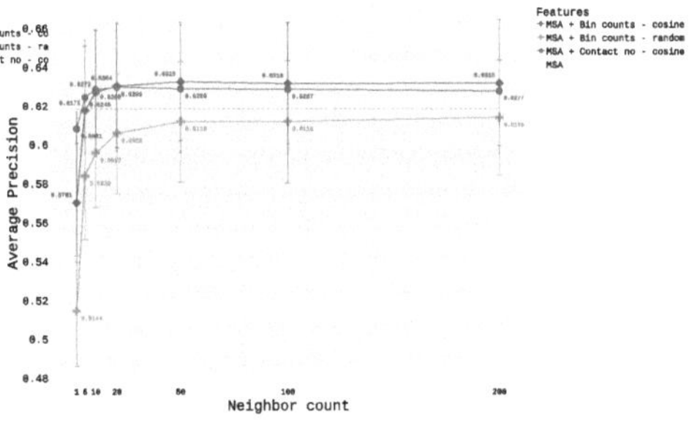

(b) S_{M_1} dataset

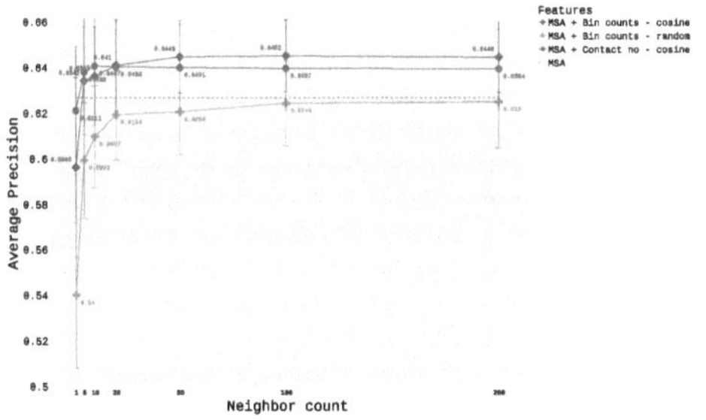

(c) S_{M_2} dataset

Fig. 4. Cross validation performance when number of neighbors is varied.

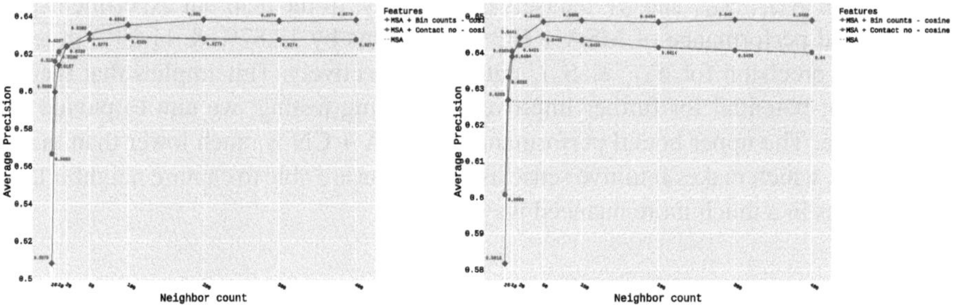

(a) S_{M_1} dataset (b) S_{M_2} dataset

Fig. 5. Held out performance when number of neighbors is varied.

Table 2. Classification results for cross validation and held out dataset experiments. For MSA + BC, MSA + CN best testing results obtained using any k neighbors are reported.

Dataset	Features	Average Precision
S_{M1}	MSA	0.6184 ± 0.0314
S_{M1}	BC	0.1588 ± 0.0275
S_{M1}	CN	0.1521 ± 0.023
S_{M1}	MSA + BC (upperbound)	0.7205 ± 0.0235
S_{M1}	MSA + CN (upperbound)	0.6564 ± 0.0293
S_{M1}	MSA + BC (cosine)	0.6325 ± 0.0305
S_{M1}	MSA + BC (random)	0.6139 ± 0.0297
S_{M1}	MSA + CN (cosine)	0.6299 ± 0.0332
S_{M2}	MSA	0.6268± 0.0189
S_{M2}	BC	0.1452 ± 0.0578
S_{M2}	CN	0.1389 ± 0.0444
S_{M2}	MSA + BC (upperbound)	0.7055 ± 0.0214
S_{M2}	MSA + CN (upperbound)	0.6625 ± 0.0197
S_{M2}	MSA + BC (cosine)	0.6452 ± 0.0158
S_{M2}	MSA + BC (random)	0.625 ± 0.0204
S_{M2}	MSA + CN (cosine)	0.6407 ± 0.0207
S_L	MSA	0.5885 ± 0.045
S_L	BC	0.2463 ± 0.0468
S_L	CN	0.2028 ± 0.0275
S_L	MSA + BC (upperbound)	0.7325 ± 0.0271
S_L	MSA + CN (upperbound)	0.6525 ± 0.0339
S_L	MSA + BC (cosine)	0.6003 ± 0.0465
S_L	MSA + BC (random)	0.5875 ± 0.0443
S_L	MSA + CN (cosine)	0.5972 ± 0.0455

(a) Classification performance using cross validation - average over 5 folds is reported

Dataset	Features	Average Precision
S_{M1}	MSA	0.6189
S_{M1}	BC	0.1545
S_{M1}	CN	0.1479
S_{M1}	MSA + BC (upperbound)	0.7476
S_{M1}	MSA + CN (upperbound)	0.6675
S_{M1}	MSA + BC (cosine)	0.6380
S_{M1}	MSA + CN (cosine)	0.6288
S_{M2}	MSA	0.6264
S_{M2}	BC	0.1532
S_{M2}	CN	0.1397
S_{M2}	MSA + BC (upperbound)	0.7423
S_{M2}	MSA + CN (upperbound)	0.6741
S_{M2}	MSA + BC (cosine)	0.6490
S_{M2}	MSA + CN (cosine)	0.6449

(b) Classification performance using held out datasets

4 Conclusion

In this work, we investigated the idea that the amount of busyness in the 3D structure neighborhood of a residue can impact whether a residue pair is in contact. We believe that inter-atomic forces - attractive or repulsive - and space constraints could cause this. We proposed two crowdedness measures - bin counts and residue contact number in terms of relative distance to a residue. We assessed the performance of these measures individually and found them to be predictive. Since unsupervised protein language models such as MSA transformer can capture residue pair relationships very well, we combined our crowdedness measure with their predictions to see if discriminative information is added. When combined with MSA transformer score, the upper bound performance increases by at least 7.87% average precision in cross validation experiments and 11.59% average precision in held out experiments, which is very promising and indicates that crowdedness in fact impacts residue contact. We developed an approach to substitute when these measures are unavailable - using ESM embeddings to identify similar residues in the training set. With this substitution approach to transfer crowdedness information, the classification performance increased by at least 1.15% average precision in cross validation experiments and by at least 1.85% in our held out experiments.

Acknowledgments. The authors thank the anonymous reviewers for their valuable suggestions. Support from the University of Delaware CBCB Bioinformatics Core Facility and use of the BIOMIX compute cluster was made possible through funding from Delaware INBRE (NIH NIGMS P20 GM103446), the State of Delaware, and the Delaware Biotechnology Institute. The work is also partly supported by grants from NSF (NSF-MCB1820103) and from Delaware Biotechnology Institute (CAT).

Disclosure of Interests. The authors have no competing interests.

References

1. AlQuraishi, M.: Machine learning in protein structure prediction. Curr. Opin. Chem. Biol. **65**, 1–8 (2021)
2. Brandes, N., Ofer, D., Peleg, Y., Rappoport, N., Linial, M.: Proteinbert: a universal deep-learning model of protein sequence and function. Bioinformatics **38**(8), 2102–2110 (2022)
3. Davis, J., Goadrich, M.: The relationship between precision-recall and roc curves. In: Proceedings of the 23rd international conference on Machine learning, pp. 233–240 (2006)
4. Dong, Q., Zhou, S., Guan, J.: Improving prediction of the contact numbers of residues in proteins from primary sequences. In: 2009 International Joint Conference on Bioinformatics, Systems Biology and Intelligent Computing, pp. 251–254. IEEE (2009)
5. Du, Z., et al.: The trRosetta server for fast and accurate protein structure prediction. Nat. Protoc. **16**(12), 5634–5651 (2021)
6. Jumper, J., et al.: Highly accurate protein structure prediction with alphafold. Nature **596**(7873), 583–589 (2021)
7. Kermani, A.A.: A guide to membrane protein x-ray crystallography. FEBS J. **288**(20), 5788–5804 (2021)
8. Kozma, D., Simon, I., Tusnady, G.E.: PDBTM: protein data bank of transmembrane proteins after 8 years. Nucleic Acids Res. **41**(D1), D524–D529 (2012)

9. Krogh, A., Larsson, B., Von Heijne, G., Sonnhammer, E.L.: Predicting transmembrane protein topology with a hidden markov model: application to complete genomes. J. Mol. Biol. **305**(3), 567–580 (2001)
10. Lagerström, M.C., Schiöth, H.B.: Structural diversity of g protein-coupled receptors and significance for drug discovery. Nat. Rev. Drug Discovery **7**(4), 339–357 (2008)
11. Lin, Z., et al.: Evolutionary-scale prediction of atomic-level protein structure with a language model. Science **379**(6637), 1123–1130 (2023)
12. Meier, J., Rao, R., Verkuil, R., Liu, J., Sercu, T., Rives, A.: Language models enable zero-shot prediction of the effects of mutations on protein function. In: Advances in Neural Information Processing Systems, vol. 34, pp. 29287–29303 (2021)
13. Paszke, A., et al.: Pytorch: an imperative style, high-performance deep learning library. In: Advances in Neural Information Processing Systems, vol. 32, pp. 8024–8035. Curran Associates, Inc. (2019)
14. Pedregosa, F., et al.: Scikit-learn: machine learning in Python. J. Mach. Learn. Res. **12**, 2825–2830 (2011)
15. Rao, R.M., et al.: MSA transformer. In: International Conference on Machine Learning, pp. 8844–8856. PMLR (2021)
16. Ren, Z., Ren, P.X., Balusu, R., Yang, X.: Transmembrane helices tilt, bend, slide, torque, and unwind between functional states of rhodopsin. Sci. Rep. **6**(1), 34129 (2016)
17. Rohl, C.A., Strauss, C.E., Misura, K.M., Baker, D.: Protein structure prediction using Rosetta. In: Methods in Enzymology, vol. 383, pp. 66–93. Elsevier (2004)
18. Sawhney, A., Li, J., Liao, L.: Improving alphafold predicted contacts for alpha-helical transmembrane proteins using structural features. Int. J. Mol. Sci. **25**(10), 5247 (2024)
19. Sheridan, R., et al.: Evfold. org: evolutionary couplings and protein 3D structure prediction. biorxiv 021022 (2015)
20. Sun, J., Frishman, D.: Deephelicon: accurate prediction of inter-helical residue contacts in transmembrane proteins by residual neural networks. J. Struct. Biol. **212**(1), 107574 (2020)
21. Wang, S., Sun, S., Li, Z., Zhang, R., Xu, J.: Accurate de novo prediction of protein contact map by ultra-deep learning model. PLoS Comput. Biol. **13**(1), e1005324 (2017)
22. Yin, H., Flynn, A.D.: Drugging membrane protein interactions. Ann. Rev. Biomed. Eng. **18**, 51–76 (2016)

Explaining Protein Folding Networks Using Integrated Gradients and Attention Mechanisms

Rukmangadh Sai Myana and Sumit Kumar Jha(✉)

Department of Computer Science, Florida International University, Miami, FL, USA
rmyan001@fiu.edu, sumit.jha@ufl.edu

Abstract. Protein folding prediction models like AlphaFold and ColabFold have revolutionized structural biology by providing accurate protein structures. However, these models present challenges when it comes to understanding how they arrive at their decisions. In this paper, we propose the application of Explainable AI (XAI) techniques, specifically Integrated Gradients and Attention Mechanisms, to elucidate the decision-making process of these complex networks. We conduct computational experiments to evaluate the effectiveness of these methods and discuss potential implications for the field.

Keywords: Protein Folding · Explainable AI · Integrated Gradients · Attention Mechanisms · AlphaFold · ColabFold

1 Introduction

Protein folding represents a central and enduring challenge within computational biology, as the elucidation of these intricate processes is paramount to deciphering the molecular underpinnings of biological function and disease etiology. The functionality of a protein is inextricably linked to its complex three-dimensional conformation, a structure uniquely determined by the physicochemical attributes inherent in its amino acid sequence. The precise prediction of protein structures is therefore of considerable importance, finding broad utility in diverse applications such as drug discovery [12], enzyme engineering [71], the study of genetic regulation [50], and the detailed analysis of molecular interactions [40].

The advent of deep learning methodologies has instigated a paradigm shift within the realm of protein structure prediction. Groundbreaking models, notably AlphaFold2 [46] and ColabFold [58], have demonstrated an unprecedented capacity to predict protein structures with accuracy approaching experimental levels [1]. These sophisticated innovations have effectively revolutionized the field of structural biology, empowering researchers to address and resolve intricate structural complexities previously deemed intractable [95].

Despite the remarkable predictive capabilities of these deep learning models, their operation as complex, 'black-box' systems presents significant challenges.

M. Alser et al. (Eds.): ICCABS 2025, LNCS 15599, pp. 195–211, 2026.
https://doi.org/10.1007/978-3-032-02489-3_15

The inherent opacity of their decision-making processes raises critical concerns regarding interpretability. Organizations encounter difficulties in discerning the rationale behind model predictions, evaluating performance consistency across diverse applications, and identifying potential systematic errors or biases embedded within the model's learning and predictive mechanisms [57,73]. This lack of transparency not only undermines trust in model outputs but also impedes the seamless integration of these powerful tools into critical domains demanding high confidence and accountability, such as clinical diagnostics and therapeutic development, necessitating the urgent address of interpretability to fully realize their transformative potential in biological research [88].

Prior to the deep learning era, protein structure prediction relied primarily on computational modeling techniques rooted in established biophysical principles. Homology modeling, a prominent approach, constructs atomic-resolution models of target proteins by exploiting sequence similarity with homologous templates [25], with widely used tools including IntFOLD [56], RaptorX [64], and Modeller [32]. Protein threading, conversely, tackles structure prediction for proteins lacking clear homologous templates by leveraging statistical relationships between known protein folds and target sequences, exemplified by tools such as I-TASSER [72], HHpred [81], and Phyre [47].

Both traditional methodologies and contemporary machine learning models are fundamentally underpinned by the physics governing protein folding [31]. The incorporation of these foundational principles into machine learning interpretability frameworks offers a promising avenue for achieving a more profound understanding of model predictions. Experimental observations have consistently demonstrated that proteins typically adopt well-defined three-dimensional structures, denoted as the native state [48,91]. Furthermore, the transition of many proteins from an unfolded state to their native state is characterized by an 'all-or-none' mechanism [5,67]. This characteristic transition imparts robustness to protein structures, thereby ensuring their requisite biological functionality, and understanding these internal mechanisms is critical for deciphering complex cellular functions, molecular interactions, and disease mechanisms [4,14,40,90].

Explainable AI (XAI) techniques offer a potent framework for interpreting the intricacies of complex models [28]. These methodologies are designed to bridge the gap between model complexity and user comprehension by providing insights into the determinants of model predictions [60]. For instance, feature attribution methods [35,36,61] can identify critical sequence motifs, physicochemical properties, or interaction patterns that exert significant influence on predicted structures. Similarly, attention mechanisms [10,55,87] within deep learning architectures can elucidate relationships between input sequence regions within a structural context. Visualization tools, including saliency maps [80] and gradient-based overlays [76], further facilitate the correlation of model outputs with underlying biological, physical, and chemical phenomena, thereby aiding hypothesis generation, experimental validation, and broader acceptance of computational predictions within high-stakes applications such as precision medicine and drug design [88]. Integrating domain-specific knowledge, particularly principles of protein physics and thermodynamics, into XAI methodologies further enhances interpretability by grounding model predictions within established biological frameworks, fostering cross-disciplinary collaboration and making computational outputs more accessible and actionable for experimental biologists and clinicians [17,19,68,74].

2 Background

Machine learning (ML) models have garnered significant attention and efficacy, broadening their impact beyond computer science into fields such as medicine, environmental science, and physics. Key innovations like the Multi-headed Attention mechanism in Transformer models [87] have transformed medical applications, enhancing diagnostics, preventive care, and medical workflow management. Specifically, Transformer-based models like ScoreNet [82], T2T-ViT [98], IL-MCAM [22], and SEViT [3] have been instrumental in histopathology image classification, while Uni4Eye [16] and LAT [85] are used for fundus image classification. Models like Chest L-Transformer [38] and FeSTA [63] demonstrate robust performance in X-ray classification.

Additionally, Transformer models have advanced *medical image segmentation* with examples like RANT [62], PCAT-UNet [20], and AMGB-Transformer [52]; *medical image detection* through models such as UTRAD [24] and STCovidNet [89]; and *medical image reconstruction* via models like SSTrans-3D [92] ReconFormer [39], and DSFormer [99]. In clinical information extraction, BioBERT [51], ClinicalBERT [42], and BlueBERT [65] have made notable contributions. Finally, applications in critical care employ models such as SimTA [94], and BENDR [49], while public health monitoring has benefited from models like BERTweet [2].

Moreover, breakthroughs in protein structure prediction include trRosetta [93], ESMFold [53], RoseTTAFold2 [9], and AlphaFold2 [46]. In pharmacology, models like TranSynergy [54], TP-DDI [97], and MolTrans [43] address drug interactions, while drug synthesis is facilitated by Molecular Transformer [75]. These advancements demonstrate the broad and transformative potential of attention based models across diverse sectors.

2.1 AlphaFold, AlphaFold2 and ColabFold

AlphaFold, a deep learning model developed by DeepMind, accurately predicts protein structures. Introduced in 2018 [77], AlphaFold generates a mean force potential for a given protein, allowing the 3D structure to be obtained via gradient descent in this potential space. This prediction is enabled by a central convolutional residual network that computes a distogram a matrix representing distances between C_β atoms of residue pairs in a protein sequence based on multiple sequence alignment (MSA) features. For each residue pair (i, j), AlphaFold predicts a discrete probability distribution over distances from 2 to 22 , split into 64 bins. Cubic spline interpolation is then used to construct the proteins distance potential from these discrete distributions. The torsion angles of side chains are modeled as Von Mises distributions, whose parameters are also predicted by AlphaFold. The final mean force potential is computed by summing the log-likelihoods of the distance distributions across all residue pairs (i, j), along with the log-likelihood of the torsion angle distributions across all residues.

In 2020, DeepMind significantly advanced their previous work with the introduction of AlphaFold2 [46], a model that set a new standard in protein structure prediction by achieving unprecedented performance in the CASP14 competition. This breakthrough was largely attributed to AlphaFold2's innovative Transformer-based architecture and the integration of novel attention mechanisms, including Axial Attention [41] and Invariant Point Attention [46]. Axial Attention generalizes the Self-Attention operation to multiple dimensions while maintaining computational efficiency, while Invariant Point Attention offers self-attention that is invariant to global euclidean transformations. Together, these mechanisms allowed AlphaFold2 to accurately model complex spatial relationships within protein structures, setting it apart from competing approaches.

To construct diverse multiple sequence alignment (MSA) features, extensive protein sequence datasets from both public and environmental databases [26,59] are queried by AlphaFold2 using advanced homology detection tools, specifically HMMer [29] and HHblits [83], which leverage hidden Markov models (HMMs). Given the considerable volume of these databases, search processes for a single protein can require several hours and demand multiple terabytes of storage. Additionally, deep neural network computations necessitate graphics processing units (GPUs) with substantial GPU RAM, particularly for proteins with larger sequence lengths. ColabFold [58] was thus introduced to accelerate single-protein predictions while maintaining a performance comparable to AlphaFold2. ColabFold achieves this by substituting AlphaFold2s homology search with the significantly faster MMseqs2 (Many-against-Many sequence searching) [84], achieving a 40 to 60-fold speed increase. For batch predictions, ColabFold reduces runtime by approximately 90-fold by eliminating recompilation and incorporating an early stopping criterion.

2.2 Explainable AI Techniques

Current interpretability methods can be broadly categorized into five approaches [7]: feature attribution, inherently interpretable models, hierarchical explanations, contrastive explanations, and counterfactual explanations.

Feature attribution methods aim to assign scores or ranks to parts of the input, quantifying their influence on a models prediction. Techniques such as LIME [69] approximate neural network decision-making by fitting an interpretable model, like a decision tree, to the local neighborhood of the input. Anchors [70] provides explanations for individual predictions of any black-box classification model by identifying IF-THEN decision rules that anchor the prediction. A rule is considered an anchor if rule-abiding changes to feature values do not affect the prediction, ensuring robustness of the explanation. Other feature attribution methods like relevance propagation [8,78], Integrated Gradients [86], and Grad-CAM [76] generate feature importance maps by back-propagating gradients through the network.

Inherently interpretable models produce a 'rationale' [6,11,18,96], a concise, human-understandable explanation of their predictions. These methods typically consist of two components: a generator that extracts the rationale and an encoder that uses it to make a prediction. Such models are often termed 'self-explainable' due to their intrinsic transparency. Simpler models like linear regression [33], decision trees [15], and RuleFit [34] also fall under this category, as they naturally lend themselves to straightforward interpretation.

Hierarchical explanations aim to elucidate model outputs at multiple levels. For example, [21] employs Shapley values to provide both feature attributions and feature interactions, while [79] extends Integrated Gradients to also capture these interactions. This multi-layered approach offers a more comprehensive understanding of the prediction process. On the other hand, contrastive explanations focus on building inherently interpretable models capable of addressing questions of the form *Why 'p'?* and *Why 'p' instead of 'q'?*. While most existing post-hoc techniques, such as gradient-based methods, are limited to addressing *Why 'p'?*, contrastive methods fill this gap by explicitly comparing alternatives. Finally, Counterfactual explanations identify minimal changes to the input data that would effectively "flip" the models prediction.

3 Methodology

3.1 Integrated Gradients

Integrated Gradients (IG) is a popular technique used in XAI to attribute the contribution of each input feature to a model's prediction. Integrated Gradients quantify how much each feature contributes to the model's output relative to a baseline. The baseline is typically chosen to represent the "absence" of information (e.g. a zero matrix of numerical inputs, a completely black image). IG then computes the cumulative gradients of the model's output with respect to the inputs, moving along a straight line from the baseline to the actual input. The integrated gradient for the i-th feature is calculated as:

$$IG_i(x) = (x_i - x'_i) \int_{\alpha=0}^{1} \frac{\partial F(x' + \alpha \cdot (x - x'))}{\partial x_i} d\alpha$$

where x is the actual input, x' is the baseline input, α is a scaling factor that interpolates between the baseline and the actual input, and F is the model's output function. Integrated Gradients (IG) offers several desirable properties for feature attribution. It ensures that features with no impact on the output receive zero attribution, maintains consistency across mathematically equivalent model implementations (model-agnostic), and satisfies the completeness property, where the sum of all feature attributions equals the difference between the model's prediction for the actual input and the baseline. In the context of protein structure prediction with ColabFold, the output function corresponds to the calculation of the model confidence metric while the baseline corresponds to absence of any input sequence (i.e. zero vectors) to the model.

Integrated Gradients (IG) has been effectively employed as a versatile interpretability technique across a range of domains, underscoring its utility in elucidating the decision-making processes of deep learning models. Within drug discovery, IG has facilitated the analysis of retrosynthetic reaction predictions [44] and the prediction of crucial drug-CYP3A4 enzyme interactions for understanding drug metabolism [45]. Furthermore, IG has been instrumental in enhancing the transparency of AI-driven diagnostics by interpreting disease detection models [27,66] and in environmental science by identifying key factors in precipitation predictions [30]. These diverse applications collectively illustrate the robustness of IG as an interpretability tool, offering valuable insights for a spectrum of scientific and practical problems.

3.2 Attention Mechanisms

Attention mechanisms, introduced in the domain of machine translation, demonstrated significant improvements in sequence-to-sequence models by dynamically focusing on relevant parts of the input during prediction [10]. With the advent and subsequent success of Transformer models, attention mechanisms have gained widespread adoption across a diverse array of tasks spanning domains such as speech recognition, natural language processing, object detection, and time series prediction illustrating their versatility and efficacy.

A notable milestone in the evolution of attention-based models is AlphaFold2, which leveraged attention mechanisms for predicting protein structures with remarkable accuracy. Specifically, AlphaFold2 employs Axial Attention, a generalized and computationally efficient variant of self-attention, designed to handle high-dimensional inputs such as protein sequences and their spatial features. The formulation of Self-Attention (SA), a foundational mechanism in modern deep learning models, is expressed mathematically as follows:

$$\mathrm{SA}(Q, K, V) = \mathrm{softmax}\left(\frac{QK^T}{\sqrt{d_k}}\right) V$$

In self-attention mechanisms, query(Q), key(K), andvalue (V) matrices, derived from input data, are utilized, with d_k representing the key vector dimensionality. Self-attention computes weighted sums of value vectors, where weights are determined by compatibility scores between query and key vectors, thereby measuring the attention one residue pays to another. Self-attention over these triplets enables the integration of contextual information across the sequence, capturing long-range dependencies crucial for accurate structure prediction. Axial Attention (AA) [41] extends the standard self-attention mechanism by restricting the computation of attention to specific axes of a high-dimensional input tensor. This approach enhances computational efficiency by avoiding the full-dimensional attention computation, which can be prohibitively expensive for large-scale inputs. When Axial Attention layers are stacked such that each layer corresponds to a different axis of the input, the combined operation is equivalent to applying standard self-attention across all axes.

4 Computational Experiments

To interpret the inter-residue relationships learned by the model, we applied Integrated Gradients (IG) to the attention, or compatibility, matrices. This approach was undertaken to extract the specific inter-residue interactions that are most salient to the model's predictions. The aim was to ascertain whether these identified interactions align with established physicochemical principles known to govern protein folding and stability, thereby providing a biologically meaningful interpretation of the model's internal representations.

4.1 Dataset Selection

Protein sequences have been obtained from protein Data Bank (PDB) [13] and Uniprot [26]. The choice of protein sequence has been determined by the availability of experimentally determined 3D structure for the protein. In this context, the B1 domain of streptococcal protein G, and Sso10a from Sulfolobus solfataricus have been explored.

B1 Domain Steptococcal Protein G [37] Protein G helps *Streptococcus* evade host defenses by binding to key host proteins. A repeating 55-residue domain in Protein G binds to the F_c region of immunoglobulin G (IgG) and α2-macroglobulin, a major protease inhibitor in human plasma. Protein G from strain GX7809 contains two such repeats, while strain GX7805 has three, with over 90% sequence identity between repeats. This domain of Protein G named as B1 domain is a significant analytical tool in immunology due to its extreme physicochemical properties and has been extensively studied. Its experimentally determined structure thus provides a benchmark for residue importance predictions by AlphaFold2.

The B1 domain (Fig. 1c) exhibits a novel topology with a four-stranded β-sheet containing a central parallel pair and a +3x crossover connecting outer strands via an α-helix, a configuration unprecedented in structural databases. Its exceptional thermal stability arises from several features: (1) 95% of its residues participate in secondary structures, contributing 45 hydrogen bonds (41 backbone-backbone, 3 side chain-backbone, and 1 side chain-side chain), which stabilize turns and helices; (2) a hydrophobic core formed by tightly packed aromatic residues, such as Trp43 and Tyr45, and hydrophobic interactions involving residues like Leu5, Leu7, and Val39; and (3) a solvent-exposed surface enriched in polar and charged residues, including Thr, Lys, Glu, and Asp. These structural features together enhance stability and functionality, making the B1 domain a remarkable model for protein studies.

Sso10a Sulfolobus Solfatarius [23] The crystal structure of Sso10a reveals an elongated dimer formed through crystallographic 2-fold rotation, with dimensions of 27 × 80 × 27 . Each monomer comprises four α-helices and three β-strands arranged in an H1B1H2H3B2B3H4 topology, with α-helices making up 76% of the structure and β-sheets contributing 13%. The protein is organized into two distinct domains: an N-terminal Winged Helix Domain (residues 1–59) that contains a DNA-binding helix-turn-helix motif and a wing structure formed by β-strands, and a C-terminal H4 Helix Domain that forms an essential antiparallel coiled-coil for dimerization.

The C-terminal domain's H4 helix (residues 60–92) forms a nine-turn structure that pairs with its counterpart to create a coiled-coil interface spanning 3.7 heptad repeats. This interface is stabilized by hydrophobic residues at the heptad's a and d positions, with an unusual feature being Asp69 occupying a typically hydrophobic d position to form a solvent-exposed ion pair with Lys86. The dimerization interface is characterized by a left-handed coiled-coil with a crossing angle of approximately 25, stabilized by hydrophobic interactions, specific ion pairs (including Glu65-Lys86), and hydrogen bonds.

5 Results

5.1 B1 Domain - Steptococcal Protein G

Despite the remarkable thermal stability of protein G, arising from its intricate structural features, including 45 hydrogen bonds (41 backbone-backbone, 3

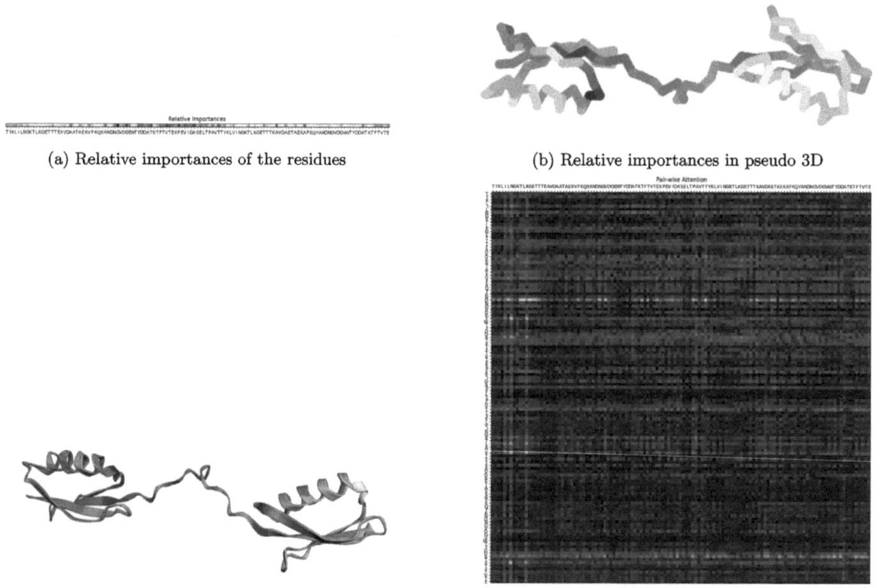

(a) Relative importances of the residues

(b) Relative importances in pseudo 3D

(c) Relative importances in 3D

(d) Attention Map

Fig. 1. *B1 Domain* of protein G

side chain-backbone, and 1 side chain-side chain) that stabilize essential structural motifs such as turns and helices; robust hydrophobic interactions involving residues like Leu5, Leu7, and Val39; and a solvent-exposed surface enriched with polar and charged residues, our analysis suggests that current interpretation techniques, such as Integrated Gradients and Attention, do not fully capture the underlying physicochemical principles in a meaningful way.

From a physiochemical perspective, the pairwise attention that each residue allocates to other residues should logically align with physiochemical interactions, such as disulfide (S-S) bonds, hydrogen bonds and other molecular interactions, which play critical roles in stabilizing protein structure. However, such relationships are conspicuously absent in the attention maps shown in Fig. 1d. For example in an α helix structure, there exists a hydrogen bond between the ith and $(i-4)^{th}$ peptides but such a pattern of attention is not visible in the figure. The misalignment between the attention matrix and fundamental biophysical interactions indicates that this matrix fails to capture the core physical principles that determine how protein residues interact with one another.

Similarly, while integrated gradients highlight certain residues in the protein structure, as depicted in Fig. 1a, and Fig. 1c, these highlighted residues do not correspond meaningfully to their importance from the perspective of physics, chemistry, or biology. For instance, residues critical for forming stabilizing interactions or participating in catalytic activity remain underemphasized or misrepresented.

5.2 Sso10a - Sulfolobus Solfatarius

Correspondingly, while the Sso10a protein exhibits a well-defined structure with four α-helices and three β-strands (comprising 76% and 13% of the structure respectively), current attempts to interpret AlphaFold2's predictions using methods like Integrated Gradients and Attention have not aligned with our understanding of these structural elements.

From a biophysical perspective, one would expect the attention matrix to reflect fundamental physicochemical interactions that stabilize protein structure, such as disulfide bonds and hydrogen bonding networks. However, the attention maps presented in Fig. 2d do not appear to clearly reflect these essential molecular relationships. The attention patterns appear disconnected from the underlying physical principles that govern residue-residue interactions in protein structures.

The Integrated Gradients analysis, visualized across multiple representations in Fig. 2a, and Fig. 2c, similarly falls short of providing biologically meaningful insights. The residues highlighted by this method do not correlate well with their known functional or structural importance. Key residues involved in structural stability or enzymatic function are often overlooked or incorrectly weighted.

6 Discussion

6.1 Implications for Structural Biology

Our analysis reveals significant challenges in applying current explainable AI methodologies to protein structure prediction networks, particularly AlphaFold2. While techniques such as Integrated Gradients and Attention mechanisms have shown promise in other domains of deep learning, their application to protein folding networks presents challenges in deriving biochemically meaningful interpretations. The disconnect between these computational outputs and established physicochemical principles underscores a fundamental limitation in our ability to decode the decision-making processes within these networks.

This interpretability gap has profound implications for structural biology. As protein structure prediction models become increasingly integrated into biological research workflows, the inability to validate their predictions against known biophysical principles poses potential risks. While these models achieve high accuracy, their black-box nature poses challenges in leveraging insights for advancing our understanding of protein folding mechanisms.

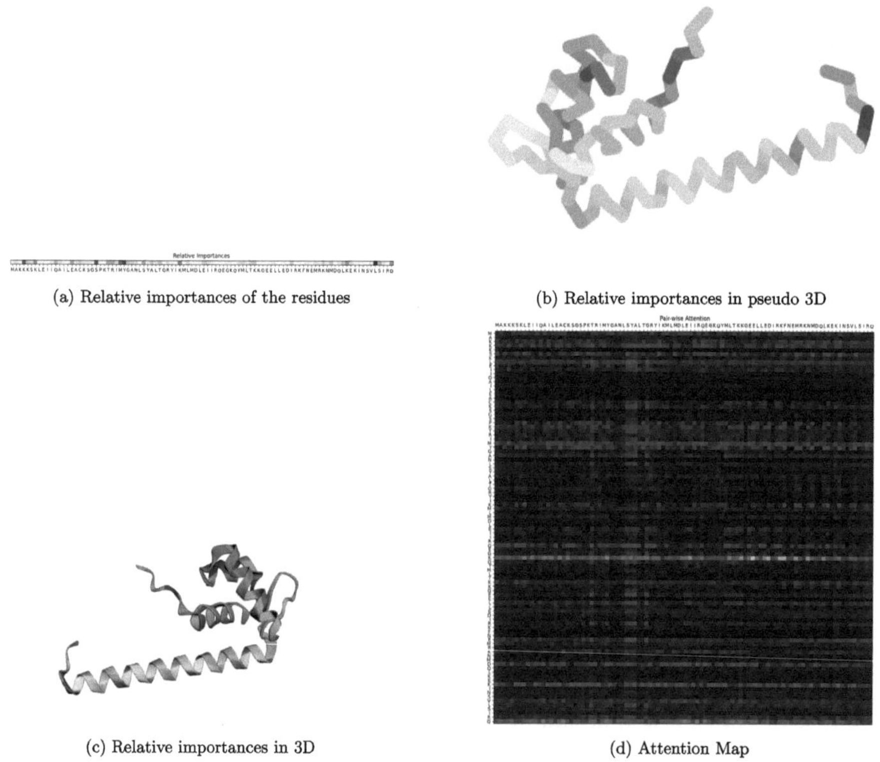

(a) Relative importances of the residues

(b) Relative importances in pseudo 3D

(c) Relative importances in 3D

(d) Attention Map

Fig. 2. *Sso10a* protein.

Following the paradigm established by Rudin [73] in other machine learning domains, we propose that the field of protein structure prediction would benefit from a fundamental shift toward inherently interpretable architectures. Rather than attempting to retrofit explanation methods onto existing models, future development should prioritize architectures that incorporate explicit representations of physical and chemical principles. Such models would ideally maintain or exceed the accuracy of current approaches while providing transparent reasoning paths that align with established biological knowledge.

6.2 Limitations and Future Directions

While our study highlights the inadequacy of current interpretability methods for AlphaFold2, several important caveats merit consideration. First, the limitations we observe may be specific to the particular combination of model architecture and explanation techniques employed.

The complexity and scale of protein folding networks, may demand novel interpretability approaches that are specifically designed for this domain. We propose that future work should focus on developing custom interpretability frameworks that explicitly incorporate biochemical and physical principles. Such frameworks might include: Physics-aware attention mechanisms that track and visualize specific types of molecular interactions, Gradient-based methods that are calibrated to known structure-stability relationships, Interpretability layers that map network activations to recognized biological motifs and interaction patterns, Hybrid approaches that combine machine learning interpretability with molecular dynamics insights. These developments would not only enhance our confidence in model predictions but also potentially provide new insights into the physical principles governing protein folding and structure.

7 Conclusion

Our investigation into explainable AI methods for protein folding networks, focusing on Integrated Gradients and Attention Mechanisms, has revealed both opportunities and significant challenges. While these methods offer initial insights into model behavior, their current implementations fall short of providing interpretations that align meaningfully with established biochemical principles and protein folding mechanisms. We plan to conduct extensive computational experiments to evaluate the robustness of proposed interpretation methods across diverse protein structures. Future work will also focus on incorporating physical principles into interpretability methods by developing attribution methods that reflect molecular interactions and creating attention mechanisms aligned with physical force fields.

While our current work has established a foundation for explaining protein folding networks, there are still notable challenges in developing interpretability methods that align closely with biochemical principles. The path forward requires close collaboration between computational scientists and structural biologists to ensure that future methods not only provide mathematical explanations but also offer genuine insights into the biological principles governing protein structure. Success in this endeavor would not only advance our understanding of protein folding networks but also contribute to the broader goal of creating more interpretable and trustworthy AI systems for scientific applications.

References

1. Ahdritz, G., et al.: OpenFold: retraining alphafold2 yields new insights into its learning mechanisms and capacity for generalization. Nat. Methods, 1–11 (2024)
2. Ahne, A., et al.: Extraction of explicit and implicit cause-effect relationships in patient-reported diabetes-related tweets from 2017 to 2021: deep learning approach. JMIR Med. Inform. **10**(7), e37201 (2022)
3. Almalik, F., Yaqub, M., Nandakumar, K.: Self-ensembling vision transformer (sevit) for robust medical image classification. In: International Conference on Medical Image Computing and Computer-Assisted Intervention, pp. 376–386. Springer (2022)
4. Ananthakrishnan, R., Ehrlicher, A.: The forces behind cell movement. Int. J. Biol. Sci. **3**(5), 303 (2007)
5. Anfinsen, C.B., Haber, E., Sela, M., White, F., Jr.: The kinetics of formation of native ribonuclease during oxidation of the reduced polypeptide chain. Proc. Natl. Acad. Sci. **47**(9), 1309–1314 (1961)
6. Antognini, D., Faltings, B.: Rationalization through concepts. arXiv preprint arXiv:2105.04837 (2021)
7. Babiker, H.K.B.: Deep interpretable modelling. Educ. Res. Arch. (2023)
8. Bach, S., Binder, A., Montavon, G., Klauschen, F., Müller, K.R., Samek, W.: On pixel-wise explanations for non-linear classifier decisions by layer-wise relevance propagation. PLoS ONE **10**(7), e0130140 (2015)
9. Baek, M., Anishchenko, I., Humphreys, I.R., Cong, Q., Baker, D., DiMaio, F.: Efficient and accurate prediction of protein structure using rosettafold2. BioRxiv 2023–05 (2023)
10. Bahdanau, D.: Neural machine translation by jointly learning to align and translate. arXiv preprint arXiv:1409.0473 (2014)
11. Bastings, J., Aziz, W., Titov, I.: Interpretable neural predictions with differentiable binary variables. arXiv preprint arXiv:1905.08160 (2019)
12. Batool, M., Ahmad, B., Choi, S.: A structure-based drug discovery paradigm. Int. J. Mol. Sci. **20**(11), 2783 (2019)
13. Berman, H.M., et al.: The protein data bank. Nucleic Acids Res. **28**(1), 235–242 (2000)
14. Blair, D.F., Dutcher, S.K.: Flagella in prokaryotes and lower eukaryotes. Curr. Opinion Genet. Dev. **2**(5), 756–767 (1992)
15. Breiman, L.: Classification and regression trees. Routledge (2017)
16. Cai, Z., Lin, L., He, H., Tang, X.: Uni4Eye: unified 2D and 3D self-supervised pre-training via masked image modeling transformer for ophthalmic image classification. In: International Conference on Medical Image Computing and Computer-Assisted Intervention, pp. 88–98. Springer (2022)

17. Caruana, R., Lou, Y., Gehrke, J., Koch, P., Sturm, M., Elhadad, N.: Intelligible models for healthcare: predicting pneumonia risk and hospital 30-day readmission. In: Proceedings of the 21th ACM SIGKDD International Conference on Knowledge Discovery and Data Mining, pp. 1721–1730 (2015)
18. Chang, S., Zhang, Y., Yu, M., Jaakkola, T.: Invariant rationalization. In: International Conference on Machine Learning, pp. 1448–1458. PMLR (2020)
19. Chen, C., Li, O., Tao, D., Barnett, A., Rudin, C., Su, J.K.: This looks like that: deep learning for interpretable image recognition. In: Advances in Neural Information Processing Systems, vol. 32 (2019)
20. Chen, D., Yang, W., Wang, L., Tan, S., Lin, J., Bu, W.: PCAT-UNet: UNet-like network fused convolution and transformer for retinal vessel segmentation. PLoS ONE **17**(1), e0262689 (2022)
21. Chen, H., Zheng, G., Ji, Y.: Generating hierarchical explanations on text classification via feature interaction detection. arXiv preprint arXiv:2004.02015 (2020)
22. Chen, H., et al.: IL-MCAM: an interactive learning and multi-channel attention mechanism-based weakly supervised colorectal histopathology image classification approach. Comput. Biol. Med. **143**, 105265 (2022)
23. Chen, L., et al.: The hyperthermophile protein sso10a is a dimer of winged helix DNA-binding domains linked by an antiparallel coiled coil rod. J. Mol. Biol. **341**(1), 73–91 (2004)
24. Chen, L., You, Z., Zhang, N., Xi, J., Le, X.: UTRAD: anomaly detection and localization with U-transformer. Neural Netw. **147**, 53–62 (2022)
25. Chothia, C., Lesk, A.M.: The relation between the divergence of sequence and structure in proteins. EMBO J. **5**(4), 823–826 (1986)
26. UniProt Consortium, Tunca Dogan: UniProt: a worldwide hub of protein knowledge. Nucleic Acids Res. **47**(D1), D506–D515 (2019)
27. DeGrave, A.J., Janizek, J.D., Lee, S.I.: AI for radiographic covid-19 detection selects shortcuts over signal. Nat. Mach. Intell. **3**(7), 610–619 (2021)
28. Doshi-Velez, F., Kim, B.: Towards a rigorous science of interpretable machine learning. arXiv preprint arXiv:1702.08608 (2017)
29. Eddy, S.R.: Accelerated profile hmm searches. PLoS Comput. Biol. **7**(10), e1002195 (2011)
30. Espeholt, L., et al.: Deep learning for twelve hour precipitation forecasts. Nat. Commun. **13**(1), 1–10 (2022)
31. Finkelstein, A., Galzitskaya, O.: Physics of protein folding. Phys. Life Rev. **1**(1), 23–56 (2004)
32. Fiser, A., Šali, A.: Modeller: generation and refinement of homology-based protein structure models. In: Methods in Enzymology, vol. 374, pp. 461–491. Elsevier (2003)
33. Freedman, D.: Statistical Models: Theory and Practice. Cambridge University Press (2005)
34. Friedman, J.H., Popescu, B.E.: Predictive learning via rule ensembles. Ann. Appl. Stat. (2008)
35. Garson, G.D.: Interpreting neural-network connection weights. AI Expert. **6**(4), 46–51 (1991)
36. Greenwell, B.M., Boehmke, B.C., McCarthy, A.J.: A simple and effective model-based variable importance measure. arXiv preprint arXiv:1805.04755 (2018)
37. Gronenborn, A.M., et al.: A novel, highly stable fold of the immunoglobulin binding domain of streptococcal protein G. Science **253**(5020), 657–661 (1991)

38. Gu, H., Wang, H., Qin, P., Wang, J.: Chest l-transformer: local features with position attention for weakly supervised chest radiograph segmentation and classification. Front. Med. **9**, 923456 (2022)
39. Guo, P., Mei, Y., Zhou, J., Jiang, S., Patel, V.M.: ReconFormer: accelerated MRI reconstruction using recurrent transformer. IEEE Trans. Med. Imaging (2023)
40. Hegyi, H., Gerstein, M.: The relationship between protein structure and function: a comprehensive survey with application to the yeast genome. J. Mol. Biol. **288**(1), 147–164 (1999)
41. Ho, J., Kalchbrenner, N., Weissenborn, D., Salimans, T.: Axial attention in multidimensional transformers. arXiv preprint arXiv:1912.12180 (2019)
42. Huang, K., Altosaar, J., Ranganath, R.: ClinicalBERT: modeling clinical notes and predicting hospital readmission. arXiv preprint arXiv:1904.05342 (2019)
43. Huang, K., Xiao, C., Glass, L.M., Sun, J.: MolTrans: molecular interaction transformer for drug-target interaction prediction. Bioinformatics **37**(6), 830–836 (2021)
44. Ishida, S., Terayama, K., Kojima, R., Takasu, K., Okuno, Y.: Prediction and interpretable visualization of retrosynthetic reactions using graph convolutional networks. J. Chem. Inf. Model. **59**(12), 5026–5033 (2019)
45. Jiménez-Luna, J., Grisoni, F., Schneider, G.: Drug discovery with explainable artificial intelligence. Nat. Mach. Intell. **2**(10), 573–584 (2020)
46. Jumper, J., et al.: Highly accurate protein structure prediction with AlphaFold. Nature **596**(7873), 583–589 (2021)
47. Kelley, L.A., Sternberg, M.J.: Protein structure prediction on the web: a case study using the Phyre server. Nat. Protoc. **4**(3), 363–371 (2009)
48. Kendrew, J.C., Bodo, G., Dintzis, H.M., Parrish, R., Wyckoff, H., Phillips, D.C.: A three-dimensional model of the myoglobin molecule obtained by X-ray analysis. Nature **181**(4610), 662–666 (1958)
49. Kostas, D., Aroca-Ouellette, S., Rudzicz, F.: BENDR: using transformers and a contrastive self-supervised learning task to learn from massive amounts of EEG data. Front. Hum. Neurosci. **15**, 653659 (2021)
50. Latchman, D.S.: Transcription factors: an overview. Int. J. Exp. Pathol. **74**(5), 417 (1993)
51. Lee, J., et al.: BioBERT: a pre-trained biomedical language representation model for biomedical text mining. Bioinformatics **36**(4), 1234–1240 (2020)
52. Li, Y., et al.: AGMB-Transformer: anatomy-guided multi-branch transformer network for automated evaluation of root canal therapy. IEEE J. Biomed. Health Inform. **26**(4), 1684–1695 (2021)
53. Lin, Z., et al.: Evolutionary-scale prediction of atomic-level protein structure with a language model. Science **379**(6637), 1123–1130 (2023)
54. Liu, Q., Xie, L.: TranSynergy: mechanism-driven interpretable deep neural network for the synergistic prediction and pathway deconvolution of drug combinations. PLoS Comput. Biol. **17**(2), e1008653 (2021)
55. Luong, M.T.: Effective approaches to attention-based neural machine translation. arXiv preprint arXiv:1508.04025 (2015)
56. McGuffin, L.J., et al.: IntFOLD: an integrated web resource for high performance protein structure and function prediction. Nucleic Acids Res. **47**(W1), W408–W413 (2019)
57. Miller, T.: Explanation in artificial intelligence: insights from the social sciences. Artif. Intell. **267**, 1–38 (2019)
58. Mirdita, M., Schütze, K., Moriwaki, Y., Heo, L., Ovchinnikov, S., Steinegger, M.: Colabfold: making protein folding accessible to all. Nat. Methods **19**(6), 679–682 (2022)

59. Mitchell, A.L., et al.: MGnify: the microbiome analysis resource in 2020. Nucleic Acids Res. **48**(D1), D570–D578 (2020)
60. Murdoch, W.J., Singh, C., Kumbier, K., Abbasi-Asl, R., Yu, B.: Definitions, methods, and applications in interpretable machine learning. Proc. Natl. Acad. Sci. **116**(44), 22071–22080 (2019)
61. Olden, J.D., Joy, M.K., Death, R.G.: An accurate comparison of methods for quantifying variable importance in artificial neural networks using simulated data. Ecol. Model. **178**(3), 389–397 (2004). https://doi.org/10.1016/j.ecolmodel.2004.03.013, https://www.sciencedirect.com/science/article/pii/S0304380004001565
62. Pan, X., Bai, W., Ma, M., Zhang, S.: RANT: a cascade reverse attention segmentation framework with hybrid transformer for laryngeal endoscope images. Biomed. Signal Process. Control **78**, 103890 (2022)
63. Park, S., Kim, G., Kim, J., Kim, B., Ye, J.C.: Federated split vision transformer for COVID-19 CXR diagnosis using task-agnostic training. arXiv preprint arXiv:2111.01338 (2021)
64. Peng, J., Xu, J.: RaptorX: exploiting structure information for protein alignment by statistical inference. Proteins: Structure, Function Bioinform. **79**(S10), 161–171 (2011)
65. Peng, Y., Yan, S., Lu, Z.: Transfer learning in biomedical natural language processing: an evaluation of BERT and ELMo on ten benchmarking datasets. In: Demner-Fushman, D., Cohen, K.B., Ananiadou, S., Tsujii, J. (eds.) Proceedings of the 18th BioNLP Workshop and Shared Task, pp. 58–65. Association for Computational Linguistics, Florence, Italy (2019). https://doi.org/10.18653/v1/W19-5006, https://aclanthology.org/W19-5006
66. Placido, D., et al.: A deep learning algorithm to predict risk of pancreatic cancer from disease trajectories. Nat. Med. **29**(5), 1113–1122 (2023)
67. Privalov, P., Khechinashvili, N.: A thermodynamic approach to the problem of stabilization of globular protein structure: a calorimetric study. J. Mol. Biol. **86**(3), 665–684 (1974)
68. Razavian, N., Blecker, S., Schmidt, A.M., Smith-McLallen, A., Nigam, S., Sontag, D.: Population-level prediction of type 2 diabetes from claims data and analysis of risk factors. Big Data **3**(4), 277–287 (2015)
69. Ribeiro, M.T., Singh, S., Guestrin, C.: Why should I trust you?: explaining the predictions of any classifier. In: Proceedings of the 22nd ACM SIGKDD International Conference on Knowledge Discovery and Data Mining, San Francisco, CA, USA, August 13-17, 2016, pp. 1135–1144 (2016)
70. Ribeiro, M.T., Singh, S., Guestrin, C.: Anchors: high-precision model-agnostic explanations. In: Proceedings of the AAAI Conference on Artificial Intelligence, vol. 32 (2018)
71. Ringe, D., Petsko, G.A.: How enzymes work. Science **320**(5882), 1428–1429 (2008)
72. Roy, A., Kucukural, A., Zhang, Y.: I-TASSER: a unified platform for automated protein structure and function prediction. Nat. Protoc. **5**(4), 725–738 (2010)
73. Rudin, C.: Stop explaining black box machine learning models for high stakes decisions and use interpretable models instead. Nat. Mach. Intell. **1**(5), 206–215 (2019)
74. Rudin, C., Ustun, B.: Optimized scoring systems: toward trust in machine learning for healthcare and criminal justice. Interfaces **48**(5), 449–466 (2018)
75. Schwaller, P., et al.: Predicting retrosynthetic pathways using transformer-based models and a hyper-graph exploration strategy. Chem. Sci. **11**(12), 3316–3325 (2020)

76. Selvaraju, R.R., Cogswell, M., Das, A., Vedantam, R., Parikh, D., Batra, D.: Grad-CAM: visual explanations from deep networks via gradient-based localization. Int. J. Comput. Vis. **128**, 336–359 (2020)
77. Senior, A.W., et al.: Improved protein structure prediction using potentials from deep learning. Nature **577**(7792), 706–710 (2020)
78. Shrikumar, A., Greenside, P., Kundaje, A.: Learning important features through propagating activation differences. In: International Conference on Machine Learning, pp. 3145–3153. PMLR (2017)
79. Sikdar, S., Bhattacharya, P., Heese, K.: Integrated directional gradients: feature interaction attribution for neural NLP models. In: Proceedings of the 59th Annual Meeting of the Association for Computational Linguistics and the 11th International Joint Conference on Natural Language Processing (Volume 1: Long Papers), pp. 865–878 (2021)
80. Simonyan, K.: Deep inside convolutional networks: visualising image classification models and saliency maps. arXiv preprint arXiv:1312.6034 (2013)
81. Söding, J., Biegert, A., Lupas, A.N.: The HHpred interactive server for protein homology detection and structure prediction. Nucleic Acids Res. **33**(suppl_2), W244–W248 (2005)
82. Stegmüller, T., Bozorgtabar, B., Spahr, A., Thiran, J.P.: ScoreNet: learning non-uniform attention and augmentation for transformer-based histopathological image classification. In: Proceedings of the IEEE/CVF Winter Conference on Applications of Computer Vision, pp. 6170–6179 (2023)
83. Steinegger, M., Meier, M., Mirdita, M., Vöhringer, H., Haunsberger, S.J., Söding, J.: HH-suite3 for fast remote homology detection and deep protein annotation. BMC Bioinform. **20**, 1–15 (2019)
84. Steinegger, M., Söding, J.: Mmseqs2 enables sensitive protein sequence searching for the analysis of massive data sets. Nat. Biotechnol. **35**(11), 1026–1028 (2017)
85. Sun, R., Li, Y., Zhang, T., Mao, Z., Wu, F., Zhang, Y.: Lesion-aware transformers for diabetic retinopathy grading. In: Proceedings of the IEEE/CVF Conference on Computer Vision and Pattern Recognition, pp. 10938–10947 (2021)
86. Sundararajan, M., Taly, A., Yan, Q.: Axiomatic attribution for deep networks. In: International Conference on Machine Learning, pp. 3319–3328. PMLR (2017)
87. Vaswani, A.: Attention is all you need. In: Advances in Neural Information Processing Systems (2017)
88. van der Velden, B.H.: Explainable AI: current status and future potential. Eur. Radiol. **34**(2), 1187–1189 (2024)
89. Wang, B., Zhang, D., Tian, Z.: STCovidNet: automatic detection model of novel coronavirus pneumonia based on swin transformer. Res. Square (2022). https://doi.org/10.21203/rs.3.rs-1401026/v1, https://europepmc.org/article/PPR/PPR465713
90. Wang, J., et al.: Exploring human diseases and biological mechanisms by protein structure prediction and modeling. Transl. Biomed. Inform. Precis. Med. Perspective, 39–61 (2016)
91. Wüthrich, K.: NMR of proteins and nucleic acids (1986)
92. Xie, H., et al.: Deep-learning-based few-angle cardiac SPECT reconstruction using transformer. IEEE Trans. Radiat. Plasma Med. Sci. **7**(1), 33–40 (2022)
93. Yang, J., Anishchenko, I., Park, H., Peng, Z., Ovchinnikov, S., Baker, D.: Improved protein structure prediction using predicted inter-residue orientations. Proc. Natl. Acad. Sci. **117**(3), 1496–1503 (2020)

94. Yang, Y., et al.: A multi-omics-based serial deep learning approach to predict clinical outcomes of single-agent anti-pd-1/pd-l1 immunotherapy in advanced stage non-small-cell lung cancer. Am. J. Transl. Res. **13**(2), 743 (2021)
95. Yang, Z., Zeng, X., Zhao, Y., Chen, R.: AlphaFold2 and its applications in the fields of biology and medicine. Signal Transduction Targeted Ther. **8**(1), 115 (2023)
96. Yu, M., Chang, S., Zhang, Y., Jaakkola, T.S.: Rethinking cooperative rationalization: introspective extraction and complement control. arXiv preprint arXiv:1910.13294 (2019)
97. Zaikis, D., Vlahavas, I.: TP-DDI: transformer-based pipeline for the extraction of drug-drug interactions. Artif. Intell. Med. **119**, 102153 (2021)
98. Zhao, C., Shuai, R., Ma, L., Liu, W., Wu, M.: Improving cervical cancer classification with imbalanced datasets combining taming transformers with T2T-ViT. Multimedia Tools Appl. **81**(17), 24265–24300 (2022)
99. Zhou, B., et al.: DSFormer: a dual-domain self-supervised transformer for accelerated multi-contrast MRI reconstruction. In: Proceedings of the IEEE/CVF Winter Conference on Applications of Computer Vision, pp. 4966–4975 (2023)

Computationally Reconstructing the Evolution of Cancer Progression Risk

Kefan Cao[1] and Russell Schwartz[2,3(✉)]

[1] Computer Science Department, Carnegie Mellon University, Pittsburgh, PA 15213, USA

[2] Ray and Stephanie Lane Computational Biology Department, Carnegie Mellon University, Pittsburgh, PA 15213, USA

[3] Department of Biological Sciences, Carnegie Mellon University, Pittsburgh, PA 15213, USA

russells@andrew.cmu.edu

Abstract. Understanding the evolution of cancer in its early stages is critical to identifying key drivers of cancer progression and developing better early diagnostics or prophylactic treatments. Early cancer is difficult to observe, though, since it is generally asymptomatic until extensive genetic damage has accumulated. In this study, we develop a computational approach to infer how once-healthy cells enter into and become committed to a pathway of aggressive cancer. We accomplish this through a strategy of using tumor phylogenetics to look backwards in time to earlier stages of tumor development combined with machine learning to infer how progression risk changes over those stages. We apply this paradigm to point mutation data from a set of cohorts from the Cancer Genome Atlas (TCGA) to formulate models of how progression risk evolves from the earliest stages of tumor growth, as well as how this evolution varies within and between cohorts. The results suggest general mechanisms by which risk develops as a cell population commits to aggressive cancer, but with significant variability between cohorts and individuals. These results imply limits to the potential for earlier diagnosis and intervention while also providing grounds for hope in extending these beyond current practice.

Keywords: Cancer · Machine Learning · Evolution · Evolution · Risk Modeling

1 Introduction

Cancer remains a major source of mortality globally [7] despite many years of intensive research into prevention and treatment. Great hope was placed in earlier screening, to identify more cancers when they are treatable. While such efforts have saved lives, they underperformed early hopes in part due to over treatment [3]: the statistical conclusion from reductions in mortality that many

M. Alser et al. (Eds.): ICCABS 2025, LNCS 15599, pp. 212–223, 2026.
https://doi.org/10.1007/978-3-032-02489-3_16

cancers detected early would never have been threatening to their hosts. Concern about the harm to patients of overtreatment in turn has led to undertreatment [12] as evolving clinical practice towards more conservative treatment of cancers judged unlikely to be aggressive can lead to failure to aggressively treat some that need it. The challenge of navigating the complementary issues of overtreatment and undertreatment has led to numerous efforts to more accurately distinguish threatening from non-threatening cancers (c.f., [25]).

Much insight into how tumors evolve and progress has come from the discipline of tumor phylogenetics [19], i.e., the study of the evolutionary history of cancer cells. Phylogenetic analysis can reveal order of mutations in a cancer cell, the timing of these mutations, and the relationships between different subclones, e.g., whether the most deadly clones arise from a single lineage or through parallel events [10]. Work on cancer evolution has suggested that cancer is not normally a primary illness but rather usually a late stage of a process of somatic hypermutability, either intrinsic or due to environmental factors [2], that could potentially predict tumor risk long before the tissue becomes phenotypically abnormal [21]. Combined with the insight of overtreatment, these observations suggest that our primary goal in early cancer screening should not be to detect cancers *per se*, but to predict which lesions of genetic damage will go on to be threatening to the patient.

Solving such prediction problems has led to a more prominent role for statistical inference and machine learning methods in personalized and precision cancer treatment. See, for example, [14] for a recent review. Machine learning has yielded ever better ability to identify cancers that will threaten patient lives and predict which tumors will respond to what treatments [15]. Nonetheless, they are limited by our ability to gather data about tumors, our imperfect understanding of their biology, and the inherent stochasticity of their progression.

Our prospects for predicting future cancer progression hinge on the question of when a cell lineage becomes committed to being a cancer, or a cancer with a bad progression outcome. If the risk of a cancer progressing is essentially constant until the moment it progresses, then the prediction task is impossible. However, if progression risk gradually increases over a cancer's history then there is hope of predicting progression well before it occurs. There has been considerable prior work to answer variants of this question of how a tissue becomes committed to being an aggressive cancer, including the classic two-hit model [11] and the more recent "bad luck" model [23], which can be conceptualized as different ways of reasoning about how we expect measures of progression risk to vary over the history of a cell lineage. Does risk of progression rise suddenly, as from a single chance mutation shifting a tissue from low-risk to high-risk, or does it increase gradually, as from a series of mutations each slightly driving aggressiveness? Do these changes tend to occur early in tumor development, long before a cancer is typically detected, or late, once it is already advanced? The answers to these questions may be of great practical importance in understanding the limits of prospects for early detection of high-risk lesions or early interventions to keep incipient cancers off of high-risk pathways.

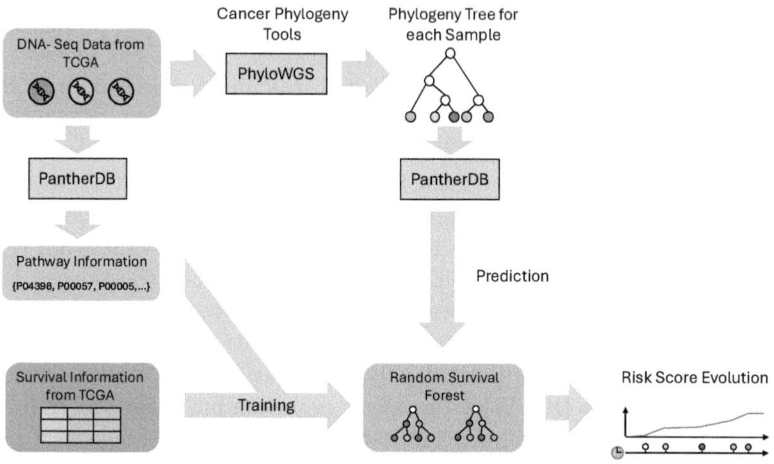

Fig. 1. Summary figure describing the overall analysis pipeline.

Our central goal in this study is to establish a model for reconstructing how progression risk develops over a tissue's history as it transitions from healthy to cancerous and potentially to lethality. We want to ask, if we could have observed the cells that eventually become cancers from their earliest development, how early could we have identified that they were on a trajectory to aggressive cancer. We emphasize that our goal here is not to provide the definitive answer to this question, but to show, first, that it is theoretically possible to ask it by computational methods applied to extant data sources and, second, that doing so will provide insight into basic cancer research and potentially actionable knowledge for improving early diagnosis and treatment. We recognize that we cannot definitively answer the question yet, largely because we do not have the ideal data for the proposed analysis, but proceed in the hope of inspiring future work to yield more definitive answers.

Our major contributions are: 1) developing a paradigm for tumor phylogenetics combined with machine learning to characterize how the landscape of cancer risk evolves over time, 2) implementing a realization of this paradigm using existing tools and data, and 3) applying that realization to a pilot study of lung and colorectal cancers to suggest how cancer risk evolves and how it can vary between tumor types and individual patients. The remainder of this paper describes how we accomplish these steps and examines the results and conclusions we can draw from them before returning in the Discussion to considering how this question might be asked better in the future.

2 Methods

2.1 Data Collection and Preprocessing

We apply our methodology to data from The Cancer Genome Atlas (TCGA) [4], which provides genomic data from large cohorts of thirty three cancer types with associated clinical and demographic metadata. We accessed TCGA data through the Genomic Data Commons (GDC) data portal [9], restricting analysis for the present study to DNA-seq single nucleotide variant (SNV) data. We also downloaded clinical data to extract survival information. We restricted our analysis to cohorts with at least 400 subjects and low class imbalance in survival outcomes due to our need for predictive regression models of censored survival. We therefore chose to focus the study on two cancers: lung adenocarcinoma (TCGA-LUAD) and colorectal adenocarcinoma (TCGA-COAD).

2.2 Phylogenetic Analysis

A limitation of TCGA data is that it normally offers only one bulk DNA-seq sample per patient, which makes clonal phylogenetic inference challenging. To reconstruct the phylogenetic trees of the cohorts, we used PhyloWGS [5], a Bayesian method for tumor phylogeny inference from bulk SNV data, which our prior experience has shown to work comparatively well on single bulk samples. We applied PhyloWGS to the SNV data of each cohort to infer the evolutionary history of the cancer cell lineages. Where PhyloWGS inferred multiple possible trees, we selected the tree with the highest posterior probability. We then used these trees to estimate the time points of key mutations in the evolution of each cell lineage, as described below. To ensure a reasonable runtime and exclude samples with insufficient mutations, we limited our analysis to samples with 45-1200 SNVs. Additionally, we omitted a small number of samples for which PhyloWGS failed to return a result. Consequently, approximately 80% of the samples from each cohort were included in the study.

2.3 Pathway and Mutation Analysis

Due to large numbers of SNV mutations ($\geq 10,000$ genes for 400 to 1100 patients), we took a systems approach to reduce the problem dimension by aggregating mutations to key pathways relevant to tumor progression [17]. We used the PANTHER [16,22] database to convert raw SNV data to affected pathways, assuming that an SNV in any gene in a pathway affects the pathway. We used random forest and Coxnet regression, via the scikit-survival [18] package, applied to the pathway data to identify key pathways that affect patient survival.

3 Results

We used a random forest classifier to derive risk scores for each patient based on their pathway-mapped mutations. Since the set of variants and affected pathways for the two cancers are different, we trained separate models for the two cancer types. Figure 2 shows how the classifier output (predicted risk) varies over time for the two cohorts, LUAD (Fig. 2a-b) and COAD (Fig. 2c-d), each separated by survival outcome. Both cohorts show a qualitatively similar portrait of slowly increasing risk over time. For both, mean risk scores are somewhat

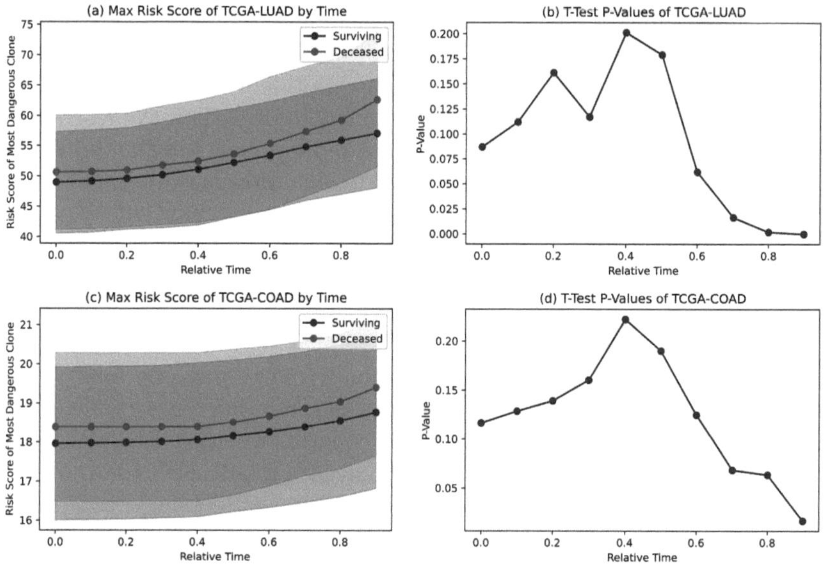

Fig. 2. Risk scores and p-values versus fraction of mutations accumulated (relative time) for LUAD and COAD cohorts. Shading areas in (a) and (c) represent the standard deviations. (a) Risk score of the most high-risk clone vs. relative time for LUAD subjects. (b) p-value for distinguishing by survival outcome in LUAD subjects. (c) Risk score of the most high-risk clone vs. relative time for COAD subjects. (d) p-value for distinguishing by survival outcome in COAD subjects.

elevated in individuals with bad versus good survival outcomes throughout the tumor history, suggesting that there are at least sometimes intrinsic differences predictive of outcome from the earliest stages of cancer development. However, variability patient-to-patient is substantially larger than the difference between good- and bad-outcome subgroups. With LUAD, there is minimal separation by outcome early on, but they diverge in the latter half of their evolutionary trajectory, indicating that a significant portion of the determination of outcome occurs late in the tumor's development. For COAD, the separation between surviving and deceased subjects is more consistent across the tumor's timeline.

For both types of cancers, risk scores are significantly different between living and deceased subjects (t-test p-value < 0.05). Our ability to distinguish patients by outcome improves sharply over the latter half of the trajectory as risk for both outcome groups grows and only becomes statistically significant late in the progression process, although this is in part a function of the cohort size among other study variables and not just an intrinsic property of the system.

To better understand what drives the evolution of the risk scores, we plotted for each tumor type the mean fraction of mutations for the five most frequently mutated driver genes of the highest-risk clones. These appear as Fig. 3 (LUAD) and Fig. 4 (COAD). Each figure provides two subplots separating surviving vs. deceased subjects over the course of the study followup. We plot risk as a

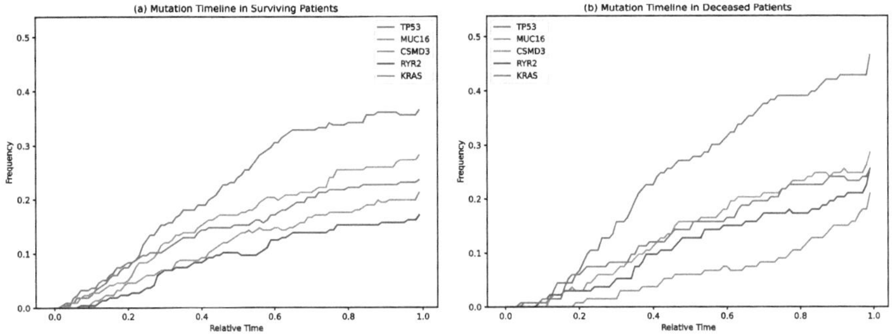

Fig. 3. Fractions of patients with key mutations in the evolution of cancer cells for LUAD. For each subfigure, the X-axis shows the fraction of mutations accumulated (relative time), and the Y-axis is the fraction of patients with the mutation. (a) Accumulation of mutations over time in surviving patients. (b) Accumulation of mutations over time in deceased patients.

function of the fraction of the total mutations accumulated, which we call the relative time. The relatively small sample sizes and high variance within each cohort preclude drawing many statistically rigorous conclusions from the data, and so we can only make suggestive qualitative observations. The figures show that while the overall risk evolution appears similar for the two cohorts, they exhibit quite different patterns of accumulation of key mutations in the evolution of cancer cells. LUAD involves a more gradual increase in inferred mutations over the course of a tumor's development. It is also less dominated by any one mutation, although TP53 emerges early as the most common mutation in each. There is little evident qualitative difference between the good and bad outcome plots aside from higher levels of TP53 mutation and lower levels of CSMD3 in bad-outcome cancers. COAD, by contrast, shows a sharper increase and higher final fraction of subjects with mutations, especially for the two most mutated genes (TP53 and APC). While mutations are predicted to accumulate in these genes from early times, there is a notable later increase in their mutation counts.

To better understand how these gene-level changes manifest as changes in predicted risk, we analyze aggregate measures of risk accumulation in Fig. 5 (LUAD) and Fig. 6 (COAD). Figure 5(a,b) show that for LUAD there are small but not statistically significant differences between good and bad outcome groups apparent even in the root node of the tree but that these rise to significance for the most high-risk clones. Figure 6(a,b) show a qualitatively similar trend for COAD but with less pronounced differences by outcome. Figure 5(c) and Fig. 6(c) show how survival time and max score relate for the two cohorts, revealing high variance patient-to-patient relative to the mean separation by outcome in either cohort. Figure 5(d) and Fig. 6(d) compare scores of root node versus the highest-risk node for each subject to show how much risk increases over the course of a cancer's history. A more pronounced change is evident for

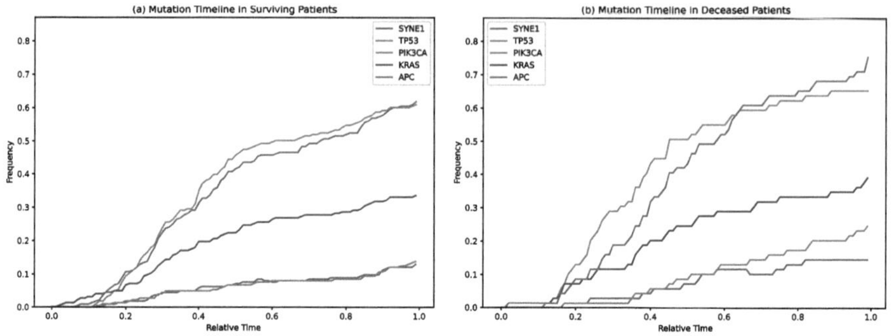

Fig. 4. Fractions of patients with key mutations in the evolution of cancer cells for COAD. For each subfigure, the X-axis shows the relative time of the mutation and the Y-axis is the fraction of patients with the mutation. (a) Accumulation of mutations over time in surviving patients. (b) Accumulation of mutations over time in deceased patients.

LUAD than COAD on average, although again with high variability subject-to-subject.

Figure 5(e-h) and Fig. 6(e-h) examine the overall accumulation of mutations for the top genes and pathways for each cancer, broken down by good and bad outcome cancers. Table 1 provides descriptors for the pathways identified by this analysis. The top-ranked genes and pathways are similar as are frequencies between cohorts. The high-ranking pathways themselves are largely consistent with prior expectations, for example in prominent representation of Wnt signaling, TP53 signaling, and inflammatory pathways. The most striking difference for LUAD is the relatively higher frequency of TP53 mutations and somewhat high frequencies of mutations across top-scoring pathways in subjects with poor outcomes. COAD shows similar results, with notably higher frequencies of mutations in TP53 and APC in deceased versus surviving groups and a qualitatively similar difference in pathways affected despite somewhat different pathways showing up as relevant. Although the p-values for individual genes do not show significant differences between the two groups, pathway-level analysis shows significant differences for some of the frequently mutated pathways of Table 1: P00059 "p53 pathway" in COAD (p-value 0.0107) and P04397 "p53 pathway by glucose deprivation" in LUAD (p-value 0.002395).

Individual genes and pathways arising in these analyses are consistent with expectations from prior literature. For example, EGFR is known to be important in early lung cancers [26] and although it does not show up as one of the five most frequently mutated genes, examining it individually shows that it is indeed predicted by our analysis to accumulate predominantly early in poor-outcome LUAD cancers (Fig. 7). Some other key mutations, such as ERBB2, occur mostly through copy number alterations (CNAs) and would not be seen in this study. Other key mutations in defining cancer risk, such as TP53 [26] have been suggested to occur more often in the later stages of cancer evolution, although our

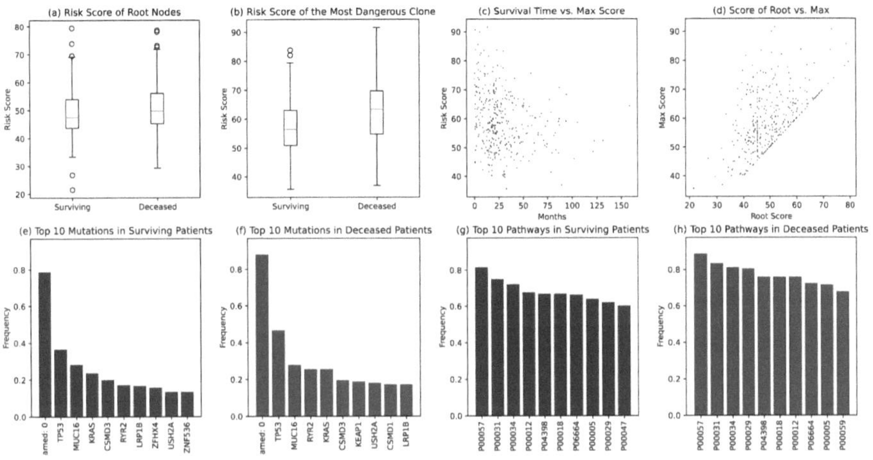

Fig. 5. Drivers of risk evolution for LUAD. In subfigures c-h, blue is used to identify data from subjects surviving as of the end of the followup and red those who were deceased as of the end of followup. (Color figure online)

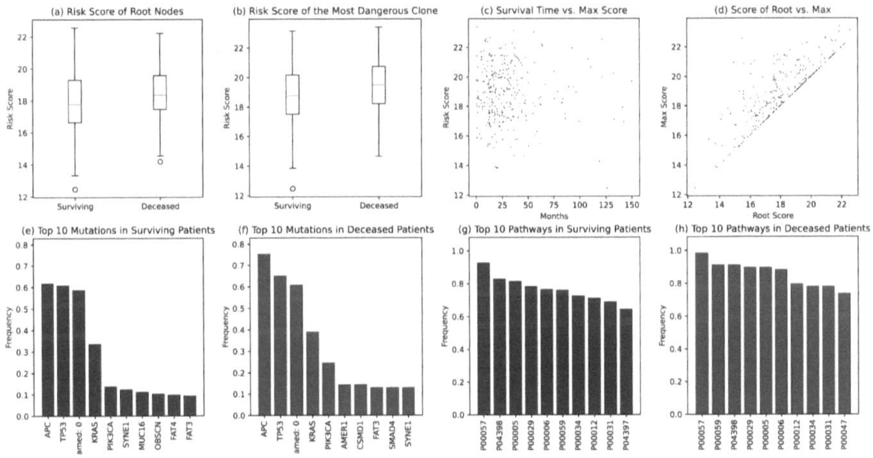

Fig. 6. Drivers of risk evolution for COAD. In subfigures c-h, blue is used to identify data from subjects surviving as of the end of the followup and red those who were deceased as of the end of followup. (Color figure online)

results suggest high patient-to-patient variability. In contrast, key risk-driving mutations in COAD [24] appear less likely to be seen in the root node of the phylogeny but rather appear to accumulate more gradually.

Table 1. Most frequently mutated PANTHER pathways in the two cohorts

Pathway ID	Pathway Name
P00057	Wnt signaling pathway
P04398	P53 pathway feedback loops 2
P00031	Inflammation mediated by chemokine and cytokine signaling pathway
P00034	Integrin signalling pathway
P00012	Cadherin signaling pathway
P00018	EGF receptor signaling pathway
P06664	Gonadotropin releasing hormone receptor pathway
P00005	Angiogenesis
P00029	Huntington disease
P00047	PDGF signaling pathway
P00059	p53 pathway
P00006	Apoptosis signaling pathway
P04397	p53 pathway by glucose deprivation

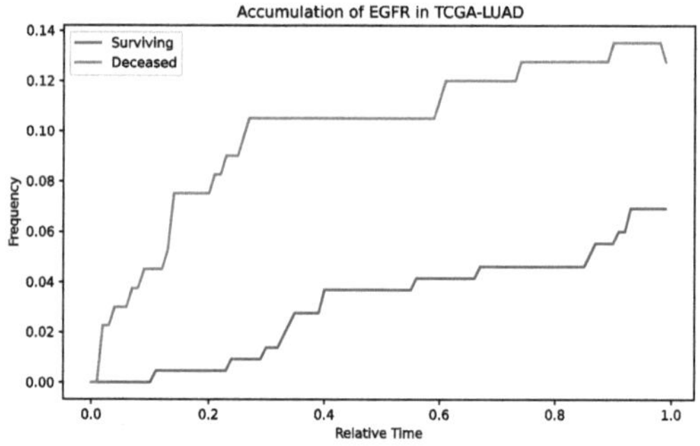

Fig. 7. Fraction of mutated EGFR versus estimated time for LUAD.

4 Conclusion and Discussion

We sought to characterize how risk of progression to aggressive cancer develops over time. We established a computational inference model for answering this question with extant cancer genomic and outcome data. We applied an implementation of that model to two cancer cohorts, revealing qualitatively similar portraits of risk accumulation although with some variability by tumor type and large variability by subject within each type. The results suggest that there are average differences predictive of outcome, captured in machine learning-derived risk scores, present from the earliest stages of tumor development although these differences are dominated by high variability within each cohort. Statistically significant separation of good and bad outcome subjects by risk score becomes possible only relatively late, in part reflecting the high variability by subject.

However, this separation might be pushed back to earlier times with larger cohorts, more comprehensive variant data, or more accurate phylogeny inferences, among other factors. The results suggest that there is potential, for at least subsets of patients, for earlier diagnostics or interventions before cancer is commonly detected—and potentially even before it is cancer—but further advances in methodology will be needed to judge the absolute limits.

As noted in the Introduction, our goal is not to definitively answer the question we pose about the evolution of cancer risk but rather to demonstrate that it is feasible and worthwhile to answer it, in the hope of inspiring future work. The biggest obstacle to answering this question better is data. The present work used TCGA data because the nature of the question requires relatively large cohorts with consistent data types, and TCGA was a landmark in making such data available. TCGA sequencing was not designed with tumor phylogenetics in mind, though, and did not have access to many technologies available to us today. We might also improve on the approach with more reliable estimates of elapsed time accounting for variability in mutation rates. We further expect that we are missing a substantial portion of risk-driving mutations by focusing only on SNVs, ignoring copy number alterations (CNAs) and structural variations (SVs) important in cancers. Methods exist to map SNV, CNA, and SV mutations to a tumor lineage tree [6,8], but they perform poorly on single-sample data such as is available with TCGA. Sufficiently large cohorts with multiple samples per patient, single-cell genomic data, longitudinal data, and protocols for profiling somatic variation in non-cancerous tissues [13] might improve the effectiveness of future studies along these lines. How to choose among many possible options in designing such a study is itself an emerging question for which tools are beginning to become available [20]. The computational tools used in this study are also older, in part because they were selected to be suitable for the data currently available; better methods for phylogenetic risk prediction might also be chosen or developed to accommodate richer data sources.

Finally, we consider what one could do with answers from a study such as is presented here. Better models of cancer risk progression could have particular value to early cancer screening and public health interventions. As we now appreciate that somatic variation is ubiquitous [1] and usually asymptomatic, it is important to develop better ways to identify when seemingly harmless variation poses a risk and better respond to problems of both overtreatment and undertreatment. Studies such as that prototyped here may also inspire better ways to respond to cancers prophylactically before they become threatening. Better characterizing the features defining high-risk lesions may also help during treatment in tailoring personalized therapies to high-risk clones or predicting when and how a tumor under treatment might recur or become more aggressive.

Acknowledgments. This work was supported by the National Human Genome Research Institute of the National Institutes of Health under award number R01HG010589. The content is solely the responsibility of the authors and does not necessarily represent the official views of the National Institutes of Health. The work was supported in part by funds from Highmark Health and UPMC Enterprises through the Center for Machine Learning and Health.

Competing Interests. No competing interest is declared.

Author Contributions Statement. This study was conceived by RS and the methodology developed by RS and KC. KC wrote the code, retrieved data, and carried out the analyses. KC and RS both contributed to data interpretation and to writing the manuscript.

References

1. Acha-Sagredo, A., Ganguli, P., Ciccarelli, F.D.: Somatic variation in normal tissues: friend or foe of cancer early detection? Ann. Oncol. **33**(12), 1239–1249 (2022)
2. Alexandrov, L.B., et al.: Signatures of mutational processes in human cancer. Nature **500**(7463), 415–421 (2013)
3. Bhatt, J.R., Klotz, L.: Overtreatment in cancer – is it a problem? Expert Opin. Pharmacother. **17**(1), 1–5 (2016). https://doi.org/10.1517/14656566.2016.1115481, pMID: 26789721
4. Cancer Genome Atlas Research Network, Weinstein, J.N., et al.: The cancer genome atlas Pan-Cancer analysis project. Nat. Genet. **45**(10), 1113–1120 (2013)
5. Deshwar, A.G., Vembu, S., Yung, C.K., Jang, G.H., Stein, L., Morris, Q.: Phylowgs: reconstructing subclonal composition and evolution from whole-genome sequencing of tumors. Genome Biol. **16**(1), 35 (2015) https://doi.org/10.1186/s13059-015-0602-8
6. Eaton, J., Wang, J., Schwartz, R.: Deconvolution and phylogeny inference of structural variations in tumor genomic samples. Bioinformatics **34**(13), i357–i365 (2018)
7. Frick, C., et al.: Quantitative estimates of preventable and treatable deaths from 36 cancers worldwide: a population-based study. Lancet Glob. Health **11**(11), e1700–e1712 (2023) https://doi.org/10.1016/S2214-109X(23)00406-0
8. Fu, X., Lei, H., Tao, Y., Schwartz, R.: Reconstructing tumor clonal lineage trees incorporating single-nucleotide variants, copy number alterations and structural variations. Bioinformatics **38**(Supplement_1), i125–i133 (2022)
9. Grossman, R.L., et al.: Toward a shared vision for cancer genomic data. N. Engl. J. Med. **375**(12), 1109–1112 (2016). https://doi.org/10.1056/NEJMp1607591
10. Hong, W.S., Shpak, M., Townsend, J.P.: Inferring the origin of metastases from cancer phylogenies. Can. Res. (2015). https://doi.org/10.1158/0008-5472.CAN-15-1889
11. Knudson, A.G., Jr.: Mutation and cancer: statistical study of retinoblastoma. Proc. Natl. Acad. Sci. **68**(4), 820–823 (1971)
12. Kocik, J.: Undertreatment or overtreatment – how far from each other in geriatric oncology? **2**(1), 18–25 (2020). https://doi.org/10.36553/WM.15
13. Martincorena, I., Campbell, P.J.: Somatic mutation in cancer and normal cells. Science **349**(6255), 1483–1489 (2015) https://doi.org/10.1126/science.aab4082
14. Masucci, M., Karlsson, C., Blomqvist, L., Ernberg, I.: Bridging the divide: a review on the implementation of personalized cancer medicine. J. Personalized Med. **14**(6) (2024). https://doi.org/10.3390/jpm14060561, https://www.mdpi.com/2075-4426/14/6/561

15. Mechahougui, H., Gutmans, J., Colarusso, G., Gouasmi, R., Friedlaender, A.: Advances in personalized oncology. Cancers (Basel) **16**(16), 2862 (2024)
16. Mi, H., Thomas, P.: PANTHER pathway: an ontology-based pathway database coupled with data analysis tools. Methods Mol. Biol. **563**, 123–140 (2009)
17. Park, Y., Shackney, S., Schwartz, R.: Network-based inference of cancer progression from microarray data. IEEE/ACM Trans. Comput. Biol. Bioinf. **6**(2), 200–212 (2008)
18. Pölsterl, S.: scikit-survival: a library for time-to-event analysis built on top of scikit-learn. J. Mach. Learn. Res. **21**(212), 1–6 (2020). http://jmlr.org/papers/v21/20-729.html
19. Schwartz, R., Schäffer, A.A.: The evolution of tumour phylogenetics: principles and practice. Nat. Rev. Genet. **18**(4), 213–229 (2017). https://doi.org/10.1038/nrg.2016.170
20. Srivatsa, A., Schwartz, R.: Optimizing design of genomics studies for clonal evolution analysis. Bioinform. Adv. **4**(1), vbae193 (2024). https://doi.org/10.1093/bioadv/vbae193
21. Tao, Y., et al.: Assessing the contribution of tumor mutational phenotypes to cancer progression risk. PLoS Comput. Biol. **17**(3), e1008777 (2021)
22. Thomas, P.D., Ebert, D., Muruganujan, A., Mushayahama, T., Albou, L.P., Mi, H.: PANTHER: making genome-scale phylogenetics accessible to all. Protein Sci. **31**(1), 8–22 (2022). https://doi.org/10.1002/pro.4218, https://onlinelibrary.wiley.com/doi/abs/10.1002/pro.4218
23. Tomasetti, C., Vogelstein, B.: Variation in cancer risk among tissues can be explained by the number of stem cell divisions. Science **347**(6217), 78–81 (2015)
24. Wu, Z., et al.: Colorectal cancer screening methods and molecular markers for early detection. Technol. Cancer Res. Treat. **19**, 1533033820980426 (2020)
25. Zhang, B., Shi, H., Wang, H.: Machine learning and AI in cancer prognosis, prediction, and treatment selection: a critical approach. J. Multidiscip. Healthc. **16**, 1779–1791 (2023)
26. Zhang, C., et al.: Genomic landscape and immune microenvironment features of preinvasive and early invasive lung adenocarcinoma. J. Thorac. Oncol. **14**(11), 1912–1923 (2019)

Cancer Diseases Classification with Sparse Neural Networks: An Information-Theoretic Approach

Zahra Jandaghi[1], Sixiang Zhang[1], Xiuzhen Huang[2], and Liming Cai[1(✉)]

[1] School of Computing, University of Georgia, Athens, GA, USA
{Sixiang.Zhang,liming}@uga.edu
[2] Department of Computational Biomedicine, Cedars-Sinai Medical Center, Los Angeles, California, USA
Xiuzhen.Huang@cshs.org

Abstract. Machine learning is indispensable for biomedical data modeling and classification. Tasks involving large, high-dimensional datasets are nevertheless computationally intensive and approximation methods are often sought to scale down the volume of raw data or model size without compromising substantial information embedded within the data. However, previous approximation methods have yielded mixed results and have yet to establish a clear framework linking feature selection and model sparsification. In this paper, we present an information-theoretic approach for cancer classification by addressing two prominent questions in data model approximation: how to identify a minimal set of critical features in cancer microarray data and how to design sparse neural networks that are effective and efficient for cancer classification. Our study highlights a key connection between these two challenges. In particular, we introduce a mutual information (MI)-based method to select a highly informative subset of genes from extensive microarray gene expression data. Each selected subset of genes, up to two orders of magnitude smaller than the original gene set, demonstrates superior performance in cancer classification compared to the full dataset. Additionally, the MI-based method enables the design of sparsified neural networks that consistently maintain or even improve classification performance compared to fully connected networks. Our test results reveal that sparsified networks selectively retain connections to the critical genes identified by the MI-based filtering method, effectively ignoring contributions from irrelevant genes.

1 Introduction

Gene expression is one of the popular data types in biology with potentials to enable effective cancer detection and classification [15]. A *mircoarray* dataset is the result of DNA microarray experiments, presenting as a massive matrix that contains numerical values for gene expression levels of a large number of genes (columns) on tested sample tissues (rows) [23,29]. The gene expression values for

M. Alser et al. (Eds.): ICCABS 2025, LNCS 15599, pp. 224–237, 2026.
https://doi.org/10.1007/978-3-032-02489-3_17

a sample may contains a patient's genetic information as being healthy or cancerous and, sometimes, different stages or types of some cancer disease. Accurately revealing such information would be extremely valuable for biomedical scientists to predict, analyze, and treat cancerous diseases [31,34]. This is a typical task in binary or multi-class classification with supervised manner when *class* labels are available. It has been assumed that cancerous genetic information in tissues is inherently different from healthy gene expression, meaning that classifiers can distinguish healthy from sick records [2,12]. Gene expression data for a tissue could also demonstrate the growing trend of those cells, which gives us clues about where cancer might materialize and how it may progress [1]

Nonetheless, a few underlying challenges make accurate classification difficult. First, most publicly available datasets often have a limited number of samples and vast amounts of genetic information (typically, tens of thousands of data points for genes). The high dimensionality of microarray data requires effective dimensionality reduction approaches that are hardly exist [12,15]. Second, samples in many datasets are often imbalanced, and the scarcity of samples introduces sampling complications and inappropriateness for gene expression information [33]. In addition, many studies have shown that microarrays contain not only noises but also redundant data, and only a relatively small portion of the genes present are critical for correct classification [36]. In summary, all these adversary scenarios may be overcome by the work of critical feature identification and benefit from processing microarray data with the emphasis on choosing the most representative features as cancer drivers [3,18].

The challenge in the classification of gene expression data is a typical issue encountered by more general research in customizing machine learning algorithms to reveal intrinsic relationships among complex data [15,35,37]. Since big data are often instances of many random variables, exact modeling them by machine learning algorithms can be extremely difficult. In particular, the task of identifying an optimal set of features from given (big) data is a computationally intractable problem [10]. Thus, approximation methods, which would scale down the amount of raw data or the model size, have been actively sought and developed [9,19,27]. To date, however, two prominent issues associated with approximation in data science and machine learning remain to be fully addressed: (1) How to select to a small set of critical features in big data; (2) How to determine a small set of pertinent connections between neurons for effective machine learning with sparse neural networks. In particular, for classification of microarray gene expression data, it is unclear whether and how these two issues may be pertinently connected [8,28].

Information theory has a great potential for investigation of intrinsic relationships among complex data given that it can offer insightful quantification on uncertainty of random data [30]. In particular, taking the advantage of mutual information between random variables has yielded some plausible methods for feature selection applications arising in data classification. Many such methods have been built upon the heuristics to identify genes one (subset) at a time in an incremental way [6,17,26,37], which nevertheless may potentially increase global irrelevance and redundancy and lead to suboptimal performance. Therefore, such methods have to incorporate ad hoc component to suppress the effect from irrelevance or/and redundancy brought forward by newly included features.

Neural networks, on the other hand, have demonstrated promising abilities to handle various complicated medical tasks, including classifications [4,38]. While neurons between two consecutive layers are fully connected, not all connections may flow useful information. Sparsifying connectivity would reduce computation cost and, in many cases, improve networks' performance [5]. Though limited in number, there have been some recent developments in pruning neural networks with mutual information. One of the studies [13] was inspired by mutual information-based feature selection in SVMs and logistic regression, with pruning from the output layer and going backward to the input layer. Another relevant research [14] trims neural connectivity based on conditional geometric MI (thus dependency) between adjacent layers, which however may suffer from inconsistence as it focuses on individual layers. Mutual information has also been applied to dropout layers in neural networks during network training to achieve better performance than simple dropout strategies [7].

In this paper, we develop an information theoretic-based approach to address the issues of feature selection and neural network sparsification. We test our findings on microarray human cancer data with the goal to establish a pertinent connection between answers that address these two issues. We first apply our method on several microarray cancer datasets and identify a significant small gene subset for each. To validate our method, we monitor classification performance while also getting biological confirmation that endorses the selected genes by our method as significant ones. In particular, through filtering tens of thousands of genes to just a few hundred, the applications of on four microarray cancer datasets with two classes (sick and healthy) show the proposed method achieves the higher efficiency and better performance compared to the state-of-the-art methods.

Second, our gene filtering method is further integrated with mutual information as a new approach for neural network sparsification. Based on the available microarray data, we deploy a neural network as the task classifier and explore pruning neural network models with the proposed mutual information method. Our work reveals sparsified neural networks improve the performance over fully-connected models on classification of cancer diseases in many cases. Moreover, our results show that the sparsified neural networks are explainable, as the input genes to the neurons in the first hidden layer closely align with the subset of genes selected by the proposed 3-MI gene selection method.

2 Backgrounds

2.1 Information Theory

In information theory, the *information content* of a specific value taken by a random variable is inversely proportional to the probability of taking this value

[30]. Typically, the information (i.e., *entropy*) of a discrete variable is measured as the expected information contained by all its values. That is, $H(X) = -\sum_x P(x)\log_2 P(x)$. This notion also leads to *conditional entropy* as well as *joint entropy* between two variables: $H(X|Y) = -\sum_{x,y} P(x,y)\log_2 P(x|y)$ and $H(X,Y) = -\sum_{x,y} P(x,y)\log_2 P(x,y)$. But most relevant to our discussion is the *mutual information* $I(X;Y)$ between two random variables X and Y that is measured by the Kullback-Leibler divergence between their joint and independent probability distributions [21,24].

$$I(X;Y) = D_{KL}\big(P(X,Y) \parallel P(X)P(Y)\big) = \sum_{x,y} P(x,y)\log_2 \frac{P(x,y)}{P(x)P(y)} \tag{1}$$

By definitions of entropy and conditional entropy, equation (refmi-definition) leads to information gain as follows.

$$I(X;Y) = H(X) - H(X|Y) \tag{2}$$

Conditional mutual information is defined as: $I(X;Y|Z) = H(X|Z) - H(X|Y,Z)$.

2.2 MI-Based Feature Selection

The critical task of feature selection for the cancer classification problem is also at the center of machine learning algorithms in general [17]. This is to select a (small) subset from a large number of features which are most relevant and meaningful for the data to be learned. Formally, let $\mathbf{X}$ be a set of random variables representing features and C be a variable representing *class*, a function in $\mathbf{X}$. Dataset D of interest contains samples for $(\mathbf{X}, C)$ and is usually represented as a matrix of numerical values where columns are features (and the class). Let d be any sample (row) in dataset D. Then $d(\mathbf{Z})$ represents the projection of row d onto any subset of columns $\mathbf{Z} \subseteq \mathbf{X} \cup \{C\}$.

A non-empty subset $\mathbf{S} \subseteq \mathbf{X}$ is called *representative set* for $\mathbf{X}$, if for every two rows d_1 and d_2 in D, $d_1(\mathbf{S}) = d_2(\mathbf{S})$ implies $d_1(C) = d_2(C)$.

Definition 1. The *minimum feature selection problem* (MIN FEATURES) is, given sample data D over variables $\mathbf{X}$ and target Y, to find a representative set $\mathbf{S}$ for $\mathbf{X}$ such that $|\mathbf{S}|$ is the minimum.

Problem MIN FEATURES has been shown computationally intractable [10]. To cope with this challenge, a number of methods [6,11,22,26,37] based on information theoretics have been proposed for feature selection on a wide range of applications. There is a compelling reason to use information theoretics to investigate feature selection. From the perspective of information gain equation (2), we conclude:

Proposition 1. Any subset of features $\mathbf{S} \subseteq \mathbf{X}$ is a representative set for $\mathbf{X}$ if and only if $I(\mathbf{S};C) = H(C)$, the entropy of class variable C.

Therefore, heuristic methods based on information theoretics usually seek a (small) subset $\mathbf{S}$ of features to maximize its mutual information with the class variable C through incrementally identifying features one (subset) at a time. As the process is very likely to increase global redundancy and irrelevance and to lead to suboptimal performance, these methods also incorporate other mutual information components to restrain or subdue the negative impacts. The following formula summarizes some common strategies used by the existing methods that evaluate a new feature variable X:

$$\phi(X) = I(X;C) - \alpha \sum_{Y \in \mathbf{S}} I(X;Y) + \beta \sum_{Y \in \mathbf{S}} I(X;Y|C) \tag{3}$$

where $\phi(X)$ computes information gain if including feature X to the existing set $\mathbf{S}$. The term that sums relevance $I(X;Y)$ over all $Y \in \mathbf{S}$ aims to cancel redundancy potentially caused by aggregation of $I(X;C)$ from multiple features X. On the other hand, to reduce irrelevance, term that sums conditional mutual information $I(X;Y|C)$ adds relevance between X and those included features conditional upon on the class variable C. However, these terms in (3) are interrelated and parameters α and β are often determined empirically. It is not unclear if the combination of these terms on the right-hand-side should be linear.

3 Methodology

In this chapter, we propose an MI-based methods for feature selection and neural network sparsification with the aim to establish relationship between the two. Next sections will demonstrate their efficacies through a series of tests on cancer microarray datasets and biological validations for cancer disease classifications.

3.1 The 3-MI Filtering Method

Our discussions will use the following known property in information theory.
Proposition 2. For any subset $\mathbf{S} \subseteq \mathbf{X}$ and $X \in \mathbf{X} \backslash \mathbf{S}$

$$I(\mathbf{S} \cup \{X\}; C) = I(\mathbf{S}; C) + I(X; C|\mathbf{S})$$

The property offers a process that one may follow to include feature X incrementally to an already chosen subset $\mathbf{S}$. However, this process, much as those previous methods, would make the ultimately yielded subset highly dependent on the order by which features are chosen. By Proposition 1, all representative sets have the same value $H(C)$, a specific small subset of features, if were to be first selected, may lead the entire feature selection process to a local optimal, possibly producing a representative subset of size much larger than the minimum.

In this work of cancer disease classification, instead of aiming to achieve a smallest representative set, we identify a representative feature set of a desired size that is most effective and efficient when applied for the classification task. In particular, we introduce a method, called *3-MI*, to filter out a large percentage of "unlikely" features (genes) by only keeping features with at least one type of mutual information having values exceeding a given threshold.

Definition 3. Let D be a dataset for feature variables $\mathbf{X}$ and class variable C. Let p be a real number between 0 and 1, inclusive. Then for any feature variable $X \in \mathbf{X}$, $I(X;C)$ is of *a high value* if and only if $I(X;C) \geq I(Y;C)$ for at least a $(1-p)$ portion of variables $Y \in \mathbf{X}$.

Likewise, we can define the conditions for $I(X;Y)$ and $I(X;Y|C)$ to be of a high value, respectively.

The *3-MI method* selects and keeps features that satisfy at least one of the following three conditions on mutual information. Given a number p, $0 \leq p \leq 1$,

1. Feature $X \in \mathbf{X}$ is selected if mutual information $I(X;C)$ is of a high value;
2. Features $X, Y \in \mathbf{X}$ are both selected if $I(X;Y)$ is of a high value;
3. Features $X, Y \in \mathbf{X}$ are both selected if $I(X;Y|C)$ is of a high value.

We point out that from the perspective cancer microarray data, the three types of mutual information $I(X;C)$, $I(X;Y)$, and $I(X;Y|C)$ bear different biological interpretations. Gene-class MI gives us hints about which gene might be correlated to a class label, whereas gene-gene MI and conditional gene-gene MI shows us which two genes are potentially similar or complementary to each other in causing disease.

We give rationales for the above strategies to choose features. First, it is obvious that if feature X has a high $I(X;C)$ value, it is $(1-p)$ more likely than other features related to the class variable and should be a part of a representative set. Second, when $I(X;Y)$ is of a high value, it is possible neither of them is of a high mutual information value with the class variable C. If that is the case, however, by the following Proposition 3, either $I(X,Y;C)$ or $I(X;Y|C)$ is of a high value. Apparently the former case justifies for the selection of both feature variables X and Y. The latter case, which also the third condition for the 3-MI method, results in a high value for $I(C \cup X;Y)$ according to Proposition 2, justifying selection of both X and Y.

Proposition 3. [20] Let D be a dataset for feature variables $\mathbf{X}$ and class variable C. If $I(X;Y)$ are of a high value, but neither $I(X;C)$ nor $I(Y;C)$ is of a high value, then either $I(X,Y;C)$ or $I(X;Y|C)$ is of a high value.

3.2 Algorithm for Gene Selection

The following figure illustrates the algorithm to select genes that consists of two phases. (1) The algorithm examines every gene and every pair of genes to identify those that satisfy conditions by the 3-MI method. In particular, genes with $1-p$ top values of 3 types of MIs are union of the subsets given by 3-MI filtering. The number of genes can be reduced by 10 to 100 fold and will be used as features in the classification problem. In next section, we will show that adjusting the magnitude of p determines our prediction's reliability level. (2) Recursive Feature Elimination (RFE) feature selection technique [16]) is then used to reduce over fitting and especially gain a stable gene subset. The pipeline of the proposed gene selection algorithm is shown in Fig. 1.

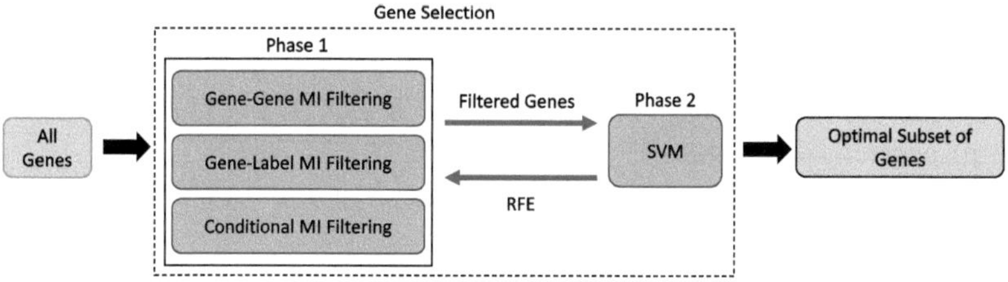

Fig. 1. Gene selection pipeline equipped with the 3-MI filtering components.

3.3 Neural Network Sparsification

Our method in neural network sparsification is to compute mutual information to prune neuron-neuron connections within neural networks. Unlike previous work [13] on MI-based sparsification of neural networks, where neurons of low MI values are removed, our method remove connections instead. In particular, mutual information $I(O_{l,i}; O_{(l+1),j})$ is computed between the output of the ith neuron on layer k and the output of the jth neuron at layer $(k+1)$ on sampled genes. The weight $W^k_{i,j}$ for the connection between these two neurons is determined by the following strategy, given a number p, $0 \le p \le 1$:

$$W^k_{i,j} = \begin{cases} W^k_{i,j} & I(O_{l,i}; O_{(l+1),j})\text{of a high value} \\ 0 & \text{otherwise.} \end{cases} \tag{4}$$

4 Performance Evaluations

Our proposed methods and algorithm have been applied to classifications of four cancer disease datasets, including lung cancer [39], breast cancer [40], colon cancer [41], and another lung cancer dataset [42]. All four are microarray data with two classes, cancerous and healthy. Table 1 gives details of numbers of tissue samples, numbers of involved genes, and whether they are balanced between the healthy and cancerous samples.

Table 1. Summary of cancer microarray data used for classification tests.

Dataset	Samples	Genes	Healthy	Cancerous	Class Distribution
Colon cancer [41]	62	2000	40	22	Imbalanced
Breast cancer [40]	86	22283	43	43	Balanced
Lung cancer [39]	107	22283	49	58	Balanced
Lung cancer [42]	156	12600	17	139	Imbalanced

4.1 Effectiveness of 3-MI-Based Gene Filtering

First, we ran our 3-MI based algorithm for gene selection and performed classifications on fully-connected neural networks on the selected genes. We compared our results with a recent study in Analysis of Variance (ANOVA) [25], which was used to measure classification performance of similar datasets containing microarray data. The performance comparisons are summarized in Table 2.

Table 2. Effectiveness comparisons between the 3-MI based gene selection and ANOVA on cancer disease classifications with microarray datasets summarized in Table 1. Classifications by both methods were performed on neural network classifiers. While both methods drastically reduce the number of genes, the 3-MI method outperforms ANOVA significantly in almost all categories of measurement.

Filtering Method	#Genes	Accuracy(%)	Precision(%)	Recall(%)	Specificity(%)	F1-score
3-MI (lung)	222	**95.28**	**95.25**	96.36	**93.77**	**0.95**
ANOVA (lung)	222	77	62.5	1	62.5	0.77
3-MI (breast)	243	**76.86**	**75.34**	81.67	**71.94**	0.78
ANOVA (breast)	243	73.73	75	**85.71**	51	0.78
3-MI (colon)	120	**70.76**	**74.97**	**85**	**43**	**0.79**
ANOVA (colon)	120	51	60	60	34	0.6

We have also confirmed that the genes filtered by the 3-MI method based algorithm are biologically significant through websites offering gene information, e.g., Pathway Browser Website [43].

Second, in order to confirm that the outstanding performance of classification with the 3-MI method in Table 2 was not due to the usage of neural networks, we tested the filtered genes with other classifiers. Table 3 gives performance comparisons between tests done by random forest and SVM classifiers on genes before and after the 3-MI based gene selection was applied. Indeed, the effectiveness of the 3-MI method for feature selection allows other machine learning methods significantly improve performance as well.

Third, we also carefully examined the individual contributions of different types of MI to the performance. Tests were conducted on our gene selection method but based on individual MI measure, gene-gene, gene-class, and conditional mutual information, respectively. Figure 2 shows individual effects on classifications of lung cancer microarray dataset [39]. It is clear that each of three types of mutual information independently contributes to the outstanding performance at different choices of parameter value p (see Sect. 3).

4.2 Effectiveness of Neural Network Sparsification

First, to evaluation our method on the effectiveness of neural network sparsification, we developed a prototype that trained a neural network model for 3 epochs

Table 3. Test results from random forest and SVM classifiers and performance comparisons between before (top) and after (bottom) the 3-MI based gene selection was applied, where SVM has parameters kernel = 'poly' or 'rbf', and degree = 3 or 6.

Classifier	Dataset	#Gene	Accuracy	Precision	Recall	F1-score
Rand Forest ($n:12$)	Lung Cancer [39]	**223**	**0.97**	**0.95**	1	**0.97**
		22283	0.94	0.90	1	0.94
Rand Forest ($n:20$)	Breast Cancer [40]	**1638**	**0.86**	**0.85**	**0.85**	**0.85**
		22283	0.79	0.78	0.78	0.78
Rand Forest ($n:12$)	Colon Cancer [41]	**120**	**0.67**	**0.67**	**0.73**	**0.7**
		2000	0.62	0.64	0.64	0.64
Rand Forest ($n:15$)	Lung Cancer [42]	**973**	**0.98**	**0.98**	**1**	**0.99**
		12600	0.96	0.96	0.99	0.98
SVM (poly, 3)	Lung Cancer [39]	**223**	**0.94**	1	**0.89**	**0.94**
		22283	0.88	1	0.78	0.87
SVM (rbf)	Breast Cancer [40]	**243**	**0.93**	**0.93**	**0.93**	**0.93**
		22283	0.86	0.92	0.78	0.87
SVM (poly, 6)	Colon Cancer [41]	**120**	**0.63**	**0.64**	**0.97**	**0.77**
		2000	0.58	0.61	0.90	0.73
SVM (poly, 3)	Lung Cancer [42]	**1951**	**0.92**	**0.92**	1	**0.96**
		12600	0.89	0.89	1	0.94

and then froze the weights. We identified which connections should be removed and fixed their weights to zero based on the mutual information between the output of incoming and outgoing neurons. Then we defined an empty model with exact architecture of the original one and loaded customized weights into the model. This same process was repeated with 14 different thresholds sharing one baseline model. Then we continued training the baseline along with sparsified models for 1024 more epochs and compared their performances. We repeated these steps 5 times and averaged all the metrics to avoid any unintentional bias.

Specifically, sparse neural networks were allowed to keep only the connections with a certain high MI values and the rest connections were fixed to have the zero weight. Their performances compared with fully connected networks, in terms of accuracy, precision, recall, specificity, and f1-measure [32], are reported in Fig. 3. In general, our sparsified models improve the fully connected baseline model in performance when more than 70% of connectivities are discarded.

4.3 Sparsified Neural Networks Are Explainable

We asked the question if the network sparsification could filter out important (relevant) genes. To answer this question, we implemented sparsification between input genes and the first hidden layer. In this setup, the rest of the network was left untouched, but the connections between the inputs and the first hidden layer

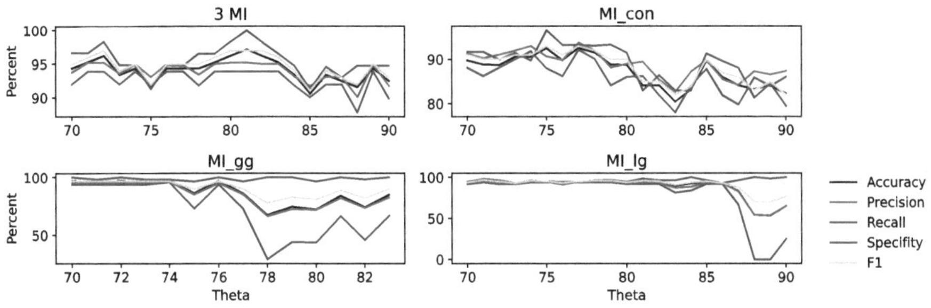

Fig. 2. Contribution of different MIs to the performance in gene selection on microarray data of lung cancer with 22283 genes [39], where MI_{gg}, MI_{lg}, and MI_{con} represent gene-gene, gene-class, and conditional mutual information, respectively, and threshold $\theta\,(Theta) = 1 - p$ (see Sect. 3).

were sparsified. We observed 70%–75% of genes from the input layer have below threshold MI levels and hence they were removed.

To find the importance of a gene, we calculated the sum of the absolute value of weights for all the outgoing connections from every input gene. This subset of genes were compared with the subset of genes obtained from the 3-MI based gene selection algorithm. Table 4 shows these two subsets largely overlap, with more than 70% to 94% genes common crossing the two subsets. This test proves that neural network sparsification based on the proposed MI-based method can highlight the important genes in an explainable manner even when only one layer in a larger neural network is considered.

Table 4. Heavy overlaps between genes selected by the 3-MI method and genes explained by sparsified neural networks, tested on three cancer microarray datasets.

Dataset	# Genes	# Genes after	# Genes in first	# Overlapping Genes
Colon cancer [41]	2000	120	500	87 (**72.5%**)
Breast cancer [40]	22283	243	6000	186 (**76%**)
Lung cancer [39]	22283	224	5000	211 (**94%**)

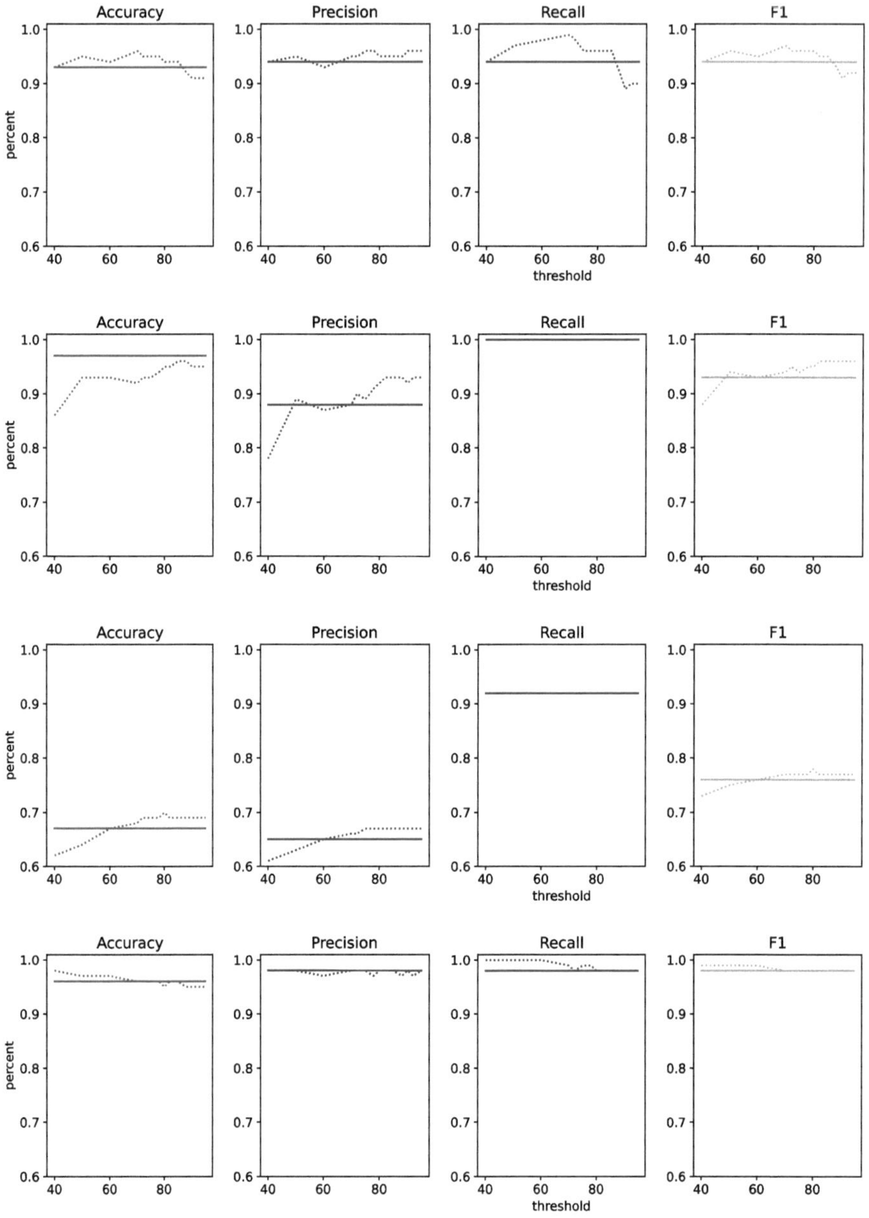

Fig. 3. Performance comparisons between sparsified and fully connected neural networks for classification of cancer microarray datasets. From the top to bottom rows: breast cancer dataset [40] with 22283 genes, lung cancer dataset [39] with 22283 genes, colon cancer dataset [41] with 2000 genes, and lung cancer dataset [42] with 12600 genes. The solid line represents the baseline model and the dotted line is the sparsified models at different thresholds. In general, the sparsification improves full connectivity in performance when more than 70% of genes are filtered.

References

1. Ahn, T., Goo, T., Lee, C.H., Kim, S.M., Han, K., Park S.: Deep learning-based identification of cancer or normal tissue using gene expression data. In: IEEE International Conference on Bioinformatics and Biomedicine (BIBM), pp. 1748–1752 (2018)
2. Alon, U., et al.: Broad patterns of gene expression revealed by clustering analysis of tumor and normal colon tissues probed by oligonucleotide arrays. Proc. Nat. Acad. Sci. **96**(12), 6745–6750 (1995)
3. Amann, J., Blasimme, A., Vayena, A., Frey, D., Madai, V.I.: Explainability for artificial intelligence in healthcare: a multidisciplinary perspective. BMC Med. Inform. Decis. Mak. **20**(1), 1–9 (2020)
4. William, G., Baxt, W.G.: Application of artificial neural networks to clinical medicine. Lancet **346**(8983), 1135–1138 (1995)
5. Blalock, D., Ortiz, J.J.G., Frankle, J., Guttag, J.: What is the state of neural network pruning?. In: Proceedings of Machine Learning and Systems (MLSys) (2020)
6. Battiti, R.: Using mutual information for selecting features in supervised neural net learning. IEEE Trans. Neural Netw. **5**(4), 537–550 (1994)
7. Chen, J., Wu, Z., Zhang, J., Li, F.: Mutual information-based dropout: learning deep relevant feature representation architectures. Neurocomputing **361**, 173–184 (2019)
8. Chen Z., et al.: Feature selection may improve deep neural networks for the bioinformatics problems. Bioinformatics **36**(5), 1542–1552 (2020)
9. Cunningham, J.P., Ghahramani, Z.: Linear dimensionality reduction: survey, insights, and generalizations. J. Mach. Learn. **16**(89), 2859–2900 (2015)
10. Davies, S., Russell, S.J.: NP-completeness of searches for smallest possible feature sets. In: AAAI Proceedings (1994)
11. Ding, C., Peng, H.: Minimum redundancy feature selection from microarray gene expression data. In: 2nd IEEE Computer Society Bioinformatics Conference Proceedings, pp. 523–529 (2003)
12. Dudoit, S., Fridlyand, J., Speed, T.P.: Comparison of discrimination methods for the classification of tumors using gene expression data. J. Am. Stat. Assoc. **97**(457), 77–87 (2002)
13. Fan C., Li, J., AoH, X., Wu, F., Meng, Y., Sun X.: Layer-wise model pruning based on mutual information. In: Proceedings of the Conference on Empirical Methods in Natural Language Processing, pp. 3079– (2021)
14. Ganesh, M.R., Corso, J.J., Sekeh, S.Y.: Mint: deep network compression via mutual information-based neuron TRIMMIN. In: 25th International Conference on Pattern Recognition (ICPR). IEEE, pp. 8251–8258 (2020)
15. Golub, T.R., Lander, E.S., et al.: Molecular classification of cancer: class discovery and class prediction by gene expression monitoring. Science **286**(5439), 531–537 (1999)
16. Guyon, I., Weston, J., Barnhill, S., Vapnik, V.: Gene selection for cancer classification using support vector machines. Mach. Learn. **46**(1), 389–422 (2002)
17. Guyon, I., Elisseeff, A.: An introduction to variable and feature selection. J. Mach. Learn. Res. **3**, 1157–1182 (2003)
18. Hambali, M., OOladele, T. Adewole, K.S.: Microarray cancer feature selection: review, challenges and research directions. Int. J. Cognit. Comput. Eng. **1**, 78–97 (2020)

19. Hinton, G.E., Salakhutdinov, R.R.: Reducing the dimensionality of data with neural networks. Science **313**(5786), 504–507 (2006)
20. Jandaghi, Z.: Mutual Information-based Machine Learning with Microarray Cancer Data, PhD Dissertation, University of Georgia (2022)
21. Kullback, S., Leibler, R.A.: On information and sufficiency. Ann. Math. Stat. **22**(1), 79–86 (1951)
22. Lin, D., Tang, X., Wang, X.: Conditional infomax learning: an integrated framework for feature extraction and fusion. In: Proceedings of the European Conference on Computer Vision, vol. 3951, pp. 68–82 (2006)
23. Lockhart, D.J., Winzeler, E.A.: Genomics, gene expression and DNA arrays. Nature **405**(6788), 827–836 (2000)
24. MacKay, D.J.C.: Information Theory, Inference, and Learning Algorithms. Cambridge University Press (2003)
25. Majumder, S., Pal, Y.V., Yadav, A., Chakrabarty, A.: Performance analysis of deep learning models for binary classification of cancer gene expression data. J. Healthc. Eng. **2022**(1), 1122536 (2022)
26. Peng, H., Long, F., Ding, C.: Feature selection based on mutual information criteria of max-Dependency, max-Relevance, and min-redundancy. IEEE Trans. Pattern Anal. Mach. Intell. **27**(8), 1226–1238 (2005)
27. Roweis, S.T., Saul, L.K.: Nonlinear dimensionality reduction by locally linear embedding. Science **290**(5500), 2323–2326 (2000)
28. Scardapane, S., Wang, D.: Randomness in neural networks: an overview. Wiley Interdiscip. Rev. Data Min. Knowl. Discov. **7**(2), e1200 (2017)
29. Schena, M., Shalon, D., Davis, R.W., Brown, P.O.: Quantitative monitoring of gene expression patterns with a complementary DNA microarray. Science **270**(5235), 467–470 (1995)
30. Shannon, C.E.: A mathematical theory of communication. Bell System Tech. J. **27**(3), 379–423 (1948)
31. Singh., D., et al.: Gene expression correlates of clinical prostate cancer behavior. Cancer Cell **1**(2), 203–209 (2002)
32. Sokolova, M., Lapalme, G.: A systematic analysis of performance measures for classification tasks. Inf. Process. Manage. **45**(4), 427–437 (2009)
33. Statnikov, A., Aliferis, C.F., Tsamardinos, I., Hardin, D., Levy, S.: A comprehensive evaluation of multicategory classification methods for microarray gene expression cancer diagnosis. Bioinformatics **21**(5), 631–643 (2005)
34. Statnikov, A., Tsamardinos, I., Dosbayev, Y., Aliferis, C.F.: GEMS: a system for automated cancer diagnosis and biomarker discovery from microarray gene expression data. Int. J. Med. Inf. **74**(7-8), 491–503 (2005)
35. Stears, R.L., Martinsky, T., Schena M.: Trends in microarray analysis. Nat. Med. **9**(1), 140–145 (2003)
36. Tan, A.C., Gilbert, D.: Ensemble machine learning on gene expression data for cancer classification. Appl. Bioinf. **2**(3), S75–S83 (2003)
37. Tang, J., Zhou, S.: A new approach for feature selection from microarray data BASEDON mutual information. IEEE/ACM Trans. Comput. Biol. Bioinf. **13**(6), 1004–1015 (2016)
38. Zhang. G.P.: Neural networks for classification: a survey. IEEE Trans. Syst. Man Cybern. Part C Appl. Rev. **30**(4), 451–462 (2000)
39. Gene expression of cigarette smoking and its role in lung adenocarcinoma development and survival (2018). https://www.ncbi.nlm.nih.gov/geo/query/acc.cgi?acc=GSE10072

40. Expression data from human breast tumors and their paired normal tissues (2009). https://www.ncbi.nlm.nih.gov/geo/query/acc.cgi?acc=GSE15852
41. Gene expression of colon tissues (1999). http://genomics-pubs.princeton.edu/oncology/affydata/index.html
42. Microarray lung cancer data. The Cancer Genome Atlas (TCGA) (2022). https://www.cancer.gov/
43. Pathway browser. https://reactome.org/PathwayBrowser

Epistatic Density of Viral Variants in Acute and Chronic HCV Patients

Alina Nemira[1,2](✉), Akshay Juyal[1,2], Pavel Skums[1,2], and Alexander Zelikovsky[2]

[1] Department of Computer Science, Georgia State University, Atlanta, GA 30302, USA
anemira1@student.gsu.edu
[2] School of Computing, University of Connecticut, Storrs, CT 06269, USA

Abstract. RNA viruses exhibit high mutation rates due to the lack of proofreading mechanisms during replication, leading to diverse intra-host viral populations. Variants with higher fitness tend to dominate the population due to enhanced transmissibility and immune escape. Fitness of viral variants depends on individual SNVs and epistatic links between pairs of SNVs as well as competition with other viral variants within the population. Recent machine learning methods have successfully predicted emerging COVID-19 variants based on epistatic SNV links, implying that SNV links contribute to fitness of viral variants.

We define the epistatic density of a viral variant as the number of positively linked SNV pairs between mutated positions in its genome. We computed epistatic density of intra-host Hepatitis C Virus (HCV) populations sampled from 85 chronic and 28 acute patients with HCV 1a genotypes. On average, epistatic density was higher in chronic patients than in acute cases. Additionally, the epistatic density distributions are more irregular and choppy in acute populations. Finally, we applied the epistatic density properties to distinguish between intra-host populations of chronic and acute HCV patients.

Keywords: Hepatitis C Virus · Evolution · Viral Variants · Coordinated Substitution Network · Epistatic Density

1 Introduction

In evolutionary biology, fitness of any organism is often quantified as the reproductive success of a genotype or phenotype compared to others in the population. Fitness of viral variant determines the ability of a virus to survive, replicate, and transmit within a host or a population [3]. Viruses demonstrate remarkable evolutionary capabilities through their ability to adapt and maintain fitness. For instance, SARS-CoV-2 variants have shown repeated epidemic surges through enhanced spreading potential. This evolutionary process involves continuous genetic modifications that can either increase or decrease fitness of viral variants [7].

The density distribution of a virus is shaped by various genetic factors, including mutations at the level of single nucleotide variants (SNVs). However, fitness is

M. Alser et al. (Eds.): ICCABS 2025, LNCS 15599, pp. 238–248, 2026.
https://doi.org/10.1007/978-3-032-02489-3_18

not solely dictated by individual mutations but also by the interactions between them - a phenomenon known as epistasis. Recently, epistasis-based approaches have shown promise in predicting emerging COVID-19 variants [10]. Epistasis refers to the phenomenon where the effect of one genetic mutation is modified or influenced by the presence of another mutation. These epistatic interactions influence viral adaptation and evolution through driven adaptive trajectories by enabling mutations to collectively enhance or decrease fitness.

While existing fitness estimations such as the log fitness ratio (LFR) [13], log relative fitness (LRF) [17], and production rate ratio (PRR) [18] have been employed to describe viral and cellular dynamic interactions and define competitive fitness indices of viral variants, they overlook the combinatorial effects of epistatic interactions. To address this limitation, we propose to estimate fitness using an epistasis-based network approach that can enhance our understanding of fitness of viral variants in rapidly evolving viruses, such as the Hepatitis C Virus (HCV).

The core idea behind a novel machine learning method is that certain pairs of SNVs appear together more frequently than expected under the assumption that they are independent. This co-occurrence is indicative of potential epistatic interactions, where the presence of one mutation influences the effect or occurrence of another. These linkages are crucial for understanding patterns of correlated mutations that may influence a pathogen's evolution, particularly traits such as fitness, pathogenicity, or drug resistance. The method represents linked SNVs in the form of an coordinated substitution network, a graphical framework where each SNV is a node (or vertex), and significant pairwise linkages between SNVs are represented as edges (or links) connecting the nodes. By constructing this network, we can visualize and analyze the relationships between mutations. A key strength of the coordinated substitution network approach is its ability to identify dense communities within the network. These dense communities often correspond to groups of mutations that collectively influence the viral phenotype, such as transmissibility, immune escape, or resistance to interventions [10].

Epistatic Density Definition and Model. We define epistatic density as the number of edges representing advantageous interactions or connections and calculate as the number of edges representing positively linked SNV pairs within the coordinated substitution network. Our fitness model assumes that emerging strains have higher relative fitness and more positive edges as dense community, in comparison, to other viral variants.

Data. We computed the epistatic density of intra-host HCV populations sampled from 85 chronic and 28 acute patients with the HCV 1a genotype, using data provided by the Centers for Disease Control and Prevention (CDC). We provide an optional step for classifying acute and chronic patients within the HCV dataset using epistatic density values and a statistical approach.

Contributions. First, we observe that chronic HCV populations exhibit higher epistatic density compared to acute cases, suggesting a more stable adaptation over time. Second, epistatic density distributions in acute cases appear more irregular and choppier than in chronic cases, indicating dynamic evolutionary pressures during early infection. Finally, we demonstrate that epistatic density metrics can serve as a discriminating factor between acute and chronic infections, offering potential applications in viral surveillance.

2 Methods

We constructed a coordinated substitution network to analyze the fitness of viral variants in patients with HCV by identifying positively linked SNV pairs using statistical tests. The coordinated substitution network was built for all HCV patients, representing SNVs as nodes and significant epistatic interactions as edges, enabling the calculation of epistatic density based on the number of edges representing pairs of positively linked SNVs within the coordinated substitution network. Epistatic density scores were computed at both the group level using aggregated patient data and the individual level by comparing each patient's mutation profile against the coordinated substitution network. To characterize viral evolution, we performed a statistical analysis of epistatic density, computing measures such as median fitness, normalized twin fitness, and normalized distinct fitness. Finally, we evaluated the model's classification accuracy using ROC analysis, demonstrating that epistatic density parameters effectively distinguish between acute and chronic HCV infections.

2.1 Research Roadmap Description

The research roadmap for the construction and analysis of coordinated substitution networks for HCV encompasses several critical steps, from data processing to network modeling, epistatic density calculation, and performance evaluation (Fig. 1). Brief descriptions of these methodological steps are provided in the following.

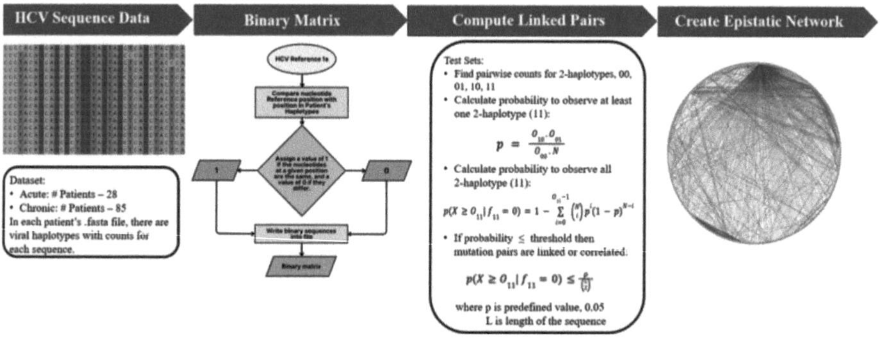

Fig. 1. Coordinated substitution network construction pipeline.

The first step involved data processing, where genetic sequences were analyzed to identify candidates for SNVs. For each viral variant, nucleotide positions were compared with the reference sequence, resulting in matching positions being assigned a value of 0 and mismatched positions receiving a value of 1. This process generated binary matrices for each patient sample, where rows represented the viral variants and columns corresponded to positions in the sequence. These matrices served as key input for subsequent analyses of mutation patterns, facilitating the computation of epistatic density.

Following data processing, coordinated substitution networks were constructed to represent epistatic interactions between SNVs. In this framework, SNVs were treated as nodes, with edges depicting significant interactions among these loci. Statistical tests were employed to quantify the likelihood of observing co-occurring mutations based on counts of four pairwise SNV combinations: ((0,0), (0,1), (1,0), (1,1)). Epistatic interactions were assessed by evaluating the statistical significance of SNV pairings, determining whether SNVs are positively linked within the coordinated substitution network.

The evaluation of epistatic density was then conducted using a defined protocol described in our developed pipeline. A fixed coordinated substitution network (CSN) was established as a global model based on aggregated data from all patients. Epistatic density scores were computed for both patient groups and individual patients, assessing how each patient's mutation profile aligned with the global network of interactions. Three statistical metrics, including median density, normalized distinct density, and twin density, were computed for each patient to characterize the distribution of epistatic densities.

For further details on the pipeline, please refer to subsection Inference of coordinated substitution networks and epistatic density calculation. Finally, the model performance was assessed through ROC analysis, allowing for classification of patients as acute or chronic based on the computed epistatic density metrics. Sensitivity and specificity metrics were computed iteratively, normalized, and used to generate ROC curves, quantifying the model's ability to accurately distinguish between patient groups.

2.2 Inference of Coordinated Substitution Networks and Epistatic Density Calculation

In real-world applications, epistatic networks cannot be directly observed. As a result, following the methodologies outlined in [1,12,14], they are approximated using coordinated substitution networks, which are derived through statistical inference applied to genomic data. An coordinated substitution network is an SNV linkage graph where vertices are SNVs and edges denote linked SNV pairs.

To identify genomic positions as potential candidates for SNVs, nucleotide positions from a reference genome were compared with corresponding positions in multiple viral variants, one at a time. Positions displaying mismatches were marked, enabling the construction of binary matrices that encapsulate viral variant data.

To advance the analysis, the inference of epistatic interactions between SNVs was performed by examining the observed counts and statistical significance of interactions between pairs of SNVs. These interactions were used to identify linked SNVs and build coordinated substitution networks representing the relationships influenced by intragenome interactions between linked SNVs. A pivotal step involves assessing the statistical significance of pairing interactions, which determines whether SNVs are positively linked within the coordinated substitution network.

For a detailed discussion of the precise mathematical framework and statistical methodology employed - including the construction and evaluation of coordinated substitution networks - interested readers are encouraged to consult the paper "Early detection of emerging viral variants through analysis of community structure of coordinated substitution networks" (Nature Communications; DOI: 10.1038/s41467-024-47304-6) [10]. This resource provides a comprehensive description of the computational framework and statistical tests used to systematically infer and validate epistatic interactions.

We developed a computational pipeline for analyzing epistatic interactions between mutations in viral genomes using coordinated substitution networks. The pipeline constructs a global network where the nodes represent the SNVs and the edges capture significant epistatic dependencies between them. Using binomial significance testing, we identify meaningful mutational interactions between HCV patients and encode these relationships in adjacency matrices. The script calculates epistatic density scores for both patient groups and individual patients by evaluating how viral sequences align with identified epistatic constraints, enabling a direct comparison of epistatic density between patients. For implementation details and access to the complete methodology for coordinated substitution networks and epistatic density calculation, please refer to our GitHub repository: https://github.com/Akshayjuyal/fitness_viral_gnome.

2.3 Epistatic Density Statistics Calculation and Data Processing

To evaluate and characterize the epistatic density distributions of viral variants across acute and chronic HCV infections, we computed a series of statistical measures derived from epistatic density values. These measures enabled epistatic density distribution comparisons between acute and chronic HCV patient populations and provided insights into viral evolutionary dynamics. The key metrics used in our analysis include normalized twin density values, normalized distinct density values, and median density values.

For each patient, epistatic density values were first sorted in ascending order to facilitate systematic computation of statistical metrics. The minimum and maximum values were extracted from the sorted dataset to establish the range. The median density values - indicative of central tendency - were calculated as the middle value of the sorted list.

Epistatic density values were sorted in ascending order. The minimum and maximum density values were extracted directly from the sorted dataset. The median was computed as the middle value of the sorted epistatic density scores. The range was defined as the difference between the maximum and minimum density values:

$$\text{Range} = \max(\text{density values}) - \min(\text{density values}) \tag{1}$$

The number of distinct values was defined as the count of unique epistatic density scores, irrespective of repetition. To quantify the diversity of epistatic density scores relative to their range, the number of distinct values was divided by the range for the normalization:

$$\text{distinct density values} = \frac{\text{Number of distinct values}}{\text{Range}} \tag{2}$$

Twin values were defined as consecutive density values differing by exactly one. This was calculated using the `np.diff()` function in Python, counting occurrences where the difference equals one. The twin values were then normalized by the range:

$$\text{Twin density values} = \frac{\text{Number of twin values}}{\text{Range}} \tag{3}$$

Data processing and statistical calculations were performed using Python (version 3.10.9) [5] with the NumPy [6], SciPy [16], Pandas [9], and statistics libraries.

The importance of specific metrics is particularly highlighted in their ability to differentiate acute and chronic infections. Chronic HCV populations demonstrated higher normalized twin values and lower normalized distinct values, reflecting tighter clustering and constrained diversity. median density values were notably higher for chronic infections, indicative of stable, optimized fitness levels. Each measure informed distinct aspects of viral evolutionary dynamics and contributed to a comprehensive characterization of epistatic density distribution profiles.

2.4 Performance Evaluation Using ROC Analysis

The statistical analysis employs sensitivity, specificity, receiver operating characteristic (ROC) curves, and area under the curve (AUC) metrics to evaluate the classification performance between acute and chronic HCV patients. We analyze the complete set of HCV patients with calculated epistatic density parameters, including median values, twin values ratios, and distinct values ratios, arranged in ascending order. The iterative process begins with zero counts for both sensitivity and specificity measures. For acute group patients, we increment the sensitivity count while maintaining the specificity count. Conversely, for chronic group patients, we increment the specificity count while maintaining the sensitivity count. The normalization procedure standardizes both sensitivity and specificity measurements. We calculate normalized sensitivity by dividing the cumulative count by the total number of acute group patients. Similarly, we compute normalized specificity by dividing the respective count by the total number of chronic group patients. The ROC curves are generated by plotting the false positive rate (1 - specificity) on the x-axis against the true positive rate (sensitivity) on the y-axis. Each plot includes a diagonal reference line representing random classifier performance. The plots are generated for three distinct scenarios: normalized twin values, normalized distinct values, and median values of epistatic density. The ROC analysis is implemented using scikit-learn (version 1.2.1) [2], a robust Python machine learning library. The AUC calculations utilize scikit-learn metrics.auc function, providing a quantitative measure of the classifier's discriminative capability. This comprehensive methodology enables objective evaluation of the classification method effectiveness in distinguishing between acute and chronic HCV patients, providing a robust framework for assessing the discrimination potential of the proposed approach.

3 Results

We analyzed intra-host HCV populations from acute (N = 98) and chronic HCV patients (N = 257), as described in [8]. From this dataset, we selected intra-host populations of the HCV 1a genotype, including samples from 28 acute (N = 28) and chronic (N = 85) patients, following the criteria outlined in [8]. HCV 1a genotype haplotypes were extracted from a mixed population of various HCV genotypes using a matching threshold of 83 % or higher to the HCV 1a reference sequence from NCBI [11]. The E1/E2 junction of the HCV 1a reference sequence was selected through pairwise alignment and trimmed from both ends using an HCV 1a sequence from a patient. The E1/E2 junction of the HCV genome (L = 263 nt each), which contains the hypervariable region 1 (HVR1), was sequenced using the GS FLX System along with the GS FLX Titanium Sequencing Kit from 454 Life Sciences, Roche, Branford, CT. Obtained sequences were processed using the Kmer Error Correction (KEC) algorithm [15] for error correction and haplotyping. All HCV sequences were aligned using MUSCLE [4]. The average number of sequences obtained per patient for acute and chronic HCV intra-host populations was n = 11,899 and n = 6,054, respectively.

To examine the evolutionary dynamics of intra-host HCV populations, we analyzed the epistatic density distributions of acute and chronic infections. epistatic density distribution histograms were generated to visualize the distribution of sequences across different density values, allowing for a comparison of selective pressures and adaptive patterns.

Distinct patterns were noted in the sequence distribution across epistatic density values within these populations, emphasizing the choppy nature of their distributions. Analyzed density histograms illustrate the distribution of sequences across different density values for intra-host HCV populations, with a focus on the choppy nature of the density distribution (Fig. 2).

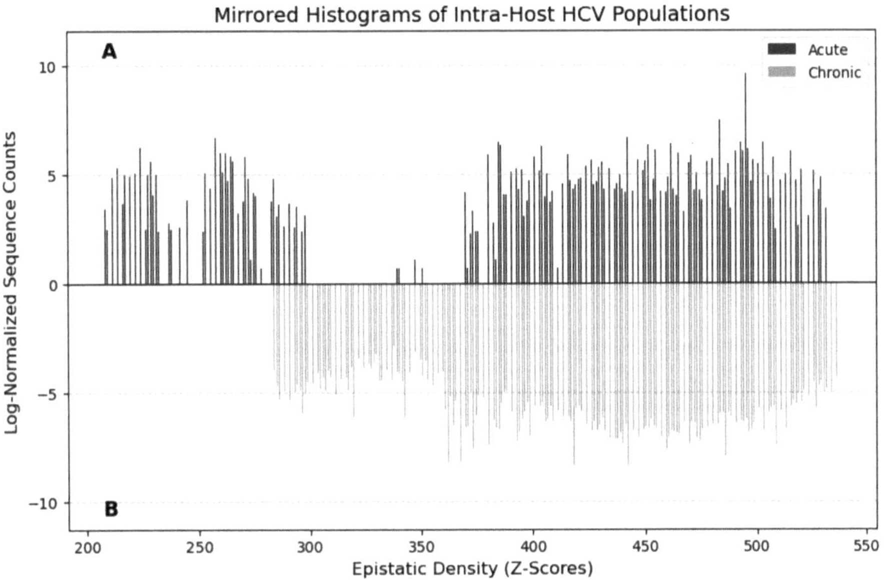

Fig. 2. Comparison of epistatic density distributions of acute (Panel A) and chronic (Panel B) intra-host HCV populations. The x-axis represents epistatic density values, which were normalized using z-score transformation, while the y-axis denotes sequence counts, which were log-normalized to improve visualization.

Epistatic density distribution for acute intra-host HCV population showed a broader range of density values from 500 to 8000 but the sequence counts are relatively sparse, indicating that only a few density values are represented. Density value peaks are less pronounced, suggesting that there may be fewer advantageous mutations or adaptations in this broader range, leading to a less diverse population in terms of fitness (Panel A, Fig. 2). In contrast, density distribution for chronic intra-host HCV population has a narrower range of values from 500 to 2250, revealing a more intricate and choppy pattern. Here, the peaks are more pronounced, indicating that certain density values are significantly more favorable, resulting in a higher concentration of sequences. This choppiness suggests that the population is experiencing strong selective pressures, favoring specific traits that enhance fitness Panel B, Fig. 2.

Overall, the comparison of these two histograms highlights the complexity of the density distribution in HCV populations. The choppy nature of the second histogram indicates a dynamic evolutionary process, where specific fitness peaks may represent adaptive strategies that are being favored in chronic versus acute infections. Understanding these patterns can provide insights into viral evolution.

We conducted a comprehensive analysis using receiver operating characteristic (ROC) curves and area under the curve (AUC) values to assess the classification performance of our model (Fig. 3).

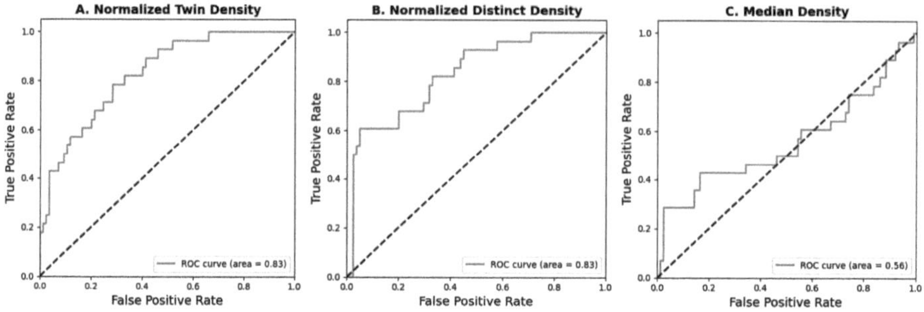

Fig. 3. Receiver Operating Characteristic (ROC) curves for classifying HCV acute and chronic patients based on normalized twin density values (**Panel A**), normalized distinct density values (**Panel B**), and median density values (**Panel C**).

Figure 3 presents three ROC curves, each evaluating the classification performance of different metrics distinguishing acute and chronic HCV patients.

Panel A, Fig. 3 the ROC curve (in orange) illustrates the classification performance based on normalized twin density values. The AUC is 0.83, indicating strong model performance in distinguishing between acute and chronic HCV cases. Higher AUC values (closer to 1.0) suggest superior discriminative ability.

Panel B, Fig. 3 displays the ROC curve for classification using normalized distinct density values. The curve follows a similar trajectory to Panel A, with an AUC of 0.83, confirming robust classification efficacy comparable to the twin ratio metric.

In contrast to the previous panels, Panel C, Fig. 3 the ROC curve for median density values exhibits significantly lower discriminative performance, with an AUC of 0.56. This value suggests poor classification capability, as the curve is closer to the diagonal reference line (dashed blue), which represents a random classifier.

Overall, the results highlight that both normalized twin and distinct density values are effective for classifying HCV infection types, whereas median density values lack sufficient discriminatory power.

4 Discussion and Conclusions

This study indicates a relation between epistatic density of viral variants increases with evolution within viral population. Our proposed epistatic density model assumes that emerging strains have higher density in the population and more positively linked edges as a dense community within the coordinated substitution network compared to other viral variants. The epistatic density is defined as the number of positively linked edges representing advantageous interactions in a genome, calculated as the number of edges representing positively linked SNV pairs within the coordinated substitution network. Additionally, our research findings suggest that analyzing coordinated substitution networks and calculating statistical parameters of epistatic density can effectively distinguish acute from chronic HCV patients. Specifically, the twin and distinct ratio values, derived from the epistatic density statistical analysis, demonstrate strong discriminatory power in classifying HCV infection types. Thus, we highlight the potential of this approach for understanding viral evolution and disease progression. These findings support the hypothesis that chronic HCV infections, due to their longer evolutionary development time, exhibit distinct epistatic density patterns compared to acute infections.

While the study presents a novel approach for classifying HCV infections using coordinated substitution networks, there are potential limitations to consider. The sample size, particularly the smaller number of acute individuals (N = 28) compared to chronic individuals (N = 85), may introduce bias and affect the generalizability of the findings. Furthermore, we restricted viral variants to the HCV 1a genotype to exclude the effects of genetic heterogeneity on classification results. We acknowledge that the reliance on specific statistical thresholds and parameters for constructing the coordinated substitution networks could influence the network structure and, consequently, the epistatic density calculations, potentially affecting the robustness of the classification.

This study introduces a novel machine learning model that leverages coordinated substitution networks for classifying HCV infection. By incorporating epistasis as a feature in the classification model, the approach captures epistatic interactions between mutations, offering valuable insights into coordinated mutation patterns that influence viral relative fitness. The model demonstrates strong discriminatory power, effectively distinguishing between acute and chronic HCV patients. This classification method has the potential to be adapted for other infectious diseases, offering a robust approach to the identification of mutation patterns and epistatic interactions that influence viral evolution. The results highlight that normalized twin and distinct density values are highly effective in classifying HCV infection types, while median density values do not provide sufficient discriminatory power, as discussed in the Results section.

Funding. Alina Nemira was supported by the GSU Brains and Behavior fellowship. Akshay Juyal was supported by the GSU Brains and Behavior fellowship. Pavel Skums has been partially supported by the NSF grants 2415564, 2415562 and 2412914. Alexander Zelikovsky has been partially supported by NSF Grant IIS-2212508, NSF Grant OISE-2412914, and by a grant of the Ministry of Research, Innovation and Digitization, under the Romania's National Recovery and Resilience Plan – Funded by EU – NextGenerationEU program, project "Metagenomics and Bioinformatics tools for Wastewater-based Genomic Surveillance of viral Pathogens for early prediction of public health risks—(MetBio-WGSP)" number 760286/27.03.2024, code 167/31.07.2023, within Pillar III, Component C9, Investment 8.

References

1. Campo, D., Dimitrova, Z., Mitchell, R.J., Lara, J., Khudyakov, Y.: Coordinated evolution of the hepatitis C virus. Proc. Natl. Acad. Sci. **105**(28), 9685–9690 (2008)

2. Claims, A.I..: Scikit-learn: machine learning in python - ACM digital library (2024). https://dl.acm.org/doi/10.5555/1953048.2078195
3. Domingo, E., Holland, J.: Rna virus mutations and fitness for survival. Annu. Rev. Microbiol. **51**(1), 151–178 (1997)
4. Edgar, R.C.: Muscle: multiple sequence alignment with high accuracy and high throughput. Nucleic Acids Res. **32**(5), 1792–1797 (2004)
5. Foundation, P.S.: Python language reference, version 3.10.9 (2022). https://www.python.org/downloads/release/python-3109/, Accessed 24 Jan 2025
6. Harris, C.R., et al.: Array programming with NumPy. Nature **585**, 357–362 (2020). https://doi.org/10.1038/s41586-020-2649-2
7. Ito, J., Strange, A.P., Liu, W., Joas, G., Lytras, S., Sato, K.: A protein language model for exploring viral fitness landscapes (2024). https://www.semanticscholar.org/paper/17d31c5da3d8181e6783cc24abe9c436a44bc3e5
8. Lara, J., Teka, M., Khudyakov, Y.: Identification of recent cases of hepatitis c virus infection using physical-chemical properties of hypervariable region 1 and a radial basis function neural network classifier. BMC Genomics **18**, 33–42 (2017)
9. McKinney, W.: Data structures for statistical computing in python. Proceedings of the 9th Python in Science Conference, pp. 56–61 (2010).https://doi.org/10.25080/Majora-92bf1922-00a
10. Mohebbi, F., Zelikovsky, A., Mangul, S., Chowell, G., Skums, P.: Early detection of emerging viral variants through analysis of community structure of coordinated substitution networks. Nat. Commun. **15**(1), 2838 (2024)
11. National Center for Biotechnology Information (NCBI): Hepatitis c virus genome, reference sequence nc_004102.1 (2025). https://www.ncbi.nlm.nih.gov/nuccore/NC_004102.1, Accessed 24 Jan 2025
12. Neverov, A.D., Fedonin, G., Popova, A., Bykova, D., Bazykin, G.: Coordinated evolution at amino acid sites of sars-cov-2 spike. Elife **12**, e82516 (2023)
13. Pantho, M.J., Bauder, L.A., Huang, S., Qin, H.: A data-driven sliding-window pairwise comparative approach for the estimation of transmission fitness of sars-cov-2 variants and construction of the evolution fitness landscape (2024)
14. Rochman, N.D., Wolf, Y.I., Faure, G., Mutz, P., Zhang, F., Koonin, E.V.: Ongoing global and regional adaptive evolution of sars-cov-2. Proc. Natl. Acad. Sci. **118**(29), e2104241118 (2021)
15. Skums, P., et al.: Efficient error correction for next-generation sequencing of viral amplicons. In: BMC Bioinformatics. vol. 13, pp. 1–13. Springer (2012)
16. Virtanen, P., et al.: SciPy 1.0: fundamental algorithms for scientific computing in python. Nat. Methods **17**, 261–272 (2020). https://doi.org/10.1038/s41592-019-0686-2
17. Wu, H., Huang, Y., Dykes, C., Liu, D., Ma, J., Perelson, A., Demeter, L.: Modeling and estimation of replication fitness of human immunodeficiency virus type 1 in vitro experiments by using a growth competition assay (2006). https://www.semanticscholar.org/paper/75d77c4ba9692c906938c33c9db19f313918680d
18. Wu, H., Huang, Y., Dykes, C., Liu, D., Ma, J., Perelson, A., Demeter, L.: Modeling and estimation of replication fitness of human immunodeficiency virus type 1 in vitro experiments by using a growth competition assay (2006). https://www.semanticscholar.org/paper/75d77c4ba9692c906938c33c9db19f313918680d

Applying Genetic Algorithm with Saltations to MAX-3SAT

Ryan Alomair(✉), Hafsa Farooq, Daniel Novikov, Akshay Juyal, and Alexander Zelikovsky

Department of Computer Science, Georgia State University, Atlanta, GA 30303, USA
ralomair1@gsu.edu

Abstract. Punctuated equilibrium, the pattern of rapid, significant mutational change, had not been observed in real time until the SARS-CoV-2 viral variants emerged with multiple mutations occurring together. Using epistasis (the circumstance in which the effect of one gene is influenced by the presence of one or more other genes) as a framework to understand this phenomenon, we can capture the relationships between different combinations of mutations, where each node is an individual mutation, and each edge represents the interaction between them, allowing us to effectively model the fitness landscape of viral variants. In exploring these relationships, it has been found that dense subgraphs within the network correspond to emerge saltation. We refer to this as an evolutionary jump and incorporate it with a genetic algorithm (GA + EJ), which can uncover high-fitness regions seemingly distant from the variant(s) from which they originally derived. We applied it to the MAX-3SAT problem and found improvement for satisfiable problem instances with 600 variables and 2550 clauses, as well as 100 variables and 429 clauses.

Keywords: MAX-3SAT Problem · Evolution · Genetic Algorithm · Epistatic network

1 Introduction

SARS-CoV-2 variants took the world by storm in November 2020, necessitating the exploration of new methodologies to predict new variants. Variants like Alpha and Omicron presented unprecedented challenges to scientists, as the virus had undergone multiple mutations that seemed to occur simultaneously before coming to fruition as Variants of Concern (VOCs) [1,2,5]. While it is not unusual for a virus to mutate multiple times before being passed on, it is uncommon for a variant to retain multiple mutations that, only when present together, translate to phenotypic effects granting higher fitness. Multiple studies have consequently tried to debunk this phenomenon, explaining it through the framework of epistasis [12,14,16–20], where interactions between multiple mutations can play a role in shaping the viral fitness landscape [10,11]. Rather than evolving through

M. Alser et al. (Eds.): ICCABS 2025, LNCS 15599, pp. 249–263, 2026.
https://doi.org/10.1007/978-3-032-02489-3_19

a gradual accumulation of independent mutations with additive effects, SARS-CoV-2 exhibits punctuated equilibrium, characterized by sudden shifts in viral fitness due to epistatic interactions [11,15].

This dynamic can be effectively modeled using an epistatic network, where each node represents a sequence containing a particular minor variant—the less frequently occurring allele at a given genomic site (referred to as a Single Nucleotide Variant (SNV)). An edge (with an edge weight) connecting a pair of nodes represents the epistatic link between the two respective SNVs. Positively linked pairs of SNVs reflect a high likelihood that they will occur in the same sequence, while a negatively linked pair reflects a low likelihood they will occur together. Clusters of positively linked SNVs form dense subgraphs in this network, where h10ighly fit viral variants can then emerge as combinations of these SNVs, resembling the pattern of an evolutionary jump [11]. Such changes contrast with traditional models of gradual evolution, where selective pressures incrementally refine individual SNVs over time. Inspired by this evolutionary phenomenon, we incorporate evolutionary jumps into the genetic algorithm.

The genetic algorithm (GA) is an optimization technique inspired by natural selection, where a population of candidate solutions evolves through iterative processes of natural selection (based on a fitness function) genetic crossover, and mutation [9]. Once an initial population of randomized solutions is produced, higher-fitness solutions are selected via tournament selection, which is then used to produce offspring through genetic crossover and mutation. The genetic algorithm (GA) is an optimization technique inspired by natural selection, where a population of candidate solutions evolves through iterative processes of selection, crossover, and mutation [9]. Starting with a randomized initial population, high-fitness solutions are chosen via tournament selection to join the mating pool, which will be used to produce a new generation of offspring solutions. Each solution's fitness is determined by a predefined fitness function, and selection is performed based on these values. Subsequently, properties (variable assignments) of higher-fitness solutions propagate over successive generations, while properties of lower fitness solutions are eliminated. As a result, each solution's properties (variable assignments) become increasingly homogeneous over time, prematurely converging to local fitness peaks without exploring potentially superior alternatives [7]. To counter this inherent fault, we use Clique-SNV to map solutions to an epistatic network and simulate evolutionary jumps. These new solutions are then inserted directly into the mating pool, allowing their properties to propagate to the next generation and effectively incorporate a tailored diversity to the mating pool [3]. By incorporating these solutions, GA can break out of its premature convergence, and explore variable assignments corresponding to solutions in other high fitness neighborhoods of the epistatic network [3].

Section 2 discusses the general process of GA and subsequent parts (k-tournament selection, genetic crossover and mutation). Section 3 goes on to discuss SARS-CoV-2/Epistasis/Punctuated equilibrium, and how these concepts manifest in our use of Clique-SNV. Section 3 then goes on to describe how we integrate Clique-SNV into simple genetic algorithm (GA), to produce the genetic algorithm with evolutionary jumps (GA+EJ). Section 3 finishes with a discussion of GA+EJ, as it is applied to the MAX-3SAT problem. Section 4 highlights specific details on our implementation and parameter selection. Section 5 covers our problem instances, results and analysis. Section 6 finishes with our concluding notes/future work.

2 Genetic Algorithm

2.1 Simple Genetic Algorithm

Simple genetic algorithm (GA) begins by initializing a randomized population of solutions for a given population size, where each solution is a binary bit string consisting of 1s and 0s. Each bit represents a boolean value that will be assigned to a corresponding variable in the problem being solved, meaning the length of each solution is equal to the number of variables in the problem instance. Each solution is then passed into a fitness function, which determines how well the solution solves the problem instance. Once a fitness value is determined for each solution, some are chosen to be part of the mating pool based on a k-tournament selection (simulating natural selection). A total of k solutions (input parameter) will be randomly selected, and the one with the highest fitness will be selected for the mating pool, shown in Fig. 1. That selection process will repeat until the mating pool reaches a predetermined size (input parameter) [9].

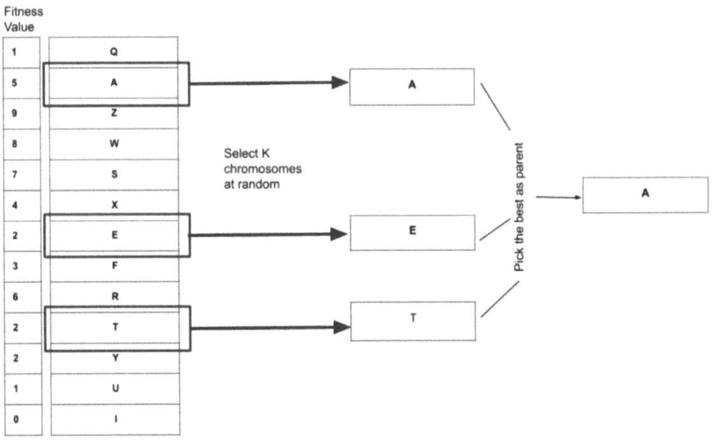

Fig. 1. Tournament selection performed for tournament size K = 3. The solution of highest fitness between the k solutions is selected for the mating pool [13].

Solutions from the mating pool are then randomly selected in pairs to produce a pair of offspring solutions. The parent solutions will first undergo a single point genetic cross-over: a random index p is chosen, and the variable assignments leading up to and including p are then taken from parent1 and given to child1, and the same for parent2 with child2, shown in Fig. 2. Then, the variable assignments beginning at p+1 until the end are taken from parent2 and given to child1, and vice versa with parent1 and child2. As a result, each child solution will have a variable complement of each parent solution, based on the randomly selected point p (note: this is often extended to a k-point crossover, where k indices are selected, and the child solutions contain alternating portions of each parent solution, still forming a complement to each other) [9].

Next, each child solution undergoes mutation, where bits are flipped at random based on a mutation probability (input parameter) as a means to help maintaining the amount of diversity within each solution, shown in Fig. 2. This process of producing two child solutions at a time is repeated until the population size is met, forming a new generation. Once the new generation of offspring is ready, each of their fitness values is calculated, and the process repeats continuously until a threshold number of generations are produced, a threshold number of generations without improvement of global fitness is reached, or a predetermined optimum global fitness is reached [9].

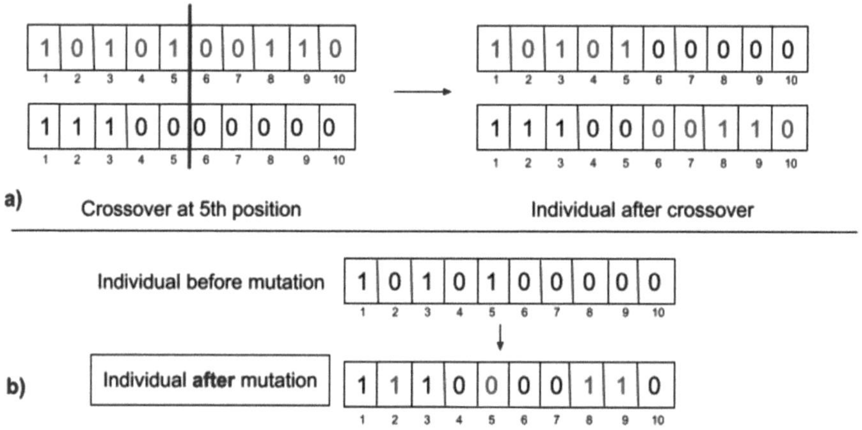

Fig. 2. a) Two individuals are performing crossover. Red line represents the point $p = 5$ of crossover. b) Mutation is performed on the 2nd, 6th and 8th genes of the individual [3]. (Color figure online)

Although the above approaches for parent selection, crossover and mutation are common for GA, these methods can be implemented in different ways, and sometimes have different approaches to producing offspring solutions altogether [9]. Ultimately, the approach must be tailored to the problem which GA solutions are being produced for. In this study, solutions are crafted for the MAX-3SAT problem, warranting a modified approach to our crossover and mutation functions, which are described in Sect. 3.4.

3 Genetic Algorithm with Evolutionary Jumps

3.1 Epistasis, Saltation and SARS-CoV-2

The evolution of SARS-CoV-2 variants can be understood through the lens of epistasis, which describes the non-additive interactions between mutations that influence viral fitness [12,14,16–20]. Epistasis constrains the set of viable genomic variants and, consequently, restricts the potential evolutionary trajectories of the virus. This constraint can be formalized using graph-theoretical frameworks, where viable genotypes form a structured subgraph within a hypercube, known as the viable space. Within this space, a genotype is considered viable if its constituent mutations collectively form a maximal clique in the underlying epistatic network. As a result, viral evolution does not proceed through a smooth and continuous exploration of all possible mutations, but is channeled along a discrete set of constrained pathways defined by these cliques instead [11].

This framework provides a powerful means to predict high-fitness variants by identifying restricted evolutionary trajectories accessible to the virus [11]. Importantly, these constraints align with saltation, in which viral lineages appear to undergo sudden drastic bursts of adaptation rather than gradual incremental change [2,15]. Such shifts may correspond to transitions between distinct regions of the viable space, where a set of epistatic interactions must be co-acquired before a new, highly fit genotype emerges. Using this network-based understanding of epistasis, it becomes possible to prioritize specific mutational combinations for functional screening, which ultimately aids in the prediction of future SARS-CoV-2 variants with high fitness potential [11].

3.2 Clique-SNV

The Clique-SNV framework can identify epistatic interactions among single nucleotide variants (SNVs) using a graph-theoretical approach. It first classifies SNV pairs as linked, forbidden, or unclassified based on their observed co-occurrence in sequencing reads. For a given genomic position, the less frequently occurring allele is referred to as the minor variant, and a pair of minor variants occurring at different position in the same sequence is referred to as a 2-haplotype. Let the number of 2-haplotypes be denoted by O_{22}. If a 2-haplotype appears significantly more often than expected by sequencing noise, it is classified as linked, meaning the variants likely co-occur in a haplotype. If its occurrence is significantly lower than expected, the pair is forbidden, indicating they are unlikely to be found together. All other pairs remain unclassified until further evaluation [6].

Next, Clique-SNV constructs an SNV graph G = (V,E) where the vertices (V) represent minor variants and the edges (E) connect the linked SNVs. This graph can capture structural constraints on co-occurring mutations. Since many isolated variants arise from sequencing errors, filtering these out refines the SNV graph and improves accuracy [6].

Finally, Clique-SNV identifies maximal cliques within the SNV graph using the BronKerbosch algorithm. A clique represents a set of SNVs that are all pairwise linked, meaning they are likely to co-occur in a viable haplotype [6]. By decomposing the SNV graph into cliques, Clique-SNV can reveal constrained evolutionary patterns, showing how certain mutations must co-occur for viability. This framework helps predict high-fitness variants and provides insight into the structured nature of viral evolution [3].

3.3 Integrating Clique-SNV into GA

Integrating Clique-SNV into the genetic algorithm (GA) allows for the detection of epistatic interactions within evolving solutions. As GA progresses, solutions to accumulate across generations and are converted to FASTA format, where 0s (reference alleles) are encoded as A's and 1s (mutations) as C's. Clique-SNV then analyzes all generated solutions, identifying statistical links between frequently co-occurring variable assignments and constructs an epistatic network.

In this network, nodes represent variable assignments (e.g., a particular position being 0 or 1), while edges form between assignments that frequently co-occur in high-fitness solutions. Note that properties of high-fitness solutions naturally persist through generations, causing them to gain greater influence in the network and reinforcing the statistical associations between their variable assignments. Strongly connected components reveal dependencies between variable states, constraining viable evolutionary pathways. The maximal cliques in this network then naturally emerge as sets of assignments that consistently appear together in viable solutions [3].

Just as dense subgraphs in SARS-CoV-2's epistatic network reveal mutations that co-occur and constrain evolutionary trajectories, Clique-SNV detects maximal cliques in the GA's epistatic network, identifying groups of variable assignments that frequently appear in high-fitness solutions.

By leveraging these maximal cliques, Clique-SNV can construct new candidate solutions by recombining linked variable assignments, simulating evolutionary jumps in the search space—just as epistatic constraints in viral evolution can similarly give rise to evoluationary jumps [3]. Thus, we use Clique-SNV to identify key epistatic interactions and facilitate punctuated equilibrium, improving the GA's ability to efficiently explore the search space beyond recombination and random mutation.

3.4 GA+EJ Applied to the MAX-3SAT Problem

The MAX-3SAT problem is a well-known NP-hard optimization problem in which the objective is to find a variable assignment that maximizes the number of satisfied clauses in a Boolean formula. The formula is expressed in conjunctive normal form (CNF), where a set of clauses are connected by logical AND statements, with each clause consisting of exactly three literals enclosed in parentheses, compared by logical ORs (each literal being a variable or its negation). It follows that a clause is satisfied if the variable assignment allows at least one literal in the clause to evaluate true. The challenge lies in determining an assignment of Boolean values (true or false) to the variables that satisfies as many clauses as possible.

In this implementation, problem instances are provided in CNF files, which specify the set of clauses and variables for a given MAX-3SAT problem. Each candidate solution in GA represents a variable assignment, where every variable in the formula is assigned either 0 (false) or 1 (true). Each solution will then consist of a binary bit string with length equal to the number of variables in the CNF formula, with the index of each bit corresponding to the variable it will be plugged into. Once these assignments are plugged into the formula, the fitness function evaluates them by simply counting the number of satisfied clauses. Higher fitness values correspond to assignments satisfying more clauses, allowing the search to tend toward optimal solutions during parent selection.

The crossover function used in the implementation for MAX-3SAT differs from the single-point crossover discussed earlier. Since the order of bits in each solution determines the variable assignment, we implement our crossover function differently than how it was described in Sect. 2. Instead of splitting the parent solutions at a fixed point, the crossover function preserves all variable assignments that are identical between the two parent solutions in both child solutions. For variables where the parents differ, one child solution inherits the assignment from one parent, while the other child takes the assignment from the other parent. This approach ensures that shared structure between high-fitness solutions is maintained while still introducing recombination between for some variable assignments.

The mutation function remains the same as previously described, applying small random changes to individual variable assignments to maintain genetic diversity and prevent premature convergence. By incorporating evolutionary jumps through Clique-SNV, the algorithm enhances its ability to escape local optima and explore new high-fitness regions of the search space efficiently.

3.5 Incorporating Evolutionary Jumps

To insert solutions corresponding to evolutionary jumps, we wait for 10 generations to be produced without any improvement in global fitness, then call Clique-SNV, which returns several EJ solutions that are inserted directly into the mating pool for the next generation. Since EJ solutions are constructed using statistically linked variable assignments in solutions that have already been produced, it is important to allow enough generations to be produced prior to calling Clique-SNV, such that these relationships have enough time to form. That said, we also must not wait for too many generations to pass, since this allows suboptimal/convergent solutions to occupy an increasingly larger share of the cumulative solution space being used to construct the epistatic network.

While our approach shares fundamental similarities with the method described in [3]—both utilizing a genetic algorithm to find optimal solutions and incorporating the Clique-SNV framework to introduce evolutionary jumps (EJs)—our implementation is distinct. The algorithm presented here introduces

EJ solutions directly into the mating pool to influence the next generation's offspring, whereas [3] replaces the lowest-fitness solutions of the current population with EJ solutions. Additionally, while [3] employs k-point crossover (described in Sect. 2.1) to generate offspring, our approach features a uniquely tailored genetic crossover specifically designed for the MAX-3SAT problem, as it described in Sect. 3.4. Lastly, our method does not incorporate any repairing mechanisms, unlike that of [3] which applies a repairing method to improve solution feasibility.

4 Implementation Specifics

4.1 Pygad as Our Base Implementation

The implementation of the genetic algorithm is built using the state-of-the-art PyGAD library, which provides a flexible and efficient framework for developing evolutionary algorithms [4]. PyGAD offers a wide array of customizable features, making it an effective tool for fine-tuning GA behavior. Its built-in mechanisms for selection, crossover, and mutation allow for seamless experimentation with different evolutionary strategies, but also give us freedom to implement these on our own. Additionally, PyGAD's callback functions, such as on_fitness and on_generation, enabled the straightforward integration of Clique-SNV into the algorithm. By leveraging these functions, we could dynamically analyze evolving solutions, construct the epistatic network, and introduce evolutionary jumps without disrupting the standard workflow of GA.

4.2 Parameters

To ensure optimal performance of the genetic algorithm, we carefully tuned various parameters to maximize the total number of clauses satisfied. The parameters adjusted during this process included population size, number of generations(gens), mating pool size and their selection, mutation type, mutation-rate and crossover method. Each of these factors played a crucial role in determining how effectively the algorithm explored the solution space and converged on high-fitness solutions. Furthermore, Clique-SNV required meticulous tuning of its own parameters to accurately capture epistatic interactions and facilitate effective evolutionary jumps. These adjustments were essential to balance exploration and exploitation in the search process. A summary of all finalized parameters is provided in Table 1, followed by a description of the tuning process for some individual parameters.

Table 1. Parameters Table

GA Parameter	Value	C-SNV Parameter	Value
Population Size	(1308),(256)	Jump Threshold	10
Num Parents	(28,38,32,50),(128)	Min. Gen. Wait	20
Gens	300	Jump Type	Global Best
Num CPUs	16	C-SNV Timeout	1200 s
Parent Type	Tournament	Threshold.Freq	0.01
K-Tournament	(35,26,41,26),(2)	Threshold.Freq+	0.01
Mutation Type	Random	Memory	96 GB
Mutation Prob.	0.01	Edge Limit	1000

1. **Population size: 1308 Solutions**
   While ensuring the population size was large enough to effectively explore a wide solution space, making it too large came at a cost to overall performance. By using a solution space too large, many generations would pass before there were enough consecutive generations without improvement. Consequently, the cumulative store of solutions became overloaded with those consisting of a suboptimal variable assignments to which the solutions were converging. As a result, EJ solutions did not introduce enough diversity in order to escape local optimums.

2. **Mating pool and tournament size**
   The mating pool size was coordinated with the tournament size to ensure a particular portion of the solution space would be explored. Tables 2 and 3 show which parameters correspond to which population sizes, shown below:

Table 2. For Population Size of 1308 solutions

Mating Pool Size	Tournament Size	% Soln. Space Explored
28	35	~53%
38	26	~54%
32	41	~64%
50	26	~64%

Table 3. For Population Size of 256 solutions

Mating Pool Size	Tournament Size	% Soln. Space Explored
128	2	~63%

3. **Clique-SNV Threshold Frequency**
   This value (denoted here as *tf*) represents the assumed lower bound on the actual occurrence rate of a given 2-haplotype (a sequence with two particular SNVs occurring simultaneously) in the dataset. It is used to control the probability of falsely identifying linked variable pairs by ensuring that observed read counts are statistically significant.
   The probability of observing at least $x \geq O_{22}$ reads (where O_{22} is the number of 2-haplotypes) given that the true frequency is at most *tf*, is constrained to be low enough to prevent false-positive associations. This threshold is set such that the expected number of false positives remains below 0.05 / $\binom{L}{2}$, where L is the haplotype length.
   We initially set *tf* to 0.01, but if the number of edges produced in the epistatic network exceeds the predefined edge limit, we increment true frequency by 0.01 and rerun Clique-SNV (the increment denoted in Table 1 as Threshold.Freq+). This adjustment controls the density of the network by filtering out weaker associations, preventing an excessive number of edges from forming. Maintaining a strict edge limit is crucial because the time required to find maximal cliques increases exponentially with the number of vertices in the network.

4. **Edge Limit**
   To manage computational feasibility, we set the edge limit to 1000, as this ensures that maximal cliques can typically be found within 20 min.

5 Results

5.1 MAX-3SAT Problem Instances

In MAX-3SAT problem instances, the number of clauses and variables plays a crucial role in determining the difficulty of finding a satisfying assignment. The challenge of solving all clauses—meaning the instance is fully satisfiable—largely depends on the clause-to-variable ratio. When this ratio is low, the problem tends to be easier because there are fewer constraints on variable assignments. Conversely, as the ratio increases, the constraints become more restrictive, making it significantly harder to find an assignment that satisfies all clauses. Research has shown that for random 3SAT instances, the satisfiability threshold occurs at a clause-to-variable ratio of approximately ∼4.267 [8]. At this critical ratio, problem instances transition from being mostly satisfiable to mostly unsatisfiable when adding more clauses, making them particularly challenging for both exact and heuristic solvers.

To maintain this complexity, we test two problem instances, one with 2,550 clauses and 600 variables and another with 429 clauses and 100 variables, both closely adhering to the 4.267 ratio. This ensures that the problem remains difficult enough to test the effectiveness of GA while still allowing room for high-fitness solutions to emerge. This is a solvable instance, meaning there exists at least one variable assignment which allows all clauses to be solved (2,550 and 429 clauses is the optimum for each respective instance). By operating near the satisfiability threshold, the instance presents a strong challenge, requiring the genetic algorithm to efficiently navigate the search space to maximize the number of satisfied clauses.

5.2 Hard MAX-3SAT Instances

Initially, we aimed to categorize problem instances into "hard" and "easy" based on their structure. However, in the context of Max-3SAT, such a distinction does not hold in a meaningful way. While some instances contain a smaller proportion of satisfiable clauses, our tests showed that both simple GA and (GA+EJ) performed relative to the optimum (the number of maximum number of satisfiable clauses) regardless of satisfiability constraints. This suggests that the presence of fewer satisfiable clauses does not inherently make an instance harder in a way that affects the comparative performance of our algorithms, rendering this distinction uninformative for our analysis.

5.3 Simple GA & GA+EJ

The following two tables contain performance scores for simple GA and GA+EJ based on 5 runs, where a score is the difference in the number of clauses satisfied by the algorithm and the optimum (2,550 and 429, since both problem instances are satisfiable). This metric was chosen because it provides the most clear measure of how close each algorithm comes to achieving the optimum. Additionally, the number of generations is reported as the generation at which each algorithm reached its optimum fitness value, ensuring that comparisons reflect the point at which the best solution was found rather than when the algorithm terminated.

Table 4. Performance scores measured by number of clauses left unsatisfied. Scores calculated over 5 runs. Runtime measured in seconds.

2550 Clauses/600 Variables				GA+EJ			Simple GA		
	Parents	Tourn. Size	Measure	Clauses Unsat.	Gens	Runtime	Clauses Unsat.	Gens	Runtime
(1)	28	35	Best	71	358	19380	72	217	2267
			Worst	89	69	2088	84	35	3231
			Average	78	217.2	8720	79.4	214.6	2550
(2)	38	26	Best	52	85	3180	58	240	2551
			Worst	64	196	5256	80	67	1671
			Average	58	258.4	9801	68	161.2	2549
(3)	32	41	Best	67	98	2497	63	106	1143
			Worst	81	102	2073	87	68	737
			Average	75.6	117.8	3319	78	187.4	2181
(4)	50	26	Best	44	323	18091	47	320	3509
			Worst	58	272	13975	62	165	1828
			Average	53.8	156.6	7231	53.8	300.2	3228

Table 5. Performance scores measured by number of clauses left unsatisfied. Scores calculated over 5 runs. Runtime measured in seconds.

429 Clauses/100 Variables			GA+EJ			Simple GA		
Parents	Tourn. Size	Measure	Clauses Unsat.	Gens	Runtime	Clauses Unsat.	Gens	Runtime
128	2	Best	3	36	12.32	5	37	19.84
		Worst	5	44	19.4	7	49	21.22
		Average	4.2	40	17.72	6	42	19.55

5.4 Simple GA Vs GA+EJ: Analysis

The results in Tables 4 and 5 demonstrate that the genetic algorithm with evolutionary jumps (GA+EJ) generally outperforms the simple genetic algorithm (GA) in solving the MAX-3SAT problem. In the larger problem instance, GA+EJ achieves a lower number of unsatisfied clauses across best, worst, and average cases in most configurations. For example, in (2) from Table 4, GA+EJ's best solution leaves 52 clauses unsatisfied, compared to 58 for the simple GA—an improvement of 10.3%. Similarly, in the worst case for this configuration, GA+EJ leaves 64 clauses unsatisfied, while the simple GA leaves 80, a 20% reduction. On average, GA+EJ leaves 14.7% fewer unsatisfied clauses (58 vs. 68).

The smaller problem instance shows an even more pronounced advantage for GA+EJ. In the best case, GA+EJ leaves only 3 clauses unsatisfied, compared to 5 for the simple GA, a 40% reduction. The worst case follows a similar trend, with GA+EJ leaving 5 clauses unsatisfied versus 7 for the simple GA (28.6% fewer). On average, GA+EJ results in 4.2 unsatisfied clauses, compared to 6 for the simple GA, marking an overall improvement of 30%.

Although, there are instances where GA+EJ does not consistently outperform simple GA. For example, in (3) from Table 4, the best-case result for GA+EJ is slightly worse (67 unsatisfied clauses) compared to the simple GA (63 unsatisfied clauses). Additionally, in the same configuration, GA+EJ's worst-case result (81 unsatisfied clauses) is only marginally better than the simple GA's worst case (87), but its average (75.6) is still close to that of the simple GA (78), suggesting less improvement in this scenario.

It is also important to mention that the runtime of the GA+EJ algorithm is often significantly greater than that of the simple GA due to the added complexity of finding maximal cliques within the epistatic network. As the number of vertices (solutions) in the network increases, the computational effort required to identify these cliques can grow exponentially (where Clique-SNV considers all solutions across all generations). The increase in complexity is particularly noticeable in the larger problem instance, highlighting the additional computational burden introduced by GA+EJ.

Additionally, GA-EJ does not consistently reduce the number of generations required to reach high-fitness solutions. In some cases, such as the settings from (1) and (3), GA+EJ requires more generations on average than the standard

GA, despite achieving slightly higher or similar fitness values. This suggests that while evolutionary jumps help maintain a higher fitness ceiling, they do not always accelerate convergence. Instead, they may introduce additional variability, allowing the algorithm to explore alternative solutions rather than prematurely converging.

However, it is important to note that both algorithms have a predefined threshold of 300 generations without improvement in global fitness before they automatically stop running. As a result, the total number of generations before termination is often larger for the genetic algorithm with evolutionary jumps. This occurs because evolutionary jumps periodically reintroduce targeted diversity into the population, sometimes causing the algorithm to run for many additional generations before reaching its final optimum. In contrast, the genetic algorithm without evolutionary jumps tends to converge prematurely; once an optimal solution is reached at a particular generation, no further improvements occur, and the algorithm simply continues for the remaining 300 stagnation generations before termination.

While the benefits of GA+EJ are sometimes configuration dependent, its overall effectiveness in producing solutions closer to the optimal is evident. These results indicate that GA+EJ generally enhances genetic diversity and helps the algorithm escape local optima, with it outperforming simple GA in 12 of the 15 scores reported, demonstrating its consistent ability to find solutions with fewer unsatisfied clauses. Additionally, in the 429-clause problem, GA+EJ outperformed the simple GA in every measured case, achieving up to a 40% reduction in unsatisfied clauses. Even in the larger 2550-clause problem, where solution space complexity increases, GA+EJ produced better results in 9 out of 12 cases. Although there were a few instances where the simple GA performed similarly or marginally better, GA+EJ's more frequent success in reducing clause violations indicates a stronger ability to approximate the optimal solution. Its advantage is most evident in cases where the standard GA requires more generations to converge, suggesting that evolutionary jumps help escape local optima and drive long-term improvements in fitness. These findings reinforce the conclusion that incorporating evolutionary jumps enhances the genetic algorithm's exploration of the solution space, ultimately leading to superior optimization performance in the MAX-3SAT problem.

6 Conclusion

The results demonstrate a marginal yet consistent improvement in the performance of the genetic algorithm when incorporating evolutionary jumps, both in achieving higher fitness and reaching high-fitness solutions in fewer generations. This enhancement aligns with the principles of punctuated equilibrium, where populations experience periods of stasis followed by sudden leaps in adaptation. By leveraging Clique-SNV to identify and exploit epistatic interactions, the algorithm introduces structured genetic changes that mimic these evolutionary jumps, reintroducing tailored diversity into the mating pool and helping escape local optima. While the improvement is sometimes limited, it underscores the potential of integrating biological principles into evolutionary algorithms to enhance their efficiency in solving complex optimization problems like MAX-3SAT.

Building on these findings, we plan to extend GA+EJ to other optimization challenges, such as the Traveling Salesman Problem, MAX-SAT instances, and the multidimensional knapsack. We also plan to implement a method that will progressively reduce the number of nodes (solutions) used in the epistatic network, since the task of finding all maximal cliques can increase exponentially with each new node. This approach aims to mitigate the drastic increase in time complexity for the GA+EJ, particularly as the size of the problem grows. By limiting the number of solutions considered in the network, we hope to improve the efficiency of the algorithm while maintaining its effectiveness in solving the MAX-3SAT problem. Additionally, we intend to further optimize performance by parallelizing the algorithm on GPUs, paving the way for even more efficient and scalable implementations.

Funding Information. Alexander Zelikovsky has been partially supported by NSF Grant IIS-2212508, NSF Grant OISE-2412914, and by a grant of the Ministry of Research, Innovation and Digitization, under the Romania's National Recovery and Resilience Plan – Funded by EU – NextGenerationEU program, project "Metagenomics and Bioinformatics tools for Wastewater-based Genomic Surveillance of viral Pathogens for early prediction of public health risks—(MetBio-WGSP)" number 760286/27.03.2024, code 167/31.07.2023, within Pillar III, Component C9, Investment 8.

References

1. Andre, M., et al.: From alpha to omicron: How different variants of concern of the sars-coronavirus-2 impacted the world. Biology (Basel) **12**(9), September 2023. https://doi.org/10.3390/biology12091267
2. Corey, L., Beyrer, C., Cohen, M.S., Michael, N.L., Bedford, T., Rolland, M.: Sars-cov-2 variants in patients with immunosuppression. N. Engl. J. Med. **385**(6), 562–566 (2021). https://doi.org/10.1056/NEJMsb2104756
3. Farooq, H., Novikov, D., Juyal, A., Zelikovsky, A.: Genetic Algorithm with Evolutionary Jumps, pp. 453–463. Springer, October 2023. https://doi.org/10.1007/978-981-99-7074-2_36
4. Gad, A.F.: Pygad: an intuitive genetic algorithm python library. Multimed. Tools Appl., 1–14 (2023)
5. Ingraham, N.E., Ingbar, D.H.: The omicron variant of sars-cov-2: understanding the known and living with unknowns. Clin. Transl. Med. **11**(12), e685 (2021). https://doi.org/10.1002/ctm2.685
6. Knyazev, S., et al.: Accurate assembly of minority viral haplotypes from next-generation sequencing through efficient noise reduction. Nucleic Acids Res. **49**(17), e102–e102 (07 2021). https://doi.org/10.1093/nar/gkab576
7. Leung, Y., Gao, Y., Xu, Z.B.: Degree of population diversity - a perspective on premature convergence in genetic algorithms and its Markov chain analysis.

IEEE Trans. Neural Networks **8**(5), 1165–1176 (1997). https://doi.org/10.1109/72.623217
8. Mézard, M., Zecchina, R.: Random k-satisfiability problem: From an analytic solution to an efficient algorithm. Phys. Rev. E **66**, 056126 (2002)
9. Mitchell, M.: An introduction to genetic algorithms. MIT press (1998)
10. Mohebbi, F., Zelikovsky, A., Mangul, S., Chowell, G., Skums, P.: Community structure and temporal dynamics of sars-cov-2 epistatic network allows for early detection of emerging variants with altered phenotypes. bioRxiv, pp. 2023–04 (2023)
11. Mohebbi, F., Zelikovsky, A., Mangul, S., Chowell, G., Skums, P.: Early detection of emerging viral variants through analysis of community structure of coordinated substitution networks. Nat. Commun. **15**(1), 2838 (2024). https://doi.org/10.1038/s41467-024-47304-6
12. Moulana, A., et al.: Compensatory epistasis maintains ace2 affinity in sars-cov-2 omicron ba.1. Nat Commun. **13**(1), 7011, November 2022. https://doi.org/10.1038/s41467-022-34506-z
13. NA: Tournament selection (2023). https://www.tutorialspoint.com/geneticalgorithms/images/tournamentselection.jpg. Accessed 4 Mar 2025
14. Neverov, A.D., Fedonin, G., Popova, A., Bykova, D., Bazykin, G.: Coordinated evolution at amino acid sites of sars-cov-2 spike. Elife **12**, February 2023. https://doi.org/10.7554/eLife.82516
15. Nielsen, B.F., et al.: Immune heterogeneity and epistasis explain punctuated evolution of sars-cov-2. medRxiv, July 2022. https://doi.org/10.1101/2022.07.27.22278129
16. Rochman, N.D., Faure, G., Wolf, Y.I., Freddolino, L., Zhang, F., Koonin, E.V.: Epistasis at the sars-cov-2 receptor-binding domain interface and the propitiously boring implications for vaccine escape. mBio **13**(2), e0013522 (2022). https://doi.org/10.1128/mbio.00135-22
17. Rochman, N.D., Wolf, Y.I., Faure, G., Mutz, P., Zhang, F., Koonin, E.V.: Ongoing global and regional adaptive evolution of sars-cov-2. Proc. Natl. Acad. Sci U S A **118**(29), July 2021. https://doi.org/10.1073/pnas.2104241118
18. Rodriguez-Rivas, J., Croce, G., Muscat, M., Weigt, M.: Epistatic models predict mutable sites in sars-cov-2 proteins and epitopes. Proc. Natl. Acad. Sci. U S A **119**(4), January 2022. https://doi.org/10.1073/pnas.2113118119
19. Zahradník, J., et al.: Sars-cov-2 variant prediction and antiviral drug design are enabled by rbd in vitro evolution. Nat. Microbiol. **6**(9), 1188–1198 (2021). https://doi.org/10.1038/s41564-021-00954-4
20. Zeng, H.L., Dichio, V., Rodríguez Horta, E., Thorell, K., Aurell, E.: Global analysis of more than 50,000 sars-cov-2 genomes reveals epistasis between eight viral genes. Proc. Natl. Acad. Sci. U S A **117**(49), 31519–31526 (2020). https://doi.org/10.1073/pnas.2012331117

Computing Gram Matrix for SMILES Strings Using RDKFingerprint and Sinkhorn-Knopp Algorithm

Sarwan Ali[1], Haris Mansoor[3], Prakash Chourasia[2(✉)], Imdad Ullah Khan[3], and Murray Patterson[2]

[1] Columbia University Irving Medical Center, New York, NY, USA
sa4559@columbia.edu
[2] Georgia State University, Atlanta, GA 30303, USA
pchourasia1@student.gsu.edu, mpatterson30@gsu.edu
[3] Lahore University of Management Sciences, Lahore, Punjab 54792, Pakistan
{16060061,imdad.khan}@lums.edu.pk

Abstract. SMILES (Simplified Molecular Input Line Entry System) strings are widely used to represent molecular structures in cheminformatics and drug discovery. However, effectively transforming these string-based representations into meaningful numerical features for machine learning remains a significant challenge due to the complex, non-Euclidean nature of molecular structures. Traditional fingerprint-based and deep learning approaches often struggle with scalability, interpretability, or computational efficiency. Our approach leverages the Morgan Fingerprint to generate molecular feature representations, followed by a pairwise kernel function to compute a structured similarity matrix. We then refine this matrix using the Sinkhorn-Knopp algorithm, ensuring it satisfies probabilistic constraints. To reduce dimensionality, we apply Kernel Principal Component Analysis (PCA), producing compact embeddings suitable for downstream machine learning tasks. We conduct a comprehensive empirical evaluation of the proposed method which is assessed for drug subcategory prediction (classification task) and solubility AlogPS "aqueous solubility and octanol/water partition coefficient" (regression task) using the benchmark SMILES string dataset. The outcomes show the proposed method outperforms baseline methods in supervised analysis and has potential uses in molecular design and drug discovery. By integrating kernel-based learning with probabilistic refinement, our method offers a promising alternative to existing cheminformatics techniques.

Keywords: Kernel Matrix · Classification · Sinkhorn Knopp Algorithm

1 Introduction

In the field of drug discovery and molecular design, the analysis of molecular structure is a fundamental task [26]. Simplified Molecular Input Line Entry Sys-

M. Alser et al. (Eds.): ICCABS 2025, LNCS 15599, pp. 264–276, 2026.
https://doi.org/10.1007/978-3-032-02489-3_20

tem (SMILES) strings have emerged as a popular choice for representing molecular structure data [25], primarily due to their simplicity and ease of use. However, modeling and analyzing molecular structures represented as SMILES strings pose several challenges [14]. These include dealing with the high dimensionality of the data and the complex non-linear relationships between the structures. Moreover, converting SMILES strings into machine-readable numerical representations is a challenging task that requires sophisticated techniques. Machine learning applications, such as domain adaptation and generalization, heavily rely on efficient low dimensional embeddings of complex data [10].

The analysis of SMILES strings has become increasingly important in the field of drug discovery and cheminformatics [4]. SMILES strings are a compact representation of a molecule's structure and have become a popular choice for encoding molecular information in machine learning models [31]. Traditional molecular fingerprinting methods e.g., Morgan Fingerprints, MACCS rely on predefined heuristics and may not fully capture structural nuances. SMILES strings represent molecules in a linear format, encoding atomic connectivity and bonding. However, molecules are graph-structured, with atoms as nodes and bonds as edges. Analyzing SMILES as sequences misses important structural relationships like functional groups, global topology, and spatially-derived physicochemical properties [14]. Deep learning-based approaches, such as graph neural networks (GNNs) and sequence-based models e.g., transformers for SMILES often require extensive labeled data, suffer from interpretability issues, and can be computationally expensive. These models are used for a range of tasks, including drug solubility [8] and subtype prediction.

Despite advances in SMILES representation learning, a scalable, interpretable, and effective kernel-based approach remains underexplored. Kernel methods offer a powerful way to capture molecular similarities [5]. There is still a need for further investigation into the effectiveness of different types of embeddings, classification, and regression models for SMILES string analysis. By converting SMILES into molecular graphs, we recover these structural relationships, enabling more expressive and accurate molecular representations. The aim of this research project is twofold: (i) to address this gap in knowledge by evaluating the performance of various embedding methods and machine learning models for classification and regression tasks using SMILES strings as input, (ii) to propose a method for SMILES string analysis. The results of this study could have significant implications for drug discovery research and help to identify the most effective methods for predicting molecular properties.

In this paper, we propose a kernel-based approach for encoding and analyzing molecular structures represented as SMILES strings. Performing two tasks (i) the drug subcategories prediction (classification task) and (ii) the solubility AlogPS (aqueous solubility and octanol/water partition coefficient) prediction (regression task). The proposed approach involves computing a kernel matrix using the Sinkhorn-Knopp algorithm [13] and using kernel principal component analysis (PCA) [9] to reduce the dimensionality of the molecular structures. The resulting low-dimensional embeddings are then used for classification and regression analysis. The proposed approach starts with converting the SMILES strings into molecular structures using the RDKit library, later we convert them into feature vectors using the RDKFingerprint function, which computes a fingerprint for each molecule. The pairwise kernels function is then used to compute the distance matrix. We present a novel method for a kernel matrix by utilizing an optimal transport matrix [17] using the SMILE string dataset. The Sinkhorn-Knopp algorithm is used to compute the final kernel matrix that satisfies the constraints of a probability distribution. Achieved iteratively by adjusting the kernel matrix until the marginal distributions of the rows and columns of the matrix match the desired marginal distributions. Kernel PCA reduces the dimensionality of the molecular structures by projecting the kernel matrix onto a lower-dimensional (LD) space. The resulting LD embeddings capture the intrinsic properties of the molecular structures, making them suitable for classification and regression analysis.

The proposed approach has several potential applications in drug discovery and molecular design. It can be used to analyze large datasets of molecular structures and to identify compounds with desirable properties. It can also be used to design new molecules with desired properties and can also be used in searching for molecules with similar properties. In summary, our contributions to this paper are the following:

1. We propose a novel kernel function-based approach for SMILES string analysis using classification and regression. Our method is based on the idea of first converting SMILES strings into molecular graphs, computing features, and using optimal transport matrix, we generate the final kernel matrix.
2. Conventional kernel approaches struggle with probabilistic constraints and computational efficiency. Our proposed kernel-based method introduces an optimal transport framework ensuring probabilistic constraints while preserving molecular similarity and balancing interpretability and scalability.
3. Using empirical analysis, we show that our kernel-based approach achieves higher performance for classification and comparable regression performance compared to baselines on benchmark SMILES string dataset.

2 Related Work

Molecular fingerprints are binary vectors used to encode a molecule structural information [20,31]. They capture substructures and are useful for predicting molecular properties. Recent studies have explored using fingerprints and embeddings to predict drug solubility [18]. Random forest regression and support vector regression outperformed other models in predicting solubility [3]. Graph convolutional neural networks achieved an R-squared value of 0.75 on a dataset of 1144 compounds [24]. More research is needed to evaluate the effectiveness of different embeddings, classification, and regression models for solubility and drug subtype prediction.

Kernel-based approaches have been widely used to capture complex relationships between data points [1,16,21]. One popular kernel method for molecular data analysis is kernel ridge regression (KRR), a supervised learning algorithm that uses a kernel matrix to represent pairwise similarities between molecular structures. The KRR algorithm has been applied to several molecular property prediction tasks, including drug solubility and toxicity prediction, and has shown promising results [7,29].

Another popular kernel method for molecular data analysis is the support vector machine (SVM) algorithm [30]. SVM is a widely used supervised learning algorithm that uses a kernel matrix to represent the pairwise similarities between data points. Several studies have applied SVM to molecular activity classification tasks, including predicting protein-ligand binding affinity [28] and identifying active compounds in high-throughput screening experiments [15].

Recently, kernel principal component analysis (PCA) has emerged as a powerful tool for dimensionality reduction and feature extraction in molecular data analysis [9,22]. Kernel PCA uses a kernel matrix to project the high-dimensional molecular structures onto a lower-dimensional space. Several studies have applied kernel PCA to molecular property prediction [9,23] and activity classification tasks, and have shown that it can significantly improve the performance of existing methods. While these methods have shown promising results, they have several limitations. For example, KRR and SVM require pre-defined kernel functions, which can be computationally expensive to compute for large datasets using their proposed methods.

3 Proposed Approach

The proposed approach takes a pair of SMILES strings S_1 and S_2 as input and computes a kernel value between the molecules represented by the SMILES strings. The kernel value is computed using the following steps described below.

3.1 Convert the SMILES Strings into Molecular Graphs

Let X_1 and X_2 be the sets of molecular graphs corresponding to the SMILES strings in S_1 and S_2, respectively. For each SMILES string, we use the MolFromSmiles function from the chem library in the RDKit library to obtain the corresponding molecular graph. This gives us the sets of molecular graphs $X_1 = x_{1,1}, x_{1,2}, \ldots, x_{1,n}$ and $X_2 = x_{2,1}, x_{2,2}, \ldots, x_{2,n}$.

3.2 Compute the Molecular Features

Next, we convert each molecular graph $x_{i,j}$ into a feature vector using the RDKit fingerprint. The fingerprint is a bit vector of a fixed length representing the presence or absence of certain molecular substructures in the molecule. The RDKit fingerprint is generated using the Morgan algorithm [18], which is a circular fingerprinting method that considers the neighborhood of each atom in the molecule up to a certain radius. The resulting fingerprint can be used to compare the structural similarity of different molecules, among other applications in cheminformatics. This fingerprint gives us the sets of feature vectors $X_1^{feat} = f_{1,1}, f_{1,2}, \ldots, f_{1,m}$ and $X_2^{feat} = f_{2,1}, f_{2,2}, \ldots, f_{2,m}$, where $m = 2048$, which is defined by Morgan Fingerprint.

3.3 Compute the Pairwise Distance Matrix

We compute the pairwise distance matrix D between the feature vectors using the Gaussian kernel of width σ:

$$D_{i,j} = exp(-\frac{||X_i^{feat} - X_j^{feat}||^2}{2\sigma^2}) \tag{1}$$

where $D_{i,j}$ is the i, j^{th} entry of the distance matrix D. Iterating over all i and j will give us a $N \times N$ matrix D, where N is the number of SMILES strings.

3.4 Compute the Kernel Matrix

We use the Sinkhorn-Knopp algorithm [13] to compute the optimal transport matrix between the two sets of molecules. Let $K_{i,j}$ be the entry in the kernel matrix K corresponding to molecules $x_{1,i}$ and $x_{2,j}$. We first normalize the pairwise distance matrix D to obtain a joint probability matrix P, kernel width σ:

$$P_{i,j} = \frac{D_{i,j}/\sigma}{\sum_{k,l} D_{k,l}} \tag{2}$$

We then use zero vectors $\overrightarrow{a}, \overrightarrow{b}$ and unit vectors $\overrightarrow{a_1}, \overrightarrow{b_1}$ and iteratively compute a bipartite graph using the probability matrix. More formally: We can define two node sets and edge weights as: $U = \{u_1, u_2, \ldots, u_m\}$ (Molecules from set 1) and $V = \{v_1, v_2, \ldots, v_n\}$ (Molecules from set 2)

$$w(u_i, v_j) = -\delta P_{i,j} + \zeta \log(\overrightarrow{a_1}[i]) + \zeta \log\left(\overrightarrow{b_1}[j]\right) \tag{3}$$

where $\zeta = 1, \delta = 10^{-10}$. Note that the ξ is the tolerance parameter. The $\overrightarrow{a_1}$ and $\overrightarrow{b_1}$ are linear sums of the rows and columns of the bipartite graph, respectively, which are then used to update $\overrightarrow{a}$ and $\overrightarrow{b}$. This process is iteratively continued until the convergence criteria are met. The convergence criteria is $max(|\overrightarrow{a_1} - \overrightarrow{a}|) < \xi$ and $max(|\overrightarrow{b_1} - \overrightarrow{b}|) < \xi$. Note that ξ is a hyperparameter whose value is the following $\xi = 10^{-6}$. Finally, we compute the kernel matrix K using the probability matrix:

$$K = a' \times P \times b' \tag{4}$$

where a' and b' are both diagonal matrix version of $\overrightarrow{a}$ and $\overrightarrow{b}$.

The overall workflow is given in Fig. 1. For a set of input SMILES strings, the first step is to convert them to molecular graphs and compute feature vectors using the Morgan algorithm (RDKFingerprint) [18], Fig. 1 (b), and (c). In the next step, the feature vectors are converted into Gaussian kernel matrix D Fig. 1 (d), and normalized to transform into a probability matrix P, Fig. 1 (e). Then we we initialize vectors a, a_1, b, b_1 and use the Sinkhorn-Knopp algorithm [13] to compute the optimal transport matrix a and b iteratively till convergence criteria are met. We use bipartite graph 3 also shown in Fig. 1 (g) and (h) to compute the matrix. After these matrix a and b returned by the Sinkhorn-Knopp algorithm are then multiplied with the probability matrix P to compute the final kernel matrix K (in Eq 4) Fig. 1 (i).

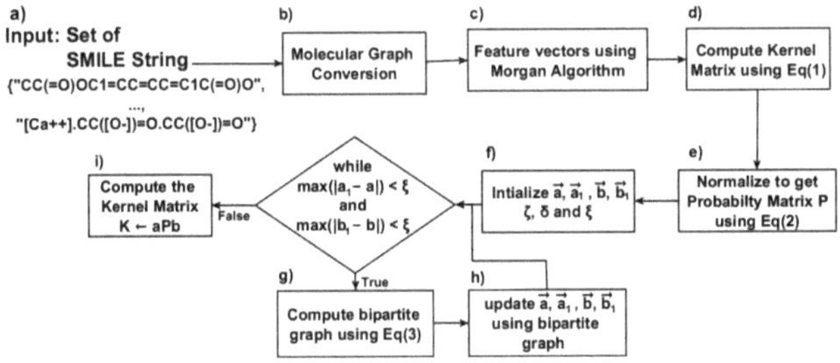

Fig. 1. Workflow of the proposed method.

3.5 Kernel PCA-Based Embeddings

A large dataset results in a large K, which can be difficult to store. Kernel-PCA addresses this by transforming K into a low-dimensional subspace. It maps the data into a higher-dimensional space and identifies a lower-dimensional manifold, capturing non-linear relationships. Using kernel-PCA, we compute the top principal components from K as feature vectors used in classifiers and regression.

4 Experimental Setup

In our drug subtype classification task, we use classifiers like SVM, Naive Bayes, MLP, KNN, Random Forest, Logistic Regression, and Decision Tree. Evaluation metrics include accuracy, precision, recall, weighted F1, macro F1, ROC-AUC, and training runtime. We employ linear regression, ridge regression, lasso regression, random forest regression, and gradient boosting regression to predict solubility AlogPS. Evaluation metrics used are mean squared error (MSE), mean absolute error (MAE), root mean squared error (RMSE), coefficient of determination (R^2), and explained variance score (EVS). We split our data into random training and test sets with a 70–30% split for both tasks, and repeat experiments 5 times. The preprocessed data, and code is publicly available at[1].

[1] https://github.com/pchourasia1/SMILE_Kernel.

Baseline Methods

- **Morgan Fingerprint**: The circular Morgan fingerprint [18] is used to encode the presence of substructures within a molecule. It generates a binary vector that indicates the presence or absence of substructures.
- **MACCS Fingerprint**: The MACCS fingerprint [6,12] is a binary fingerprint that uses predefined substructures based on functional groups and ring systems commonly found in organic molecules.
- **k -mers**: This method is a sequence-based embedding that encodes the frequencies of overlapping sub-sequences of length k [11] in the SMILES string.
- **Weighted k -mers**: To enhance the k-mers-based embedding's quality, we use a weighted version that employs Inverse Document Frequency (IDF) to assign weights to each k-mer within the embedding [19].

Dataset Statistics: We used 2 datasets for experimentation. First, a set of 6299 SMILES strings from the DrugBank dataset [27] is used. To classify the drugs, we assigned drug subtypes (totaling 188 distinct subcategories) as target labels. For regression analysis, we used solubility AlogPS. The top 10 drug subcategories, extracted from the Food and Drug Administration (FDA) website[2], are presented in Table 1. The second dataset consists of a set of 16395 SMILES strings from the ChEMBL [2] dataset where we classify these sequences for 51 Standard Type and regression task for AlogP value. The top 10 Standard Types and their count, Maximin, and minimum string length are presented in Table 1.

Table 1. Drug subtypes (top 10) extracted from FDA website for **DrugBank dataset** and Standard type (top 10) extracted from **ChEMBL dataset**.

Drug Bank Dataset					ChEMBL Dataset				
Drug Subcategory		String Length Statistics					String Length Statistics		
	Count	Min.	Max.	Avg.	Standard Type	Count	Min.	Max.	Avg.
Others	6299	2	569	55.4448	IC50	4876	2	248	53.3169
Barbiturate	54	16	136	51.2407	Activity	2373	10	169	56.7821
Amide Local Anesthetic	53	9	149	39.1886	AC50	2201	7	234	50.4371
Non-Standardized Plant Allergenic Extract	30	10	255	66.8965	RBA	1421	23	155	53.9078
Sulfonylurea	17	22	148	59.7647	Ki	1390	2	248	53.8388
Corticosteroid	16	57	123	95.4375	EC50	1306	19	114	49.1256
Nonsteroidal Anti-inflammatory Drug	15	29	169	53.6000	Potency	766	4	248	42.6802
Nucleoside Metabolic Inhibitor	11	16	145	59.9090	Efficacy	749	28	107	54.2390
Nitroimidazole Antimicrobial	10	27	147	103.800	Inhibition	456	22	103	53.7478
Muscle Relaxant	10	9	82	49.8000	Emax	172	23	96	61.5698

5 Results and Discussion

This section reports the classification and regression results for baselines and the proposed kernel approach for the SMILES string analysis.

[2] https://www.fda.gov/.

Classification Results: Table 2 displays the classification results for the DrugBank dataset and ChEMBL dataset. For the DrugBank dataset results indicate that our proposed kernel-based approach outperforms other embedding methods and classifiers in terms of average accuracy, precision, recall, weighted F1 score, and ROC-AUC. The weighted k-mers method performs better than other methods in terms of Macro F1. These results indicate that our kernel approach, by projecting data into a higher-dimensional space, enhances the ability of underlying classifiers to differentiate between various types of SMILES strings. Although the classification results for the ChEMBL dataset show Morgan Fingerprint outperforms for accuracy, precision, recall, and weighted F1 score. But we can see for ROC-AUC our proposed method performs better.

Table 2. Classification results (of 5 runs) for different methods using different evaluation metrics on **DrugBank dataset** and **ChEMBL dataset**. The best values are shown in bold.

		Drug Bank Dataset							ChEMBL Dataset						
Embed.	Algo.	Acc. ↑	Prec. ↑	Recall ↑	F1 (Wt.) ↑	F1 (Mac.) ↑	ROC-AUC ↑	Train Time (Sec.) ↓	Acc. ↑	Prec. ↑	Recall ↑	F1 (Wt.) ↑	F1 (Mac.) ↑	ROC-AUC ↑	Train Time (Sec.) ↓
Morgan Fingerprint	SVM	0.8838	0.8577	0.8838	0.8696	0.0591	0.5383	17.6993	0.4427	0.4367	0.4427	0.4369	0.2474	0.6174	216.976
	NB	0.8969	0.8454	0.8969	0.8697	0.0275	0.5068	3.5027	0.2794	0.4307	0.2794	0.2679	0.2086	0.6765	10.2868
	MLP	0.8297	0.8493	0.8297	0.8390	0.0245	0.5239	17.4977	0.4300	0.4271	0.4300	0.4259	0.1800	0.5834	76.8557
	KNN	0.9129	0.8543	0.9129	0.8795	0.0374	0.5130	0.2560	0.4829	0.4812	0.4829	0.4748	0.2606	0.6229	4.1277
	RF	0.9109	0.8499	0.9109	0.8764	0.0258	0.5088	3.4253	0.4764	0.4686	0.4764	0.4695	0.2460	0.6103	55.3410
	LR	0.9131	0.8520	0.9131	0.8784	0.0378	0.5148	2.8179	**0.4934**	**0.4868**	**0.4934**	**0.4870**	0.2689	0.6215	10.2591
	DT	0.8569	0.8512	0.8569	0.8534	0.0333	0.5286	1.2680	0.4321	0.4298	0.4321	0.4251	0.2263	0.6074	4.3639
MACCS Fingerprint	SVM	0.8705	0.8539	0.8705	0.8613	0.0520	0.5441	3.1812	0.4676	0.4626	0.4676	0.4418	0.2594	0.6368	76.7914
	NB	0.2458	0.8473	0.2458	0.3698	0.0359	0.5224	0.5048	0.0842	0.3224	0.0842	0.0915	0.1396	0.7046	0.7156
	MLP	0.8659	0.8444	0.8659	0.8547	0.0220	0.5175	21.0636	0.4638	0.4415	0.4638	0.4399	0.1893	0.5949	25.2533
	KNN	0.9076	0.8447	0.9076	0.8741	0.0305	0.5107	0.0903	0.4816	0.4771	0.4816	0.4711	0.2352	0.6191	**0.3670**
	RF	0.9057	0.8499	0.9057	0.8749	0.0344	0.5149	1.1254	0.4819	0.4721	0.4819	0.4751	**0.2724**	0.6380	6.6008
	LR	0.9126	0.8331	0.9126	0.8710	0.0100	0.5000	3.2345	0.4426	0.4328	0.4426	0.4186	0.2251	0.6016	8.9684
	DT	0.8227	0.8522	0.8227	0.8363	0.0457	0.5436	**0.1100**	0.4444	0.4408	0.4444	0.4368	0.2308	0.6254	0.2999
k-mers	SVM	0.8190	0.8514	0.8190	0.8341	0.0413	0.5487	11640.03	0.4271	0.4164	0.4271	0.4154	0.2265	0.6235	764.380
	NB	0.7325	0.8425	0.7325	0.7816	0.0247	0.5149	2348.88	0.0799	0.2705	0.0799	0.0696	0.1095	0.6881	0.9319
	MLP	0.8397	0.8465	0.8397	0.8426	0.0270	0.5311	7092.26	0.3821	0.3603	0.3821	0.3651	0.1176	0.5545	78.6732
	KNN	0.9101	0.8480	0.9101	0.8766	0.0429	0.5167	68.50	0.4334	0.4284	0.4334	0.4216	0.1657	0.5913	1.6151
	RF	0.4461	0.4362	0.4461	0.4351	0.1938	0.5874	14.1722	0.4819	0.4721	0.4819	0.4751	**0.2724**	0.6380	6.6008
	LR	0.8885	0.8423	0.8885	0.8642	0.0461	0.5286	1995.11	0.4211	0.4057	0.4211	0.4065	0.2015	0.5962	87.3623
	DT	0.8429	0.8490	0.8429	0.8455	0.0397	0.5361	211.38	0.3883	0.3837	0.3883	0.3813	0.1690	0.5809	1.5972
Weighted k-mers	SVM	0.8219	0.8355	0.8219	0.8368	0.0451	0.5490	9926.76	0.4698	0.4629	0.4698	0.4604	0.2658	0.6409	405.782
	NB	0.7490	0.8475	0.7490	0.7931	0.0360	0.5221	2564.96	0.1929	0.2973	0.1929	0.2008	0.1940	0.6925	1.3470
	MLP	0.8288	0.8511	0.8288	0.8392	0.0270	0.5345	7306.79	0.4397	0.4224	0.4397	0.4259	0.1587	0.5835	84.402
	KNN	0.9122	0.8473	0.9122	0.8728	0.0307	0.5091	53.06	0.4626	0.4619	0.4626	0.4536	0.2096	0.6098	1.597
	RF	0.9135	0.8455	0.9135	0.8758	0.0245	0.5067	619.65	0.4340	0.4313	0.4340	0.4250	0.1987	0.5867	29.458
	LR	0.8928	0.8492	0.8928	0.8697	**0.0595**	0.5293	1788.37	0.4786	0.4670	0.4786	0.4674	0.2566	0.6260	103.276
	DT	0.8420	0.8518	0.8420	0.8461	0.0445	0.5347	147.47	0.3666	0.3648	0.3666	0.3605	0.1541	0.5738	8.838
SMILES Kernel	SVM	0.8430	0.8554	0.843	0.8478	0.0519	0.5375	32.3892	0.4436	0.4405	0.4436	0.4398	0.2474	0.6257	535.8723
	NB	0.6256	**0.8624**	0.6256	0.7209	0.0755	0.5412	4.1092	0.2246	0.3327	0.2246	0.2303	0.2290	**0.7053**	4.8780
	MLP	0.8222	0.8437	0.8222	0.8326	0.0204	0.5078	34.097	0.4241	0.4178	0.4241	0.4188	0.1616	0.5751	54.1374
	KNN	0.9116	0.8501	0.9116	0.8783	0.0442	0.5147	0.5929	0.4899	0.4861	0.4899	0.4808	0.2324	0.6104	1.8517
	RF	0.9145	0.8580	0.9145	**0.8801**	0.0324	0.5118	91.06	0.4671	0.4597	0.4671	0.4597	0.2331	0.6067	103.7672
	LR	**0.915**	0.8372	**0.915**	0.8744	0.0112	0.5001	56.4993	0.2971	0.0882	0.2971	0.1361	0.0108	0.5000	31.9301
	DT	0.828	0.8499	0.828	0.8381	0.043	**0.5733**	59.5459	0.4110	0.4099	0.4110	0.4048	0.1985	0.5968	22.0352

Regression Results: Table 3 presents the regression results for the DrugBank dataset, which indicate that the random forest regression model with MACCS fingerprint outperforms all other embedding methods and regression models. Although our proposed kernel-based approach did not perform comparitively, it is still able to achieve results comparable to those obtained using the MACCS fingerprint in combination with random forest regression. Table 4 presents the regression results for the ChEMBL dataset. We can see Linear regression and Ridge regression models with weighted k-mer for our proposed embeddings.

Table 3. Regression results for different models and evaluation metrics on **DrugBank dataset**. The best values are shown in bold.

Embedding	Algo.	MAE ↓	MSE ↓	RMSE ↓	R^2 ↑	EVS ↑
Morgan Fingerprint	Linear Regression	63.2345	11601.2046	107.7088	0.3139	0.3143
	Ridge Regression	62.6110	11529.2733	107.3744	0.3182	0.3185
	Lasso Regression	53.4116	11043.7095	105.0890	0.3469	0.3474
	Random Forest Regression	24.0881	7722.9372	87.8802	0.5433	0.5439
	Gradient Boosting Regression	32.4982	8853.8418	94.0948	0.4764	0.4768
2.1cmMACCS Fingerprint	Linear Regression	55.7719	11202.9967	105.8442	0.3375	0.3378
	Ridge Regression	55.5289	11167.1285	105.6746	0.3396	0.3399
	Lasso Regression	54.1349	11189.4825	105.7803	0.3383	0.3385
	Random Forest Regression	**17.8092**	**3711.9790**	**60.9260**	**0.7804**	**0.7809**
	Gradient Boosting Regression	31.4769	7308.5600	85.4901	0.5678	0.5678
k-mers	Linear Regression	8.3616e+10	4.6111e+23	6.7905e+11	−2.72674e+19	−2.72670e+19
	Ridge Regression	59.1402	12955.0398	113.8202	0.2339	0.2339
	Lasso Regression	51.7842	12608.1103	112.2858	0.2544	0.2545
	Random Forest Regression	23.2473	6073.5836	77.9331	0.6408	0.6420
	Gradient Boosting Regression	32.3582	8709.4397	93.3243	0.4849	0.4855
Weighted k-mers	Linear Regression	1.3608e+11	1.6509e+24	1.2848e+12	−9.7624e+19	−9.7527e+19
	Ridge Regression	62.8535	13187.9852	114.8389	0.2201	0.2202
	Lasso Regression	55.5155	12241.4725	110.6411	0.2761	0.2762
	Random Forest Regression	24.0294	6224.7174	78.8968	0.6319	0.6330
	Gradient Boosting Regression	33.0856	9066.1662	95.2164	0.4638	0.4644
SMILES kernel	Linear Regression	55.4431	10084.3079	100.4206	0.40368	0.40418
	Ridge Regression	50.5369	16914.8699	130.0571	−0.00023	0.000006
	Lasso Regression	50.5372	16914.9748	130.0575	−0.00023	0.0
	Random Forest Regression	23.0957	5056.7441	71.1107	0.7009	0.7023
	Gradient Boosting Regression	25.4830	5642.1382	75.1141	0.6663	0.6664

Inter-class Embedding Interaction: We use heat maps to assess our kernel's ability to distinguish classes by averaging similarity values and computing pairwise cosine similarity between class embeddings. The heat map is further normalized between [0–1] to the identity pattern. The heatmaps for the baseline (i.e., Morgan Fingerprint) and its comparison with the proposed SMILES kernel-based embeddings are reported in Fig. 2. We can observe that in the case of the Morgan fingerprint, the embeddings for different labels are similar. This shows that it is difficult to distinguish between classes due to high pairwise similarities among their vectors. The proposed SMILES kernel-based embeddings show clear pairwise similarity within the same class and distinct separation between different classes, demonstrating their effectiveness in distinguishing class relationships. We define Inter-Class Embedding Interaction as the difference in kernel similarity scores between embeddings:

$$\Delta K = K_{\text{Gaussian}}\ (\mathbf{z}_1, \mathbf{z}_2) - K_{\text{SMILES}}\ (\mathbf{z}_1, \mathbf{z}_2) \tag{5}$$

where, $\Delta K > 0$, means the Gaussian kernel overestimates similarity, failing to capture class-specific variations. $\Delta K < 0$, means a better separates classes. $\mathbf{z}_1, \mathbf{z}_2 \in \mathbb{R}^{d'}$ is low-dimensional projections of $\mathbf{x}_1, \mathbf{x}_2$, where $d' = 100$.

The strings belonging to different classes (i.e., drug subcategories) are shown in Fig. 3, where, using morgan fingerprint-based embeddings (a benchmark

Table 4. Regression results for **ChEMBL dataset** for different models and evaluation metrics. The best values are shown in bold.

Embedding	Algo.	MAE ↓	MSE ↓	RMSE ↓	R^2 ↑	EVS ↑
Morgan Fingerprint	Linear Regression	0.4782	0.5745	0.7580	0.8793	0.8794
	Ridge Regression	0.4701	0.5583	0.7472	0.8827	0.8828
	Lasso Regression	1.2661	3.0094	1.7347	0.3679	0.3681
	Random Forest Regression	0.2927	0.4280	0.6542	0.9101	0.9103
	Gradient Boosting Regression	0.7672	1.1591	1.0766	0.7565	0.7567
MACCS Fingerprint	Linear Regression	2.9302e+06	2.8153e+16	1.6779e+08	−5.9133e+15	−5.9115e+15
	Ridge Regression	0.8614	1.4766	1.2152	0.6898	0.6901
	Lasso Regression	1.1829	2.6943	1.6414	0.4341	0.4343
	Random Forest Regression	0.2939	0.4034	0.6351	0.9153	0.9153
	Gradient Boosting Regression	0.7530	1.1529	1.0737	0.7578	0.7581
k-mers	Linear Regression	0.4294	0.3504	0.5920	0.9264	0.9264
	Ridge Regression	0.4292	0.3501	0.5917	0.9265	0.9265
	Lasso Regression	0.5544	0.5588	0.7475	0.8826	0.8826
	Random Forest Regression	0.2133	0.2067	0.4546	0.9566	0.9566
	Gradient Boosting Regression	0.4287	0.3730	0.6108	0.9217	0.9217
Weighted k-mers	Linear Regression	**0.2862**	**0.1521**	**0.3899**	**0.9681**	**0.9681**
	Ridge Regression	**0.2862**	0.1520	**0.3899**	**0.9681**	**0.9681**
	Lasso Regression	0.7781	1.0640	1.0315	0.7765	0.7765
	Random Forest Regression	0.3257	0.4030	0.6348	0.9153	0.9154
	Gradient Boosting Regression	0.6523	0.8361	0.9144	0.8244	0.8244
SMILES kernel	Linear Regression	0.6138	0.9016	0.9495	0.8106	0.8110
	Ridge Regression	1.6340	4.7613	2.1820	−0.0001	0.0000
	Lasso Regression	1.6340	4.7613	2.1820	−0.0001	0.0000
	Random Forest Regression	0.3335	0.5013	0.7080	0.8947	0.8948
	Gradient Boosting Regression	1.0104	1.9996	1.4141	0.5800	0.5803

embedding method from the literature) for two random SMILES string samples (i.e., SMILES strings belonging to classes "Anticoagulant" and "Calcineurin Inhibitor Immunosuppressant") as input, we compute kernel value between the embeddings using the typical Gaussian kernel and the proposed SMILES kernel. To get an effective representation, we used kernel PCA and reduced the data dimensionality of the embeddings to 100 (getting top principal components) before computing the kernel value. Our proposed SMILES kernel, which gave us a smaller value of 0.17, can capture differences between classes more effectively compared to the typical Gaussian kernel, which gives us a larger kernel value of 0.21 (smaller kernel value is better). The visualizations show sharp spikes, especially in the lower dimensions (closer to 0 on the x-axis). This indicates that there is significant variance in those dimensions, and they are crucial for classification and SMILES kernel can identify and prioritize the variance captured in these initial dimensions better than the Gaussian kernel. The later dimensions (closer to 100) have relatively low variance. Better results of our proposed kernel indicate that it can handle noise effectively. The Gaussian kernel assumes a certain shape and structure to the data. It might not always be the best fit for all datasets. Our proposed kernel seems to be more attuned to the characteristics and nuances of the data, leading to better performance.

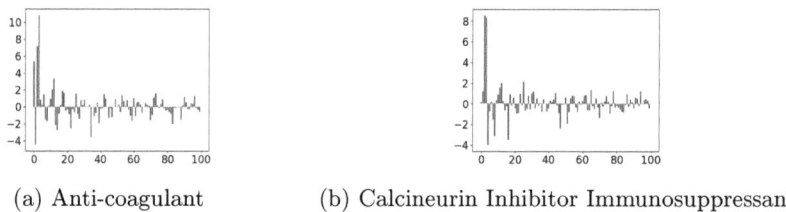

(a) Morgan Fingerprint (b) Smiles Kernel

Fig. 2. Heatmap for classes in **DrugBank dataset** for different drug subtypes. The figure is best seen in color. (Color figure online)

(a) Anti-coagulant (b) Calcineurin Inhibitor Immunosuppressant

Fig. 3. Comparing two pairs of classes. (a) and (b) belong to different classes. The Gaussian kernel for (a) and (b) is 0.21 while for the proposed method is 0.17 (a smaller value is better) on **DrugBank dataset**. Bar plot where we used kernel PCA with k = 100 (x-axis) and respective values (y-axis).

Discussion: For the DrugBank dataset, our kernel-based method slightly outperformed other embeddings in accuracy, precision, recall, F1 score, and ROC-AUC, while weighted k-mers excelled in Macro F1 score. For solubility prediction, the MACCS fingerprint with random forest regression performed best in RMSE, MAE, and MSE. In the ChEMBL dataset, our kernel-based method showed superior ROC-AUC but comparable performance in other metrics. For solubility prediction, weighted k-mers with linear/ridge regression achieved the best results in RMSE, MAE, and MSE. Although our methods did not outperform baselines, they provided comparable results and improved segregation through heatmaps, kernel values, and other visualizations.

6 Conclusion

We proposed a kernel-based approach for encoding and analyzing molecular structures represented as SMILES strings. We evaluated the proposed approach using the SMILES string dataset for molecular property prediction and activity classification. The proposed kernel-based approach represents a promising direction for the analysis and design of molecular structures using kernel methods. Further research can explore the use of other types of kernels and the application of the proposed approach to other areas of chemistry and material science. We believe that our proposed approach will contribute to the development of new drugs and materials with desirable properties, leading to significant advancements in healthcare and technology.

References

1. Boukerche, A., Wang, J.: Machine learning-based traffic prediction models for intelligent transportation systems. Comput. Netw. **181**, 107530 (2020)
2. ChEMBL Website. https://www.ebi.ac.uk/chembl/
3. Chen, H., Engkvist, O., et al.: The rise of deep learning in drug discovery. Drug Discov. Today **23**(6), 1241–1250 (2018)
4. Chen, H., Kogej, T., Engkvist, O.: Cheminformatics in drug discovery, an industrial perspective. Mol. Inf. **37**(9–10), 1800041 (2018)
5. Dührkop, K., Shen, H., Meusel, M., Rousu, J., Böcker, S.: Searching molecular structure databases with tandem mass spectra using CSI: Fingerid. Proc. Natl. Acad. Sci. **112**(41), 12580–12585 (2015)
6. Durant, J.L., et al.: Reoptimization of mdl keys for use in drug discovery. J. Chem. Inf. Comput. Sci. **42**(6), 1273–1280 (2002)
7. Fabregat, R., et al.: Metric learning for kernel ridge regression: assessment of molecular similarity. Mach. Learn. Sci. Technol. **3**(3), 035015 (2022)
8. Francoeur, P.G., Koes, D.R.: Soltrannet-a machine learning tool for fast aqueous solubility prediction. J. Chem. Inf. Model. **61**(6), 2530–2536 (2021)
9. Fu, G.H., Cao, D.S., et al.: Combination of kernel PCA and linear support vector machine for modeling a nonlinear relationship between bioactivity and molecular descriptors. J. Chemom. **25**(2), 92–99 (2011)
10. Glorot, X., Bordes, A., Bengio, Y.: Domain adaptation for large-scale sentiment classification: a deep learning approach. In: ICML pp. 513–520 (2011)
11. Kang, J.L., Chiu, C.T., Huang, J.S., Wong, D.S.H.: A surrogate model of sigma profile and COSMOSAC activity coefficient predictions of using transformer with smiles input. Digit. Chem. Eng. **2**, 100016 (2022)
12. Keys, M.S.: MDL information systems Inc. San Leandro, CA (2005)
13. Knight, P.A.: The Sinkhorn-Knopp algorithm: convergence and applications. SIAM J. Matrix Anal. App **30**(1), 261–275 (2008)
14. Krenn, M., et al.: Self-referencing embedded strings (selfies): A 100% robust molecular string representation. Mach. Learn. Sci. Technol. **1**(4), 045024 (2020)
15. Li, Q., Wang, Y., Bryant, S.H.: A novel method for mining highly imbalanced high-throughput screening data in PubCHEM. Bioinformatics **25**(24), 3310–3316 (2009)
16. Li, X., Chen, H.: Recommendation as link prediction in bipartite graphs: a graph kernel-based machine learning approach. DSS **54**(2), 880–890 (2013)
17. Mialon, G., et al.: A trainable optimal transport embedding for feature aggregation and its relationship to attention. arXiv preprint arXiv:2006.12065 (2020)
18. Nakajima, M., Nemoto, T.: Machine learning enabling prediction of the bond dissociation enthalpy of hypervalent iodine from smiles. Sci. Rep. **11**(1), 20207 (2021)
19. Öztürk, H., et al.: Exploring chemical space using natural language processing methodologies for drug discovery. Drug Discov. Today **25**(4), 689–705 (2020)
20. Probst, D., Reymond, J.-L.: A probabilistic molecular fingerprint for big data settings. J. Cheminf. **10**(1), 1–12 (2018). https://doi.org/10.1186/s13321-018-0321-8

21. Qiu, J., Wu, Q., et al.: A survey of machine learning for big data processing. EURASIP J. Adv. Sig. Process. **2016**, 1–16 (2016)
22. Rensi, S., Altman, R.B.: Flexible analog search with kernel PCA embedded molecule vectors. Comput. Struct. Biotechnol. J. **15**, 320–327 (2017)
23. Rensi, S.E., Altman, R.B.: Shallow representation learning via kernel PCA improves QSAR modelability. J. Chem. Inf. Model. **57**(8), 1859–1867 (2017)
24. Rupp, M., et al.: Fast and accurate modeling of molecular atomization energies with machine learning. Phys. Rev. Lett. **108**(5), 058301 (2012)
25. Schwaller, P., et al.: Machine intelligence for chemical reaction space. Wiley Interdiscip. Rev. Comput. Mol. Sci. **12**(5), e1604 (2022)
26. Sellwood, M.A., Ahmed, M., et al.: Artificial intelligence in drug discovery (2018)
27. Shamay, Y., Shah, J., Işık, M., et al.: Quantitative self-assembly prediction yields targeted nanomedicines. Nat. Mater. **17**(4), 361–368 (2018)
28. Shen, C., Ding, J., et al.: From machine learning to deep learning: Advances in scoring functions for protein-ligand docking. Wiley Interdiscip. Rev. Comput. Mol. Sci. **10**(1), e1429 (2020)
29. Stuke, A., et al.: Chemical diversity in molecular orbital energy predictions with kernel ridge regression. J. Chem. Phys. **150**(20), 204121 (2019)
30. Tkachev, V., Sorokin, M., et al.: Floating-window projective separator (flowps): a data trimming tool for support vector machines (SVM) to improve robustness of the classifier. Front. Genet. **9**, 717 (2019)
31. Wigh, D.S., Goodman, J.M., Lapkin, A.A.: A review of molecular representation in the age of machine learning. Wiley Interdiscip. Rev. Comput. Mol. Sci. **12**(5), e1603 (2022)

Enhancing Privacy Preservation and Reducing Analysis Time with Federated Transfer Learning in Digital Twins-Based Computed Tomography Scan Analysis

Avais Jan, Qasim Zia, and Murray Patterson(✉)

Georgia State University, Atlanta, GA 30303, USA
{ajan3,qzia1}@student.gsu.edu, mpatterson30@gsu.edu

Abstract. The application of Digital Twin (DT) technology and Federated Learning (FL) has great potential to change the field of biomedical image analysis, particularly for Computed Tomography (CT) scans. This paper presents Federated Transfer Learning (FTL) as a new Digital Twin-based CT scan analysis paradigm. FTL uses pre-trained models and knowledge transfer between peer nodes to solve problems such as data privacy, limited computing resources, and data heterogeneity. The proposed framework allows real-time collaboration between cloud servers and Digital Twin-enabled CT scanners while protecting patient identity.

We apply the FTL method to a heterogeneous CT scan dataset and assess model performance using convergence time, model accuracy, precision, recall, F1 score, and confusion matrix. It has been shown to perform better than conventional FL and Clustered Federated Learning (CFL) methods with better precision, accuracy, recall, and F1-score. The technique is beneficial in settings where the data is not independently and identically distributed (non-IID), and it offers reliable, efficient, and secure solutions for medical diagnosis. These findings highlight the possibility of using FTL to improve decision-making in digital twin-based CT scan analysis, secure and efficient medical image analysis, promote privacy, and open new possibilities for applying precision medicine and smart healthcare systems.

Keywords: Digital Twin · Federated Learning · Federated Transfer Learning · Computed Tomography (CT) Scan · Privacy Preservation

1 Introduction

Medical imaging is one of the most essential parts of the modern healthcare system and helps doctors make the right decisions regarding the treatment of the patient and his or her further therapy. Among all modalities, computed tomography (CT) scans produce cross-sectional images that help diagnose diseases,

M. Alser et al. (Eds.): ICCABS 2025, LNCS 15599, pp. 277–289, 2026.
https://doi.org/10.1007/978-3-032-02489-3_21

fractures, and infections [1]. However, the centralized nature of traditional CT scan interpretation poses several challenges, such as data privacy, computational cost, and lack of collaboration between different institutions [2].

Federated Learning (FL) is a form of machine learning where many clients for example hospitals, medical institutions, or edge devices can train a common model without exchanging their data locally. Federated learning has emerged as an effective technique to overcome challenges such as these through the training of multiple medical institutions in machine learning models without exchanging patient raw data [3]. This decentralized framework improves privacy by keeping data in the client's digital profile and yet improves collective intelligence [4]. However, conventional FL approaches are sensitive to domain shifts between different institutions, resulting in performance deterioration. To this end, Federated Transfer Learning (FTL) extends the FL framework with transfer learning techniques to enable knowledge transfer from pre-trained models and then fine-tunes them for the given local data distributions [5].

A Digital Twin is a real-time virtual copy of a system, process, or entity that continuously receives data from its physical counterpart. Digital twins (DT) are becoming popular for monitoring and proactive care in healthcare. The digital twin of a patient's CT scan can model the disease evolution, support individualized care, and shape decision-making [6]. Thus, the application of FTL to DT-based CT scan analysis makes it possible to improve the generalization of models, respect data privacy, and save computing resources.

This research designs an FTL-based digital twin framework for CT scan analysis to balance privacy, time, and accuracy. The main contributions of this work are:

1. Proposing an FTL framework for CT scan analysis in DT environments for private and efficient model training.
2. Optimization techniques to improve analysis speed without sacrificing diagnosis level.
3. Comparisons between FTL, Clustered FL, and conventional FL highlight FTL's accuracy and performance benefits.

This paper is structured as follows: related works are described in Sect. 2. We suggest an architecture of an FTL-based DT-CT scan model in Sect. 3. The comprehensive experiments are then carried out and assessed in Sect. 4 to determine the effectiveness and efficiency of our suggested architecture. We finally arrive at our conclusions in Sect. 5.

2 Related Works

Federated Learning(FL) is a well-known model for privacy-preserving decentralized training without the exchange of sensitive data. Various studies have been conducted on FL in medical imaging, especially for analyzing computed tomography (CT) scans. For example, FL's feasibility was shown by Sheller et al. [7]

in brain tumor segmentation while preserving data privacy to the level of central training. Using a priority algorithm, Zia et al. [8] improve response time. Likewise, Dou et al. [10] investigated FL for the detection of lung disease on CT scans and pointed out its benefits in data security and collaborative learning in hospitals. Zia et al. [9] discussed optimized decision-making with the help of EfficientNet in DT-VANET.

However, conventional FL experiences challenges in domain change when deployed in heterogeneous medical datasets. To resolve this, Federated Transfer Learning (FTL) has become an effective approach to expand on pre-trained models and then fine-tune them for target domains. Yin et al. [11] introduced an FTL framework for cross-institutional medical image analysis, enhancing the generalization capacity of the model on unseen data sets. Zia et al. [12] improve communication using a priority algorithm in a network based on Digital Twin (DT). However, Irfan et al. [13] have integrated FTL with an attention mechanism to improve feature extraction in chest CT scan classification, thus reducing misclassification rates. Zia et al. [14] have discussed the use of FTL in DT-VANET.

However, several challenges in such FTL approaches related to computational efficiency and convergence time are inherent. These challenges arise from the high communication overhead during model aggregation, the heterogeneous data distributions across clients that slow convergence, and the varying computational resources at different nodes. Zhang et al. [15] had also proposed optimized model aggregation techniques to reduce communication overhead in FL, but their application to real-time medical imaging has been limited. Zia et al. [16] discuss different protocols necessary for network communication. Furthermore, recent work in digital twin (DT)–based healthcare systems, including those of Saha et al. [17], has explored DT models for personalized patient monitoring. Therefore, integrating FTL into DT frameworks offers a promising direction to enhance privacy and computational efficiency in medical imaging. A DT provides a continuous virtual model of the physical imaging system and continuous real-time data synchronization and predictive analytics. This integration leads to more adaptive model updates and more efficient resource allocation that can help alleviate the computational burdens and convergence delays typical of traditional FL systems. Consequently, by taking advantage of the real-time and personalization of DT monitoring functionality, the overall framework can improve privacy and reduce the analysis time in CT scan analysis.

Based on these previous works, our research aims to optimize real-time decision-making by combining DT with FTL in the context of CT scan analysis. Our approach also addresses privacy issues and reduces the analysis time, to help overcome the accuracy-feasibility gap in federated medical imaging.

3 Proposed Approach

This study presents a Federated Transfer Learning (FTL) framework to support the analysis of Digital Twin-based Computed Tomography (DT-Scan). The

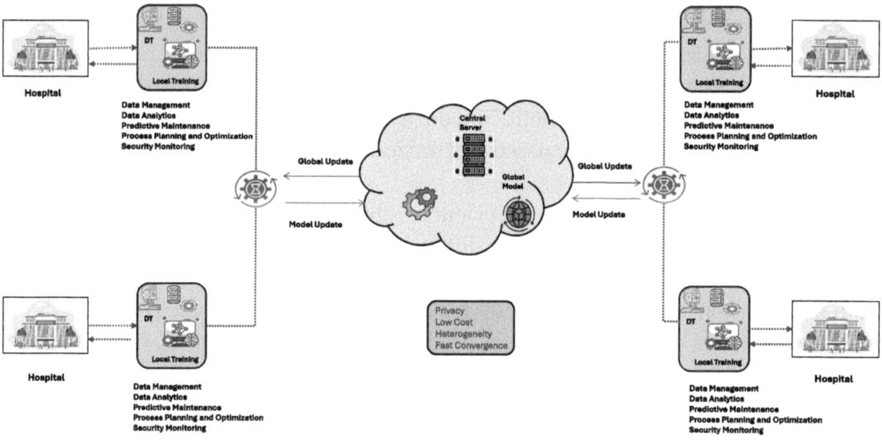

Fig. 1. High-level architecture of Federated Transfer Learning for digital twin-based Computed Tomography scan.

proposed model is intended to preserve privacy while significantly reducing the analysis time for medical imaging applications. In contrast to conventional centralized approaches, FTL enables several medical healthcare centers to jointly train deep learning models without exchanging raw patient data to comply with data privacy regulations such as HIPAA and GDPR.

3.1 FTL Method Architecture

This section provides a high-level architecture for FTL in DT-Scan, as seen in Fig. 1. The proposed DT-based FTL method consists of three main components:

Medical Institutions (Clients). Every participating hospital and diagnostic center maintains its local CT image data set. A local model such as DenseNet-121 is deployed and trained on local patient data. Raw images are not shared, as gradients representing local model updates are exchanged between institutions.

Federated Server (Global Aggregator). The server gathers model updates from various institutions and then applies Federated Averaging (FedAvg). The aggregator uses transfer learning methods to join data sources while accommodating differences in CT scan features. The globally updated model is returned to local nodes as a reference for additional refinement.

Digital Twin for CT-Scan Analysis. A virtual digital twin monitors the progression of the disease by assigning each patient's CT scan to this model. The FTL model provides real-time analysis with optimized inference speed for diagnosis. Digital Twin models are improved through continuous federated updates, which leads to growing accuracy over time.

This architecture has a few advantages. First, since raw medical images are not exchanged, we have a reduced risk of data breaches. Second, because of its transfer learning capabilities, DenseNet-121 accelerates model convergence during analysis and reduces analysis time. Furthermore, improved model generalization improves diagnostic accuracy by aggregating knowledge from various sources. Lastly, scalability and adaptability can extend the FTL framework to multiple healthcare institutions with varying data distributions.

3.2 Cloud Server Federated Transfer Learning

This algorithm represents a federated transfer learning process in which the central server updates its global model parameters W_{global} by collaborating with multiple hospitals (CT scanners in our case). First, the model parameters of the federated transfer learning process are already on the central cloud server. The group of hospitals $\mathcal{H}_n$ that have not yet taken part in Federated Transfer Learning is initialized. The process repeats several update rounds. The global model parameters W_{global} are sent to a Digital Twin of a selected hospital in each round. Using the parameters received, denoted by $l = \text{LOCAL-UPDATE}(W_{\text{global}^-})$, the hospital then performs a local update. A weighting function F_{weight} (W_{global^+}) is then used to update the global model which is explained further in detail next section. After the update, the selected hospital is removed from the list of participating hospitals since every hospital should participate in the training process. This is done until all updates are made and the optimized global model parameters W_{global} are acquired.

Algorithm 1. Weighted Cloud Server Cycling Model Update

The central server has existing model parameters W_{global};
Initialize $\mathcal{H}_n$ the set of participating hospitals;
for each update $j = 0, 1, 2, \ldots, j-1$ **do**
  $W_{\text{global}-} = W_{\text{global}}$;
  Select hospital i from $\mathcal{H}_n$;
  Send $W_{\text{global}-}$ to the Digital Twin of hospital i;
  $l = \text{LOCAL-UPDATE}(W_{\text{global}-})$;
  Calculate F_{weight} $(W_{\text{global}+})$;
  Update W_{global} ;
  Eliminate i from the remaining hospitals H_n;
end for
Output the final global model parameters W_{global};

3.3 Hospital Local Update Federated Transfer Learning

This algorithm describes the hospital's local update process in Federated Transfer Learning. This algorithm aims to fine-tune the parameters of the global model W_{global^-} at the hospital level before sending them back to the central server. The process is divided into two main parts and briefed as follows: the operations carried out by the hospital server and its digital twin(CT scanner in our case). The hospital server receives the global model parameters $W_{\text{global}-}$ from the central server and then forwards it to its local digital twin system. The digital twin (of CT scanner in our case) then fine-tunes the model using its local dataset D_i and produces the updated model parameters $W_{\text{global}+}$. The fine-tuned parameters are then forwarded to the hospital server, which sends them to the central server. This process helps to improve the global model with hospital-level information before the model makes its contribution to the federated transfer learning system.

Algorithm 2. Hospital Local Update Federated Transfer Learning

Input : The model parameters of the hospital that need to be fine-tuned $W_{\text{global}-}$;
Output : The fine-tuned model parameters $W_{\text{global}+}$;

procedure LOCAL-UPDATE
  (I) For the hospital server:
  Receive $W_{\text{global}-}$ from the central server;
  Send $W_{\text{global}-}$ to the local digital twin system;
  Receive $W_{\text{global}+}$ from the local digital twin system;
  Send $W_{\text{global}+}$ to the central server;
  (II) For the hospital's digital twin:
  Receive $W_{\text{global}-}$ from the hospital server;
  Fine-tune the model:

$$W_{\text{global}+} = \text{FINE-TUNE}(W_{\text{global}-}, D_i);$$

  Send $W_{\text{global}+}$ to the hospital server;
end procedure

4 Performance Evaluation

This section uses various performance evaluations to assess our proposed framework using Federated Transfer Learning.

4.1 Dataset

We used the Chest CT-Scan images[1] [18] dataset for our simulations. This is one of the most commonly used datasets for CT scan images. Three distinct forms of chest cancer (adenocarcinoma, large cell carcinoma, squamous cell carcinoma), as well as normal CT-Scan images, are included in this dataset, as mentioned in Fig. 2, in which the red circles indicate the affected regions. These CT scan images reveal differences in cancer types, imaging protocols, resolution, and contrast across different hospitals while still being useful to build a single, robust global model. Through these various cancer types appearing in CT scans, the visuals further highlight the problem of accurate tumor detection and classification. The features of the dataset are mentioned in the Table 1.

[1] https://www.kaggle.com/datasets/mohamedhanyyy/chest-ctscan-images?datasetId=839140.

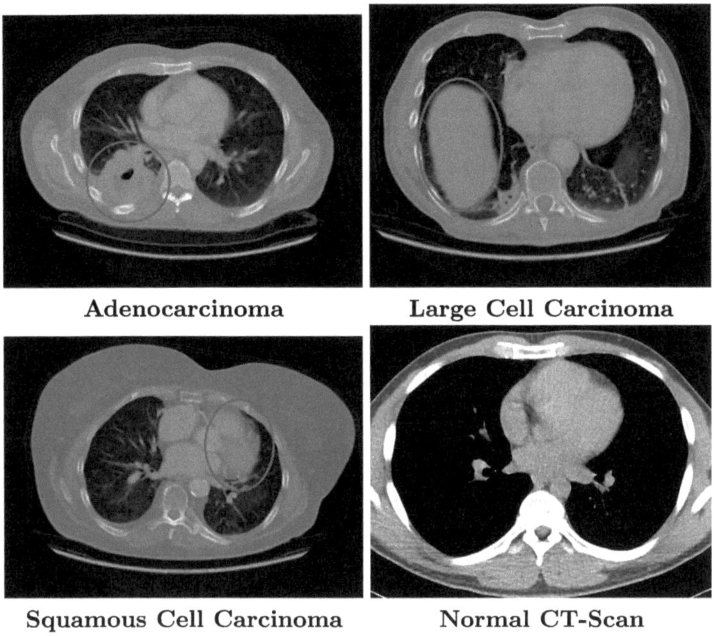

Fig. 2. Sample of computed tomography scan images from the dataset [18]

Table 1. Features of the Chest Computed Tomography Scan Dataset

Cancer Types	Origin	Grade/Severity	Symptoms
Adenocarcinoma	Outer regions of lungs	Varies	Persistent cough, hoarseness, weight loss, weakness
Large Cell Carcinoma	Anywhere in lungs	High (aggressive)	Rapid progression, large masses, necrosis
Squamous Cell Carcinoma	Central lung, near bronchi	Varies	Cough, airway obstruction, cavitation, weight loss

4.2 Baseline Studies

We will compare techniques such as FL, CFL, and FTL. We chose FL and CFL for comparison because they are closely related to FTL. We have used the DenseNet-121 model because it is pre-trained on the medical dataset CheXpert. So, there is no need to simulate data partition among hospitals. As our analysis is focused on how these three approaches perform and the convergence time when have data at the client (hospital) side. So, we are also not considering any latency issues caused during communication between the cloud and hospital server.

4.3 Performance Metrics

The efficiency of the suggested FTL framework for Digital Twins-based CT scan analysis will be assessed using various critical performance criteria.

Accuracy. Accuracy represents the proportion of correct classifications of CT scan images out of the total samples used in the model. It is computed as:

$$\text{Accuracy} = \frac{\text{TP} + \text{TN}}{\text{TP} + \text{TN} + \text{FP} + \text{FN}} \tag{1}$$

TN, TP, FN, and FP represent true negatives, true positives, false negatives, and false positives, respectively.

Precision. Precision calculates the percentage of actual positive instances correctly identified among all the cases predicted as positive. It is defined as:

$$\text{Precision} = \frac{\text{TP}}{\text{TP} + \text{FP}} \tag{2}$$

A higher precision represents fewer false positives.

Recall (Sensitivity). Recall measures how many actual positives were correctly classified. It is calculated as:

$$\text{Recall} = \frac{\text{TP}}{\text{TP} + \text{FN}} \tag{3}$$

The high recall value shows the model's ability to effectively identify positive cases.

F1-Score. We need a metric that can capture precision and recall to understand how well a model performs in classification comprehensively. This is where the F1 score comes in. It is calculated as:

$$\text{F1-Score} = \frac{2 \times \text{Precision} \times \text{Recall}}{\text{Precision} + \text{Recall}} \tag{4}$$

The model achieved a higher F1 score, providing optimal precision and recall trade-off.

$$\text{Overall Accuracy} = \frac{\sum \text{True Positives}}{\text{Total Samples}} = \frac{\sum_{i=1}^{n} \text{TP}_i}{\sum_{i=1}^{n} \text{Total Samples}_i} \tag{5}$$

where: True Positives$_i$ is the number of correct predictions of class i (diagonal elements of the confusion matrix),

Total Samples_i is the total amount of class i samples (sum of each row in the confusion matrix),

n is the number of classes.

The convergence time is known as the time when the FTL model converges to an optimal value during training. This metric allows us to compare the efficiency of the proposed approach with federated learning and clustered federated learning paradigms. The model is considered to have converged when any improvement in accuracy is less than a predetermined threshold and when the change in loss between consecutive rounds is less than a minimal threshold for several rounds in sequence.

The confusion matrix is also calculated based on TP, TN, FP, and FN. These metrics will allow the study to show that Federated Transfer Learning improves privacy, reduces analysis time, and still provides high classification performance in CT scan analysis.

4.4 Experimental and Performance Analysis

This section will discuss the performance of the proposed FTL approach in Digital Twins-based CT scan analysis. To this end, we design experiments to assess the time taken for convergence and the model's accuracy. We will compare our FTL-based method with traditional FL and CFL approaches.

FTL's performance is compared using various performance metrics that are very important for reducing analysis time and improving privacy, which includes test accuracy, recall, F1 score, and precision in Table 2. The overall accuracy of Federated Learning is 0.8000, clustered Federated Learning is 0.8476, and Federated Transfer Learning is 0.8730. The evaluation of these methods for different cancer types demonstrates robust performance, particularly for Adenocarcinoma and Squamous Cell Carcinoma, indicating that these models effectively distinguish between different cancer types. Notably, FTL exhibits the highest accuracy and F1-score for Adenocarcinoma and Squamous Cell Carcinoma, suggesting that transfer learning enhances the model's ability to generalize from pre-trained knowledge, leading to improved classification performance. This improvement is likely due to the leveraging of pre-trained models, which provide a strong foundation for learning, thereby reducing the convergence time and enhancing efficiency. The classification metrics are assessed using the confusion matrix as shown in Fig. 3.

However, the performance varies for Large Cell Carcinoma and Normal classes, where the precision and recall are relatively lower compared to the other classes. This discrepancy may be attributed to the smaller sample size for these classes, which can affect the model's ability to learn distinctive features. Despite this, CFL shows promise in improving the recall for Large Cell Carcinoma, indicating that clustering similar data points can help better capture the underlying patterns. Overall, the results demonstrate the potential of federated learning techniques in medical imaging, particularly in scenarios where data privacy is

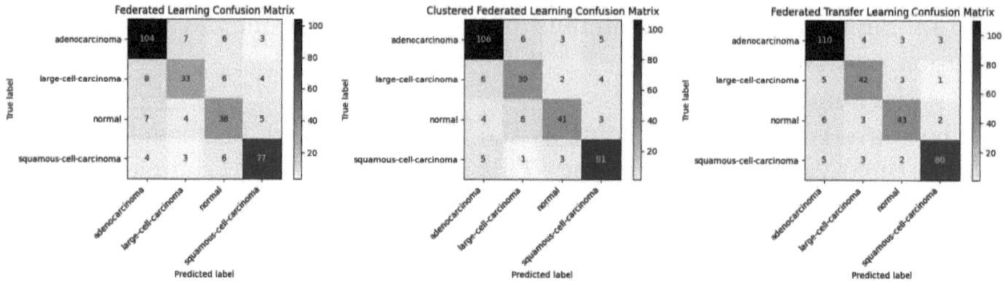

Fig. 3. Comparison of confusion matrices for Federated Learning, Clustered Federated Learning, and Federated Transfer Learning

paramount. The convergence behavior, as illustrated in the graph, further supports the efficiency of FTL, which outperforms standard FL and CFL in terms of training time, making it a viable approach for real-world applications.

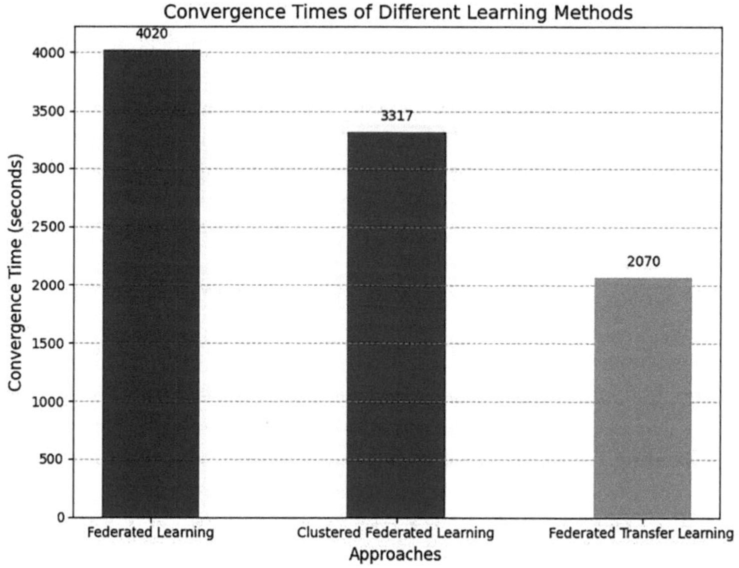

Fig. 4. Comparison of convergence time for Federated Learning, Centralized Federated Learning, and Federated Transfer Learning for classification of different cancer types from computed tomography scans

Table 2. Comparison of evaluation metrics for lung cancer type classification using Federated Learning, Clustered Federated Learning, and Federated Transfer Learning

Method	Classes	# of Images	Precision	Recall	F1-Score	Accuracy
FL	Adenocarcinoma	120	0.8455	0.8667	0.8560	0.8667
	Large Cell Carcinoma	51	0.7021	0.6471	0.6735	0.6471
	Normal	54	0.6786	0.7037	0.6909	0.7037
	Squamous Cell Carcinoma	90	0.8652	0.8556	0.8603	0.8556
CFL	Adenocarcinoma	120	0.8760	0.8833	0.8797	0.8833
	Large Cell Carcinoma	51	0.7500	0.7647	0.7573	0.7647
	Normal	54	0.8367	0.7593	0.7961	0.7593
	Squamous Cell Carcinoma	90	0.8710	0.9000	0.8852	0.9000
FTL	Adenocarcinoma	120	0.8730	0.9167	0.8943	0.9167
	Large Cell Carcinoma	51	0.8077	0.8235	0.8155	0.8235
	Normal	54	0.8431	0.7963	0.8190	0.7963
	Squamous Cell Carcinoma	90	0.9302	0.8889	0.9091	0.8889

5 Conclusion

This research paper aims to explore whether FTL can improve privacy protection and speed up the analysis in the application of Digital Twins-based Computed Tomography (CT) scan analysis. The results show that FTL can preserve patients' privacy by using decentralized learning on several local datasets to train the model and still achieve high accuracy in CT scan analysis. As a result, no sensitive data needs to be shared between institutions, thus fulfilling privacy policies like GDPR and HIPAA.

Additionally, the application of transfer learning can help the model learn quickly and with less computational time than the traditional federated and clustered federated learning processes. Digital Twin technology and system integration improve real-time simulation and monitoring of CT scan processes, thus improving accuracy and efficiency.

Therefore, in conclusion, FTL can be regarded as a potential improvement in protecting privacy in medical imaging and reducing the time taken to analyze the images. Future work should include improving the size model and checking its effectiveness in clinical settings to confirm its efficacy and robustness. This research used real-time simulation by updating the digital model with current data at regular intervals from the CT scanners. The digital twin of the CT scanner is updated in real time to replicate the actual process, thus enabling the monitoring of the system for proactive maintenance. Digital Twin technology integrated into our Federated Transfer Learning framework enables online updates of the model based on real-time data. The results of the experiments show that the performance metrics and the model convergence are improved, which are the immediate advantages of having a system that can simulate and observe the CT scan processes online.

Acknowledgements. The icons in the figures of this research chapter are taken from the Freepik website.

References

1. Hussain, S., et al.: Modern diagnostic imaging technique applications and risk factors in the medical field: a review. Biomed. Res. Int. **2022**, 5164970 (2022)
2. Shakor, M., Khaleel, M.: Recent advances in big medical image data analysis through deep learning and cloud computing. Electronics **13**, 4860 (2024)
3. Nguyen, D., Pham, Q., Pathirana, P., Ding, M., Seneviratne, A., Lin, Z., Dobre, O., Hwang, W.: Federated learning for smart healthcare: a survey. ACM Comput. Surv. (Csur). **55**, 1–37 (2022)
4. Cirillo, F., De Santis, M., Esposito, C.: Applications of Solid Platform and Federated Learning for Decentralized Health Data Management. In:Artificial Intelligence Techniques For Analysing Sensitive Data In Medical Cyber-Physical Systems: System Protection And Data Analysis, pp. 95-111 (2025)
5. Dai, S., Meng, F.: Addressing modern and practical challenges in machine learning: a survey of online federated and transfer learning. Appl. Intell. **53**, 11045–11072 (2023)
6. Subasi, A., Subasi, M.: Digital twins in healthcare and biomedicine. In: Artificial Intelligence, Big Data, Blockchain And 5G for the Digital Transformation of the Healthcare Industry, pp. 365–401 (2024)
7. Sheller, M., Reina, G., Edwards, B., Martin, J., Bakas, S.: Multi-institutional deep learning modeling without sharing patient data: a feasibility study on brain tumor segmentation. In: Brainlesion: Glioma, Multiple Sclerosis, Stroke And Traumatic Brain Injuries: 4th International Workshop, BrainLes 2018, Held In Conjunction With MICCAI 2018, Granada, Spain, September 16, 2018, Revised Selected Papers, Part I 4, pp. 92-104 (2019)
8. Zia, Q., Farooq, M., Abid, A.: Improving Response Time of Vehicular Ad hoc NETworks (VANET) (2016)
9. Zia, Q., Jan, A., Yang, D., Zhang, H., Li, Y.: Optimized real-time decision making with EfficientNet in digital twin-based vehicular networks. Electronics **14**, 1084 (2025)
10. Dou, Q., et al.: Federated deep learning for detecting COVID-19 lung abnormalities in CT: a privacy-preserving multinational validation study. NPJ Digital Med. **4**, 60 (2021)
11. Yin, Z., Wang, H., Chen, B., Zhang, X., Lin, X., Sun, H., Li, A., Zhou, C.: Federated semi-supervised representation augmentation with cross-institutional knowledge transfer for healthcare collaboration. Knowl.-Based Syst. **300**, 112208 (2024)
12. Zia, Q., Wang, C., Zhu, S., Li, Y.: Priority based inter-twin communication in vehicular digital twin networks, pp. 1–16. Int. J. Parallel Emergent Distribut. Syst. (2024)
13. Irfan, M., Malik, K., Muhammad, K.: Federated fusion learning with attention mechanism for multi-client medical image analysis. Inf. Fusion **108**, 102364 (2024)
14. Zia, Q., Zhu, S., Wang, H., Iqbal, Z., Li, Y.: Hierarchical Federated Transfer Learning in digital twin-based vehicular networks. High-Confidence Comput., 100303 (2025)

15. Zhang, W., Yang, D., Wu, W., Peng, H., Zhang, N., Zhang, H., Shen, X.: Optimizing federated learning in distributed industrial IoT: A multi-agent approach. IEEE J. Sel. Areas Commun. **39**, 3688–3703 (2021)
16. Zia, Q.: A survey of data-centric protocols for wireless sensor networks. Comput. Sci. Syst. Biol. OMICS Publishing Group. **8**, 127–131 (2015)
17. Sahal, R., Alsamhi, S., Brown, K.: Personal digital twin: a close look into the present and a step towards the future of personalised healthcare industry. Sensors. **22**, 5918 (2022)
18. Hany, M.: Chest CT-scan images dataset. Kaggle **13**, 2022 (2020)

Improved Graph-Based Antibody-Aware Epitope Prediction with Protein Language Model-Based Embeddings

Mansoor Ahmed[1], Sarwan Ali[2], Avais Jan[1], Imdad Ullah Khan[3], and Murray Patterson[1(✉)]

[1] Georgia State University, Atlanta, GA, USA
{mahmed76,ajan3}@student.gsu.edu, mpatterson30@gsu.edu
[2] Columbia University Irving Medical Center, New York, NY, USA
sa4559@columbia.edu
[3] Lahore University of Management Sciences, Lahore, Pakistan
imdad.khan@lums.edu.pk

Abstract. The accurate identification of B-cell epitopes is critical in antibody design, diagnostics, and immunotherapies. Many *in silico* approaches have recently been proposed to predict epitopes, but these approaches struggle primarily because of the variational and conformational nature of epitopes. However, deep learning-based approaches have recently shown great promise in achieving better performance at the epitope prediction task. In this paper, we employ a graph convolutional network (GCN) coupled with pre-trained protein language model (PLM)-based embeddings for epitope prediction on a benchmark antibody-specific epitope prediction (AsEP) dataset. We explore the use of different PLM-embedding methods on the epitope prediction task and show that the choice of PLM embeddings impacts the performance. Specifically, we find that antibody-specific PLMs such as AntiBERTy and general PLMs such as ProtTrans and ESM-2 for antigens provide improved epitope prediction performance with an AUCROC of 0.65, precision of 0.28, and recall of 0.46. The source code is available at: https://github.com/mansoor181/walle-pp.git

Keywords: Antibody Design · Protein Language Models · Epitope Prediction · Antigen-Antibody Interaction · Artificial Intelligence · Deep Learning

1 Introduction

Antibodies are large, Y-shaped proteins produced by B-cells that play a critical role in the immune system by identifying and neutralizing foreign substances such as toxins, bacteria, and viruses, collectively known as antigens. Antibodies are currently known to be the largest class of biotherapeutics, where five of the current top 10 blockbuster drugs are monoclonal antibodies [1]. Antibodies also have wide applications in diagnostics and biological research and are traditionally produced *in vivo* by immunizing animals [2,3]. However, these approaches are time-consuming, expensive, and laborious and make it imperative to develop *in silico* methods for antibody design by exploiting recent advances in artificial intelligence, which can potentially reduce the antibody design search space by producing cost-effective antibodies [4–6].

M. Alser et al. (Eds.): ICCABS 2025, LNCS 15599, pp. 290–302, 2026.
https://doi.org/10.1007/978-3-032-02489-3_22

A critical step in antibody design is antigen binding site or epitope prediction, which involves identifying the residues on the surface of an antigen that are recognized and bound by an antibody [7]. Accurate epitope prediction is essential for understanding antibody-antigen interactions and designing antibodies with high binding affinity [5]. However, this problem is inherently challenging due to several factors [8,9] . First, epitopes are often non-linear and conformationally diverse, making it difficult to capture their spatial relationships using sequence-based methods alone [6]. Second, a single antigen can have multiple epitopes, which further complicates the prediction task [1]. Finally, most of the current approaches rely on sequence-based data for epitope prediction, which fail to model the complex spatial arrangements of antigen binding sites [2].

Current approaches to epitope prediction can be broadly categorized into sequence-based and structure-based methods [10]. Sequence-based methods rely on amino acid sequences to predict epitopes, while structure-based methods leverage the 3D structure of antigens to identify potential binding sites [11]. Additionally, these methods can be classified as antibody-agnostic or antibody-aware [12]. Antibody-agnostic methods predict epitopes without considering the specific antibody, whereas antibody-aware methods incorporate information about the antibody's structure and binding interaction [3]. The latter approaches have proved to be more accurate as they enable the identification of specific epitopes on the antigen surface that are recognized by a particular antibody [1]. Graph-based approaches, which represent antigens and antibodies as graphs, have shown further promise due to their ability to model complex spatial relationships [7, 10]. However, the choice of embedding methods for encoding residues in these graphs is critical for achieving accurate predictions [13].

In this work, we explore the use of protein language models (PLMs) for embedding residues in a graph-based antibody-aware epitope prediction setup. Specifically, we evaluate the performance of pre-trained PLMs such as ESM-2 [14], AntiBERTy [15], ProtTrans [16], and ESM-IF [17], alongside classical embedding methods like one-hot encoding and BLOSUM62. We extend the WALLE [13] framework by integrating these embeddings into a graph convolutional network (GCN) to predict epitope residues on antigen surfaces. We pose the epitope prediction problem as a link prediction task: for each node in the antibody and antigen graphs, predict whether there is an edge between them or not. The nodes in the antigen graph participating in the binding are classified as epitopes. Our main contributions are as follows:

1. We provide a comprehensive comparison of different PLM-based embedding methods for the epitope prediction problem.
2. We propose the use of AntiBERTy for antibody residue embedding and ProtTrans for antigen embedding, along with the GCN model for improved epitope prediction.

The remainder of this paper is organized as follows. Section 2 discusses related literature on epitope prediction and graph-based methods. Section 3 presents the proposed approach with the problem formulation, data representation, and model architecture. Section 4 discusses the dataset and model training details. Section 5 discusses the experimental results and comparisons with baseline methods and outlines future research directions. Finally, Sect. 6 concludes the paper.

2 Related Work

Antibody-agnostic methods often rely on sequential, structural, or a combination of both features but lack the specificity required for antibody-specific applications [11]. For instance, *epitope1D* [3], *GraphBepi* [7], and *EpiGraph* [10] are prominent B-cell epitope prediction methods using graph-based representations that do not incorporate antibody-specific information. On the other hand, antibody-aware methods such as *EpiScan* [8], *PECAN* [18], and *EPMP* [11] explicitly incorporate information about the antibody's structure or sequence to predict epitopes.

Graph-based approaches have emerged as a powerful paradigm for epitope prediction, capturing the inherent spatial and sequential relationships in protein structures. Methods like *PECAN* [18], *PInet* [19], and [20] uses graph neural networks (GNNs) to predict protein-protein interaction interfaces, including antibody-antigen binding sites by employing attention mechanisms. *EPMP* [11] is a neural message-passing framework that uses asymmetrical architectures for paratope-epitope prediction, while *EpiPred* [12] combines conformational matching and antibody-antigen scoring to improve epitope prediction results. Moreover, [21] uses GNN to capture information on spatial neighbors of a target residue and attention-based didirectional long short-term memory (Att-BLSTM) networks to extract global information from the whole antigen sequence.

Some recent approaches have combined PLM-based embeddings and graph-based architectures for epitope prediction. *EpiGraph* [10] uses graph attention networks (GAT) with ESM-2 and ESM-IF embeddings to predict epitopes, achieving state-of-the-art performance. Similarly, [13] introduced *AsEP*, a benchmark dataset and a graph-based method (*WALLE*) that employs embeddings from ESM-2 and AntiBERTy in a GNN setup for antibody-specific epitope prediction. *GraphBepi* [7] is another graph-based model that leveraged ESM-2 embeddings for epitope prediction. *DeepProSite* [9] uses *ESMFold* [14] for protein structure prediction and employs embeddings from pre-trained ProtTrans to predict protein binding sites using graph transformers, while ESM-Bind [22] finetunes complex ESM-2 models ranging from 8M to 650 parameters on an annotated protein binding sites dataset for protein binding site prediction task.

3 Proposed Approach

This section presents the problem formulation of the epitope prediction task, the data representation strategies employed, and the overall architecture of the proposed method.

3.1 Problem Formulation

The interaction between antibodies and antigens can be framed as predicting links between the vertices of two disjoint undirected graphs: an antibody graph $G_A = (V_A, E_A)$ and an antigen graph $G_B = (V_B, E_B)$, *i.e.*, $V_A \cap V_B = \emptyset$. Here, V_A and V_B represent the sets of vertices (residues) for the antibody and antigen graphs, respectively, while E_A and E_B represent their sets of edges based on residue proximity. Each vertex is encoded into a vector using a function $h_x : V_x \to \mathbb{R}^{D_x}$ where $x \in \{A, B\}$ and h can be any encoding function such as one-hot encoding or pre-trained embeddings from a PLM. The adjacency matrix $E_x \in \{0,1\}^{|V_x| \times |V_x|}$ is derived from the distance matrix of the residues of antigen or antibody. Each entry e_{ij} indicates the proximity between residue i and residue j; $e_{ij} = 1$ if the Euclidean distance between any non-hydrogen atoms of residue i and residue j is less than 4.5Å, and $e_{ij} = 0$ otherwise. In order to make the computations faster, the antibody graph G_A is constructed only from the CDR residues of the antibody's heavy and light chains, while the antigen graph G_B is constructed from the surface residues of the antigen. We address the following two key tasks in this setup:

Epitope Node Prediction. This task involves identifying antigen residues that interact with the antibody. A residue $v \in V_B$ is labeled as an epitope (1) if it is within 4.5Å of any residue in V_A; otherwise, it is labeled as a non-epitope (0). Formally, the task is a binary node classification problem on the antigen graph G_B, conditioned on the antibody graph G_A. The classifier $f : V_B \to \{0,1\}$ is defined as:

$$f(v; G_B, G_A) = \begin{cases} 1 & \text{if } v \text{ is an epitope,} \\ 0 & \text{otherwise.} \end{cases}$$

Link Prediction. This task predicts interactions between antibody and antigen residues, forming a bipartite graph $K_{m,n} = (V_A, V_B, E)$, where $m = |V_A|$, $n = |V_B|$, and E represents potential inter-graph edges. An edge $e \in E$ is labeled as 1 if the corresponding residues are within 4.5Å of each other and 0 otherwise. The task is formulated as a binary edge classification problem, with the classifier $g : K_{m,n} \to \{0,1\}$ defined as:

$$g(v_a, v_b; K_{m,n}) = \begin{cases} 1 & \text{if } v_a \text{ and } v_b \text{ are in contact,} \\ 0 & \text{otherwise.} \end{cases}$$

The problem formulation is derived partially from [10, 13, 18].

3.2 Graph Representation

Each antibody-antigen complex is represented as a graph, where protein residues are modeled as vertices. Edges are drawn between residues if their non-hydrogen atoms are within 4.5Å of each other. We focus on surface residues, excluding buried residues with zero solvent-accessible surface area from the antigen graph, and non-CDR residues from the antibody graph. In our study, we employ various embedding methods to represent protein residues as continuous vectors. These methods include both classical techniques and state-of-the-art PLMs. Specifically, we use AntiBERTy[1], an antibody-specific transformer language model pre-trained on 558M natural antibody sequences, which produces embeddings of fixed size $L \times 512$ where L represents the length of the sequence. Additionally, we utilize ESM-2[2], a PLM that captures the evolutionary

[1] https://github.com/jeffreyruffolo/AntiBERTy.git.
[2] https://github.com/facebookresearch/esm.git.

features of the protein by generating embeddings of size $L \times 1280$ using the model checkpoint ESM-2_t33_650M_UR50D. Conversely, WALLE [13] used a smaller ESM-2 model ESM-2_t12_35M_UR50D to produce 480-dimensional sequence embeddings. We also employ ESM-IF1, an inverse folding model, to produce 512-dimensional structure embedding vectors with the model esm_if1_gvp4_t16_142M_UR50. ProtTrans[3] is a self-supervised transformer-based autoencoder fine-tuned on more than 500 million protein sequences and is used to generate 1024-dimensional embeddings with the model prot_t5_xl_uniref50. For comparison, we also include classical sequence embedding methods such as BLOSUM62 and One-Hot Encoding.

3.3 Model Architecture

The overall architecture of our graph-based framework for antibody-aware epitope prediction is shown in Fig. 1. The model takes as input an antibody-antigen graph pair, constructed as detailed in Sect. 4.1, and performs node- and edge-level predictions.

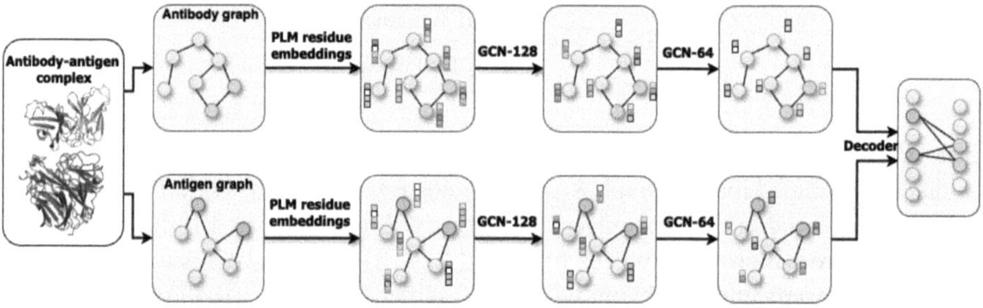

Fig. 1. The schematic of representing an antibody-antigen complex structure as a graph pair, employing pre-trained protein language model (PLM)-based embeddings and the graph convolutional network (GCN) model architecture. The yellow-colored nodes represent non-binding residues, while green and red-colored nodes represent paratopes and epitopes, respectively. (Color figure online)

We incorporate graph modules that process the input graphs of antibody and antigen structures separately, inspired by WALLE [13], PECAN [18], and EPMP [11]. The antibody graph is represented by node embeddings $X_A \in \mathbb{R}^{m \times D_A}$ and an adjacency matrix E_A, while the antigen graph is described by node embeddings $X_B \in \mathbb{R}^{n \times D_B}$ and its corresponding adjacency matrix E_B. The embedding vector sizes D_A and D_B vary depending on the embedding method used, such as AntiBERTy, ESM-2, or other methods.

Both antibody and antigen graph nodes are first projected into a dimensionality of 128 using fully connected layers. The resulting embeddings are then passed through two GCN modules consecutively to refine the features and yield updated node embeddings X'_A and X'_B with a reduced dimensionality of $m \times 64$ and $n \times 64$, respectively. The output from the first GCN layer is passed through a ReLU activation function while outputs from the second GCN layer are directly fed into the decoder module. These GCNs

[3] https://github.com/agemagician/ProtTrans.git.

operate independently, each with its own parameters, ensuring that the learned representations are specific to the antibody or the antigen. The use of separate GCN modules for the antibody and antigen allows for the capture of unique structural and functional characteristics pertinent to each molecule before any interaction analysis. This design choice aligns with the understanding that the antibody and antigen have distinct roles in their interactions, and their molecular features should be processed separately.

A decoder is employed to predict the binary labels of edges between the antibody and antigen graphs. The decoder takes a pair of node embeddings output by the graph modules, X'_A and X'_B, as input and constructs a bipartite adjacency matrix of size $m \times n$. Then, it predicts the probability of each edge: an edge is assigned a binary label of 1 if the predicted probability is greater than 0.5 or 0 otherwise. For the epitope prediction task, edge-level predictions are converted to node-level by summing the predicted probabilities of all edges connected to an antigen node. An antigen node is assigned a label of 1 if the number of connected edges is greater than a threshold or 0 otherwise. The threshold is treated as a hyperparameter and is optimized in the experiments.

4 Experimental Setup

In this section, we provide the dataset description, accompanied by an exploratory data analysis and dataset split strategies, and the implementation details of our framework.

4.1 Dataset

We utilized the AsEP dataset [13], a novel benchmark dataset of antibody-antigen complexes designed specifically for epitope prediction tasks. The dataset was derived from the Antibody Database (AbDb) and the Protein Data Bank (PDB), comprising 1,723 unique antibody-antigen complexes after filtering and deduplication. Each complex includes a single-chain protein antigen paired with a conventional antibody containing both heavy and light chains. Non-canonical residues and unresolved CDR regions were excluded to ensure data quality.

Exploratory Data Analysis (EDA). Our EDA revealed several key insights into the dataset and are shown in Fig. 2. The distribution of epitope residues showed a mean of 19 ± 4.7, while the antigen surface residues numbered in the hundreds. The contact distribution between residues in the bipartite graph had a mean of 43.7 contacts with a standard deviation of 12.8. Additionally, the dataset includes 641 unique antigens and 973 epitope groups, highlighting the diversity and complexity of the antibody-antigen interactions captured in the AsEP dataset.

Dataset Split. We employed two dataset splits for the model training and evaluation:

1. **Epitope to Antigen Surface Ratio:** This split ensures a similar distribution of epitope to non-epitope nodes across the training, validation, and test sets, with 1383 complexes for training and 170 complexes each for validation and testing.

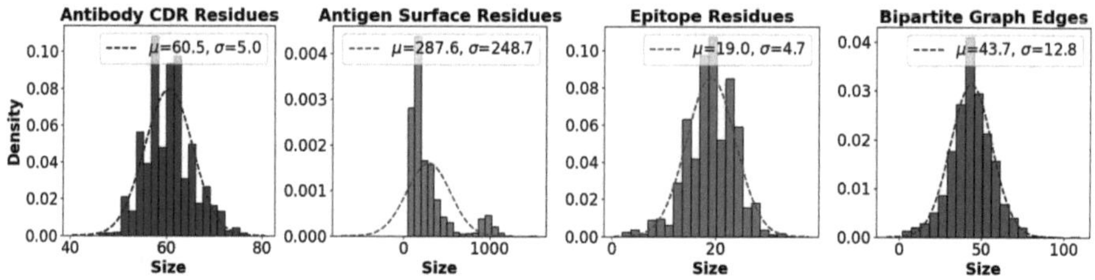

Fig. 2. (left to right) The distribution of the size of antibody CDR residues, antigen surface residues, epitope residues, and bipartite graph edges (antibody-antigen residue-residue contacts) in the dataset. The dotted lines represent the fitted normal distribution.

2. **Epitope Groups:** This split challenges the model to generalize to unseen epitope groups, with 641 unique antigens and 973 epitope groups having multi-epitope antigens, each treated as an antigen-antibody pair. The dataset was divided into training, validation, and test sets with an 80/10/10 split, resulting in 1383 complexes for training and 170 complexes each for validation and testing.

4.2 Implementation Details

The model implementation details were followed in part from WALLE [13]. We used the PyTorch Geometric framework to build the graph modules in the proposed model. We used a weighted loss function consisting of a sum of binary cross-entropy loss for the bipartite graph link reconstruction and a regularizer for the number of positive edges in the reconstructed bipartite graph to train the model. We used the mean of edges in the antibody-antigen bipartite graph (as shown in Fig. 2) to set the regularizer in the loss function to constrain the model not to predict too many or too few positive edges. We used two configurations for the decoder: a fully connected linear layer (GCN-L) and an inner product decoder (GCN). We used the same set of hyperparameters and loss functions for both dataset split settings. We evaluated the performance of our model using the standard classification metrics of Matthews correlation coefficient (MCC), the area under the precision-recall (AUPRC) curve, the area under the ROC curve (AUCROC), precision, recall, F1 score, and balanced accuracy (BACC). All experiments were run in parallel on a 40-CPU research server with Ubuntu 16.04, and it took an average of 1.5 h for embedding generation and 50 min for training on one configuration.

5 Results and Discussion

In this section, we report and discuss the results of our experiments on epitope node classification and bipartite link prediction tasks under epitope-to-antigen surface ratio and epitope group split settings using standard classification metrics.

5.1 Epitope Node Prediction

For epitope node classification, PLM embeddings demonstrated superior performance compared to classical encoding methods across both split strategies. Under epitope ratio dataset split (Table 1 - group a), AntiBERTy/ProtTrans embeddings configuration for an antibody-antigen pair achieved the highest MCC of 0.263 and AUCROC of 0.65, outperforming other embedding configurations. These results were consistent for both configurations of the decoder in our model, *i.e.*, GCN and GCN with a fully connected linear layer. AntiBERTy consistently provided improved performance for antibody embeddings on the metrics of MCC, AUPRC, precision, and F1 score, while ProtTrans showed better performance for antigen embeddings across different combinations on these metrics. These results are consistent with [9,23], suggesting that ProtTrans embeddings provide a rich context for binding site prediction. It was also observed that ESM-IF embeddings for antigens demonstrated very poor performance on all metrics except recall, while it provided average performance in the case of antibodies.

The more challenging epitope group-based split (Table 1 - group b) revealed interesting tradeoffs. The results were not consistent for a specific configuration of embeddings on the majority of evaluation metrics. In general, ESM-2/ProtTrans led to the best results on the metrics of MCC, AUPRC, precision, F1 score, and balanced accuracy (BACC). Similar to the observations in the case of the epitope-ratio split, employing AntiBERTy for antibodies consistently led to better overall results on all metrics in the epitope group split. Consistent with the observations in [13], we also observed that the model struggled with the epitope group split setting compared to the epitope ratio because it could not generalize well on unseen epitopes and showed biases towards the training set. While our model with ESM-2 embeddings attained the best results overall on these metrics, classical one-hot encoding surprisingly achieved the highest recall (0.556), suggesting simpler methods may capture broader structural patterns beneficial for unseen epitope detection. This dichotomy highlights the need for hybrid approaches combining PLM-derived semantics with geometric features [8,19].

5.2 Link Prediction

The link prediction results (Table 2) demonstrated the task's inherent complexity, with maximum MCC scores of 0.126 and 0.054 for ratio and group splits, respectively. The performance gap between tasks suggests node classification benefits more from local residue features, while link prediction requires finer-grained interaction patterns that current GCN architectures struggle to capture. Consistent with the epitope prediction task, the overall results for bipartite link prediction demonstrated that the AntiBERTy/ProtTrans embedding configuration achieves the best performance across almost all metrics for epitope-ratio split. Contrary to epitope node classification, we observed very high AUCROC scores of 0.852 for bipartite link prediction, where we were able to get a maximum AUCROC of 0.65. It was also observed that classical embedding methods such as BLOSUM62 and one-hot encoding showed degraded performance for both dataset splits and GCN decoder configurations on all metrics except AUCROC. Overall, these results suggest that GCN modules with a simple decoder performing the inner product of antibody and antigen embeddings provide consistent and best results compared to GCN with a linear layer as the decoder for both epitope node prediction and bipartite link prediction tasks. Consistent with the investigations in [10,13], our results demonstrated that the choice of the PLM embedding method impacts the model performance on the epitope prediction and bipartite link prediction tasks.

Table 1. Performance of our model on the test set for epitope node prediction with epitope to antigen surface ratio (epitope ratio) and epitope group split strategies. The best values are highlighted in bold.

Algo.	Embedding (Ab/Ag)	(a) Epitope Ratio Split							(b) Epitope Group Split						
		MCC	AUPRC	AUCROC	Prec.	Recall	F1	BACC	MCC	AUPRC	AUCROC	Prec.	Recall	F1	BACC
GCN	AntiBERTy/ESM-IF	0.000	0.101	0.500	0.074	0.211	0.110	0.510	0.000	0.098	0.500	0.076	0.275	0.119	0.521
	AntiBERTy/ESM-2	0.245	0.226	0.649	0.247	0.488	0.328	0.690	0.112	0.149	0.566	0.158	0.264	0.197	0.583
	AntiBERTy/ProtTrans	**0.263**	**0.251**	**0.650**	**0.281**	0.457	**0.348**	0.686	0.112	**0.153**	0.564	0.167	0.240	0.197	0.578
	ESM-IF/ESM-IF	0.000	0.101	0.500	0.068	1.000	0.127	0.500	0.000	0.098	0.500	0.065	1.000	0.122	0.500
	ESM-IF/ESM-2	0.219	0.210	0.633	0.243	0.466	0.320	0.680	0.117	0.148	**0.573**	0.165	0.291	0.211	**0.594**
	ESM-IF/ProtTrans	0.232	0.225	0.640	0.253	0.459	0.326	0.680	0.107	0.152	0.560	0.162	0.231	0.190	0.574
	ESM-2/ESM-IF	0.000	0.101	0.500	0.042	0.094	0.058	0.468	0.000	0.098	0.500	0.063	0.025	0.036	0.500
	ESM-2/ESM-2	0.228	0.211	0.643	0.231	0.504	0.317	0.691	0.112	0.151	0.566	0.157	0.252	0.193	0.579
	ESM-2/ProtTrans	0.228	0.205	0.655	0.218	0.553	0.313	**0.705**	**0.117**	0.152	0.567	**0.181**	0.255	**0.212**	0.587
	ProtTrans/ESM-IF	0.000	0.101	0.500	0.068	**0.956**	0.127	0.501	0.000	0.098	0.500	0.061	0.445	0.108	0.485
	ProtTrans/ESM-2	0.222	0.215	0.631	0.249	0.441	0.319	0.672	0.098	0.145	0.551	0.175	0.188	0.181	0.563
	ProtTrans/ProtTrans	0.236	0.225	0.642	0.259	0.461	0.332	0.682	0.094	0.144	0.552	0.172	0.195	0.183	0.565
	BLOSUM62/BLOSUM62	0.048	0.113	0.541	0.081	0.465	0.139	0.541	0.045	0.112	0.537	0.084	0.412	0.139	0.541
	One-Hot/One-Hot	0.049	0.113	0.543	0.079	0.533	0.138	0.542	0.044	0.108	0.542	0.076	**0.556**	0.134	0.543
GCN-L	AntiBERTy/ESM-IF	0.000	0.101	0.500	0.074	0.242	0.113	0.510	0.000	0.098	0.500	0.079	0.289	0.124	0.527
	AntiBERTy/ESM-2	**0.256**	0.247	**0.650**	0.264	0.468	0.338	**0.687**	**0.123**	0.147	**0.569**	0.162	0.280	**0.205**	**0.590**
	AntiBERTy/ProtTrans	0.252	**0.251**	0.642	**0.283**	0.426	**0.340**	0.674	0.112	**0.153**	0.564	0.167	0.240	0.197	0.578
	ESM-IF/ESM-IF	0.000	0.101	0.500	0.068	**1.000**	0.127	0.500	0.000	0.098	0.500	0.065	1.000	0.122	0.500
	ESM-IF/ESM-2	0.000	0.101	0.500	0.068	**1.000**	0.127	0.500	0.000	0.098	0.500	0.060	0.535	0.109	0.478
	ESM-IF/ProtTrans	0.219	0.218	0.634	0.240	0.472	0.318	0.682	0.099	0.150	0.557	0.155	0.239	0.188	0.574
	ESM-2/ESM-IF	0.000	0.101	0.500	0.045	0.146	0.069	0.461	0.000	0.098	0.500	0.053	0.183	0.082	0.478
	ESM-2/ESM-2	0.220	0.203	0.641	0.222	**0.521**	0.311	0.694	0.060	0.071	0.430	0.032	0.024	0.074	0.038
	ESM-2/ProtTrans	0.244	0.224	0.653	0.244	0.511	0.330	0.698	0.095	0.144	0.549	0.173	0.186	0.179	0.562
	ProtTrans/ESM-IF	0.000	0.101	0.500	0.056	0.293	0.094	0.466	0.000	0.098	0.500	0.060	0.535	0.109	0.478
	ProtTrans/ESM-2	0.234	0.217	0.640	0.253	0.470	0.329	0.684	0.095	0.144	0.549	0.173	0.186	0.179	0.562
	ProtTrans/ProtTrans	0.216	0.204	0.634	0.238	0.480	0.318	0.684	0.115	0.151	0.564	**0.183**	0.234	0.205	0.581
	BLOSUM62/BLOSUM62	0.048	0.113	0.541	0.082	0.457	0.140	0.543	0.050	0.113	0.543	0.080	0.490	0.138	0.540
	One-Hot/One-Hot	0.045	0.112	0.539	0.078	0.509	0.136	0.537	0.046	0.109	0.544	0.076	0.542	0.134	0.543

5.3 Baselines Comparison

We compared our proposed approach with established baselines (WALLE[4] [13], EpiPred [24], MaSIT-site [25], and ESMBind [22]) and revealed significant performance improvements (Table 3). Readers are referred to the related work (Sect. 2) for further description of these different baseline methods. For the epitope ratio split, our proposed approach achieved an MCC of 0.263, outperforming WALLE [13] and the next best model, EpiPred [12]. Notably, our model's superior performance extended across all metrics, with a precision of 0.281 and recall of 0.457, indicating a balanced ability to identify true epitopes while minimizing false positives. The AUC-ROC of 0.65 further confirmed our approach's robust epitope predictive capability, significantly surpassing ESMBind's 0.506. PECAN [18] originally employed a GNN architecture for epitope prediction and achieved a precision of 0.154 without PLM embeddings. This improvement suggests that the integration of GCN with different PLMs captures critical

[4] The code for the preprint version https://arxiv.org/abs/2407.18184v1 was the latest available for comparison at the time of publication.

Table 2. Performance of our proposed model on bipartite link prediction task. The best values are highlighted in bold.

Algo.	(a) Epitope Ratio Split								(b) Epitope Group Split						
	Embedding (Ab/Ag)	MCC	AUPRC	AUCROC	Prec.	Recall	F1	BACC	MCC	AUPRC	AUCROC	Prec.	Recall	F1	BACC
GCN	AntiBERTy/ESM-IF	0.000	0.013	0.752	0.000	0.000	0.000	0.500	0.000	0.012	0.733	0.000	0.000	0.000	0.500
	AntiBERTy/ESM-2	0.114	0.086	**0.852**	0.058	0.298	0.098	0.643	**0.054**	**0.038**	**0.796**	0.031	0.124	0.050	0.557
	AntiBERTy/ProtTrans	**0.126**	**0.100**	0.840	**0.063**	0.302	**0.104**	0.645	0.050	0.034	0.779	0.027	0.128	0.045	0.558
	ESM-IF/ESM-IF	0.000	0.005	0.496	0.000	0.000	0.000	0.500	0.000	0.004	0.510	0.000	0.000	0.000	0.500
	ESM-IF/ESM-2	0.036	0.022	0.723	0.027	0.128	0.045	0.558	0.017	0.013	0.648	0.015	0.060	0.024	0.525
	ESM-IF/ProtTrans	0.042	0.023	0.713	0.028	0.164	0.047	0.574	0.015	0.012	0.644	0.012	0.071	0.021	0.528
	ESM-2/ESM-IF	0.000	0.013	0.761	0.000	0.000	0.000	0.500	0.000	0.013	0.745	0.000	0.000	0.000	0.500
	ESM-2/ESM-2	0.100	0.064	0.843	0.048	0.308	0.083	0.646	0.051	0.032	0.779	0.027	**0.136**	0.045	**0.561**
	ESM-2/ProtTrans	0.102	0.070	0.833	0.039	**0.373**	0.070	**0.674**	0.046	0.031	0.789	**0.034**	0.111	**0.052**	0.551
	ProtTrans/ESM-IF	0.000	0.013	0.763	0.000	0.000	0.000	0.500	0.000	0.014	0.750	0.000	0.000	0.000	0.500
	ProtTrans/ESM-2	0.099	0.065	0.840	0.054	0.247	0.089	0.618	0.045	0.034	0.770	0.029	0.100	0.045	0.546
	ProtTrans/ProtTrans	0.113	0.078	0.827	0.053	0.312	0.090	0.649	0.043	0.028	0.768	0.030	0.108	0.047	0.549
	BLOSUM62/BLOSUM62	0.000	0.011	0.688	0.000	0.000	0.000	0.500	0.000	0.011	0.667	0.000	0.000	0.000	0.500
	One-Hot/One-Hot	0.000	0.011	0.687	0.000	0.000	0.000	0.500	0.000	0.011	0.668	0.000	0.000	0.000	0.500
GCN-L	AntiBERTy/ESM-IF	0.000	0.012	0.752	0.000	0.000	0.000	0.500	0.000	0.012	0.731	0.000	0.000	0.000	0.500
	AntiBERTy/ESM-2	**0.121**	0.089	0.843	0.057	0.315	0.096	0.651	**0.057**	**0.037**	**0.807**	**0.037**	0.121	**0.056**	0.556
	AntiBERTy/ProtTrans	0.115	**0.090**	0.833	**0.059**	0.297	**0.098**	0.642	0.050	0.034	0.779	0.027	0.128	0.045	0.558
	ESM-IF/ESM-IF	0.000	0.004	0.504	0.000	0.000	0.000	0.500	0.000	0.004	0.490	0.000	0.000	0.000	0.500
	ESM-IF/ESM-2	0.001	0.014	0.676	0.023	0.003	0.005	0.501	0.000	0.011	0.686	0.000	0.000	0.000	0.500
	ESM-IF/ProtTrans	0.045	0.023	0.708	0.027	0.178	0.046	0.580	0.020	0.013	0.641	0.014	0.083	0.024	0.534
	ESM-2/ESM-IF	0.000	0.013	0.761	0.000	0.000	0.000	0.500	0.000	0.013	0.743	0.000	0.000	0.000	0.500
	ESM-2/ESM-2	0.091	0.054	0.845	0.045	0.278	0.077	0.631	0.045	0.028	0.767	0.029	0.106	0.046	0.548
	ESM-2/ProtTrans	0.109	0.083	0.838	0.046	**0.344**	0.081	**0.662**	0.045	0.028	0.767	0.029	0.106	0.046	0.548
	ProtTrans/ESM-IF	0.000	0.013	0.765	0.000	0.000	0.000	0.500	0.000	0.014	0.749	0.000	0.000	0.000	0.500
	ProtTrans/ESM-2	0.105	0.072	**0.850**	0.058	0.274	0.096	0.631	0.000	0.011	0.686	0.000	0.000	0.000	0.500
	ProtTrans/ProtTrans	0.098	0.067	0.835	0.049	0.288	0.083	0.637	0.053	0.033	0.771	0.031	**0.129**	0.050	**0.559**
	BLOSUM62/BLOSUM62	0.000	0.011	0.684	0.000	0.000	0.000	0.500	0.000	0.011	0.686	0.000	0.000	0.000	0.500
	One-Hot/One-Hot	0.000	0.011	0.680	0.000	0.000	0.000	0.500	0.000	0.012	0.669	0.000	0.000	0.000	0.500

structural and sequential information about antigens and antibodies that were previously not explored in detail. Moreover, the performance gap between our proposed model and sequence-based methods like ESMBind (MCC of 0.016) underscores the importance of incorporating structural information for accurate epitope prediction [7, 10].

Though the performance dropped in the challenging epitope groups split across all methods, many of which struggled to achieve positive MCC values, our model maintained its lead with an MCC of 0.077. MaSIF-site [25] showed competitive performance, suggesting that its geometric deep learning approach captures some generalizable epitope features. This suggests that current approaches, including WALLE, may be overfitting to epitope patterns seen in training data, emphasizing the need for more diverse training sets and potentially more sophisticated regularization techniques.

5.4 Limitations and Future Research Directions

Though our results demonstrated performance improvement for epitope prediction, there are limitations to our approach, and hence we note some potential future research directions. The authors in [10] demonstrate that the concatenation of ESM-2 and

Table 3. Performance comparison of our proposed model with different baselines on two dataset split strategies. The values in bold represent the best values, while the dash represents that these metrics were not computed.

	Epitope Ratio Split					Epitope Group Split				
Algorithm	MCC	Prec.	Recall	AUCROC	F1	MCC	Prec.	Recall	AUCROC	F1
ESMBind	0.016	0.106	0.090	0.506	0.064	0.002	0.082	0.076	0.500	0.064
EpiPred	0.029	0.122	0.142	—	0.112	-0.006	0.089	0.158	—	0.112
MaSIF-site	0.037	0.125	0.114	—	0.128	0.046	**0.164**	0.174	—	0.128
WALLE	0.210	0.235	0.258	0.635	0.145	0.077	0.143	0.266	0.544	0.145
Our Approach	**0.263**	**0.281**	**0.457**	**0.650**	**0.348**	**0.123**	0.162	**0.280**	**0.569**	**0.205**

ESM-IF1 embeddings improved the epitope prediction performance. In the future, we will explore the integration of different PLM-based embedding methods, such as ESM-2 with ProtTrans, for antibody-aware epitope prediction using separate GAT modules. Other deep learning models, such as bi-directional long short-term memory (BiLSTM [7]), have shown potential for predicting binding sites and need to be explored with concatenated PLM embeddings for the epitope prediction task. There is also immense potential in exploring other geometric deep learning models, such as diffusion-convolutional neural networks (DCNN [26]) for epitope prediction. We will also explore other protein representation approaches, such as point clouds and meshes [25], that have demonstrated success for protein binding site prediction [19].

6 Conclusion

We extended a graph-based approach (WALLE) for antibody-aware epitope prediction on a novel benchmark dataset by exploring different pre-trained PLMs and classical embedding methods. We show that the use of PLM-based embeddings integrated with the graph-based models provides improved epitope prediction performance. Specifically, our experiments with different state-of-the-art PLM-based embedding methods show that using AntiBERTy for antibody embedding and ProtTrans for antigen embedding improves the epitope classification accuracy. We also discuss potential future research directions for improved epitope prediction.

References

1. Norman, R., Ambrosetti, F., Bonvin, A., Colwell, L., Kelm, S., Kumar, S., Krawczyk, K.: Computational approaches to therapeutic antibody design: established methods and emerging trends. Brief. Bioinform. **21**(5), 1549–1567 (2020)
2. Joubbi, S., et al.: Antibody design using deep learning: from sequence and structure design to affinity maturation. Briefings Bioinform. **25**(4), bbae307 (2024)
3. Silva, B., Ascher, D., Pires, D.: epitope1d: accurate taxonomy-aware B-cell linear epitope prediction. Briefings Bioinform. **24**(3), bbad114 (2023)

4. Fischman, S., Ofran, Y.: Computational design of antibodies. Curr. Opin. Struct. Biol. **51**, 156–162 (2018)
5. Krishnan, S., et al.: rAbDesFlow: a novel workflow for computational recombinant antibody design for healthcare engineering. Antibody Therapeutics **7**(3), 256–265 (2024)
6. Hummer, A., Abanades, B., Deane, C.: Advances in computational structure-based antibody design. Curr. Opin. Struct. Biol. **74**, 102379 (2022)
7. Zeng, Y., et al.: Identifying B-cell epitopes using AlphaFold2 predicted structures and pretrained language model. Bioinformatics **39**(4), btad187 (2023)
8. Wang, C., Wang, J., Song, W., Luo, G., Jiang, T.: EpiScan: accurate high-throughput mapping of antibody-specific epitopes using sequence information. NPJ Syst. Biol. Appl. **10**(1), 101 (2024)
9. Fang, Y., Jiang, Y., Wei, L., Ma, Q., Ren, Z., Yuan, Q., Wei, D.: DeepProSite: structure-aware protein binding site prediction using ESMFold and pretrained language model. Bioinformatics **39**(12), btad718 (2023)
10. Choi, S., Kim, D.: B cell epitope prediction by capturing spatial clustering property of the epitopes using graph attention network. Sci. Rep. **14**(1), 27496 (2024)
11. Vecchio, A., Deac, A., Liò, P., Veličković, P.: Neural message passing for joint paratope-epitope prediction. arXiv preprint arXiv:2106.00757 (2021)
12. Krawczyk, K., Liu, X., Baker, T., Shi, J., Deane, C.: Improving B-cell epitope prediction and its application to global antibody-antigen docking. Bioinformatics **30**(16), 2288–2294 (2014)
13. Liu, C., Denzler, L., Chen, Y., Martin, A., Paige, B.: AsEP: benchmarking deep learning methods for antibody-specific epitope prediction. arXiv preprint arXiv:2407.18184v1 (2024)
14. Lin, Z., Akin, H., Rao, R., Hie, B., et al.: Language models of protein sequences at the scale of evolution enable accurate structure prediction. BioRxiv **2022**, 500902 (2022)
15. Ruffolo, J., Gray, J., Sulam, J.: Deciphering antibody affinity maturation with language models and weakly supervised learning. arXiv preprint arXiv:2112.07782 (2021)
16. Elnaggar, A., Heinzinger, M., Dallago, C., Rehawi, G., et al.: ProtTrans: Toward understanding the language of life through self-supervised learning. IEEE Trans. Pattern Anal. Mach. Intell. **44**(10), 7112–7127 (2021)
17. Hsu, C., et al.: Learning inverse folding from millions of predicted structures. In: International Conference on Machine Learning, pp. 8946–8970 (2022)
18. Pittala, S., Bailey-Kellogg, C.: Learning context-aware structural representations to predict antigen and antibody binding interfaces. Bioinformatics **36**(13), 3996–4003 (2020)
19. Dai, B., Bailey-Kellogg, C.: Protein interaction interface region prediction by geometric deep learning. Bioinformatics **37**, 2580–2588 (2021)
20. Jha, K., Saha, S., Singh, H.: Prediction of protein-protein interaction using graph neural networks. Sci. Rep. **12**(1), 8360 (2022)
21. Lu, S., Li, Y., Ma, Q., Nan, X., Zhang, S.: A structure-based B-cell epitope prediction model through combing local and global features. Front. Immunol. **13**, 890943 (2022)
22. Schreiber, A.: ESMBind and QBind: LoRA, QLoRA, and ESM-2 for predicting binding sites and post translational modification. BioRxiv, pp. 2023–11 (2023)
23. Kalemati, M., Noroozi, A., Shahbakhsh, A., Koohi, S.: ParaAntiProt provides paratope prediction using antibody and protein language models. Sci. Rep. **14**(1), 29141 (2024)
24. Kulytė, P., Vargas, F., Mathis, S.V., Wang, Y.G., Hernández-Lobato, J.M., Liò, P.: Improving antibody design with force-guided sampling in diffusion models. arXiv preprint arXiv:2406.05832 (2024)

25. Gainza, P., Sverrisson, F., Monti, F., Rodola, E., Boscaini, D., Bronstein, M., Correia, B.: Deciphering interaction fingerprints from protein molecular surfaces using geometric deep learning. Nat. Methods **17**(2), 184–192 (2020)
26. Atwood, J., Towsley, D.: Diffusion-convolutional neural networks. Advances in neural information processing systems, vol. 29 (2016)

Leveraging RNA LLMs for 3D Structure Prediction via Data Augmentation

Sixiang Zhang, Harish Anand, and Liming Cai(✉)

School of Computing, University of Georgia, Athens, GA, USA
{Sixiang.Zhang,Harish.Anand,liming}@uga.edu

Abstract. Ribonucleic acid (RNA) is a complex macromolecule essential for living organisms to function in cells. Understanding its three-dimensional (3D) structure is critical for elucidating its cellular roles. However, computational prediction of RNA 3D structures remains a significant challenge due to the vast conformational space that RNA molecules can adopt. Although machine learning, particularly deep learning-based methods, has recently gained traction, the lack of a large dataset of native RNA structures for training has limited these methods from achieving desired performance. In this study, we leverage pre-trained RNA large language models to predict RNA 3D conformations directly from input RNA sequences. Specifically, we introduce data augmentation techniques to address the issue of data scarcity in RNA 3D structures. This present paper focuses on predicting backbone conformations to evaluate the effectiveness of our method. Preliminary results demonstrate promising accuracy, with predicted structures achieving an average RMSD of 3.85Å against native 3D structures in the PDB—a 50% reduction in performance error compared to predictions made without the data augmentation method.

1 Introduction

RNA (ribonucleic acid) is a fundamental molecule in biology and a cornerstone of biomedical research, particularly in recent high-profile advancements in disease diagnosis and treatment [12]. RNA plays crucial functional roles in cells, including protein synthesis, gene expression and regulation, as well as catalytic and structural contributions to various cellular processes [18]. Since RNA molecules exert their functions through spatial interactions with other biomolecules, understanding their three-dimensional (3D) structure is essential for unraveling their biological roles.

Structure prediction via computer algorithms have been active research that can complement the slow and expensive experimental elucidation of RNA structure [31]. However, compared with the similar task in the protein counterpart, 3D structure prediction of RNA may still be in its infancy, specifically due to a large discrepancy in investment of resources for structure elucidation between protein and RNA [22]. In the past decades, efforts for computational prediction of RNA 3D structure have resulted in a number of methods [2,20,38], each with its

M. Alser et al. (Eds.): ICCABS 2025, LNCS 15599, pp. 303–316, 2026.
https://doi.org/10.1007/978-3-032-02489-3_23

own strengths and limitations nevertheless. In particular, the most accurate are homology-based methods that predict the structure of a target RNA sequence through the use of known 3D structures of similar RNA sequences [23–26], which however can be difficult to obtain, especial for novel RNAs. Folding algorithms generate 3D structure from secondary structure through transformation and refinement [15,25,44], yet accuracy of prediction of secondary structure is usually in question. Fragment-based methods builds the 3D structure by selecting and assembling known small 3D motif fragments [6,8,23,24], a computationally intensive task that nevertheless can struggle with large or very novel RNA. Yet another very computationally intensive method is physics-based, using molecular dynamics or other energy minimization techniques to model RNA 3D structures [3,4,43].

Recently, deep learning has become increasingly prominent in RNA structure prediction [28,35,39], with transformer architectures leveraging self-attention to model complex sequence dependencies. Once trained on sufficiently large datasets of native RNA 3D structures, these models can capture both local and global folding patterns, leading to improved accuracy. In particular these methods are developed to work with RNA large language models (RNA-LLMs) that have been pre-trained over millions of RNA sequences. Such models can learn well patterns of nucleotide arrangements and functional motifs. They are suitable (and some have yielded outstanding performance) in predicting splicing sites, functional elements, RNA secondary structure, and binding affinities for RNA-binding proteins. In particular, the work in [28] predicts tertiary structures via secondary-structure-based intermediate representations. While these methods can indeed learn rich sequence-to-structure mappings, the limited availability of high-quality RNA 3D data continues to constrain their performance.

This paper introduces our work in 3D structure prediction with RNAErnie [39], an RNA large language model, through a proposed method to address the data scarcity issue in RNA 3D structure. RNAErnie is structured similarly to general-purpose LLMs but with the incorporation of specialized knowledge of RNA motifs. It has demonstrated its utility by offering robust starting points for downstream tasks such as RNA secondary structure prediction and RNA (long) sequence classification. This present work utilizes output embeddings of RNAErnie, which encompass not only long-window context but also capture diverse aspects of RNA molecules. In particular, to overcome the issue of limited availability of RNA 3D structure data, we propose data augmentation techniques and design a training pipeline tailored for leveraging the output embeddings of the RNA-LLM and transforming high-dimensional per-position data into 3D coordinates. Evaluation of our method on 147 experimentally validated RNA 3D models from the Protein Data Bank (PDB) demonstrates the effectiveness of the proposed data augmentation approach. Specifically, when trained on 90% of the RNA models, predictions for the remaining 10% achieved an average RMSD of 3.85Å against native 3D structures in the PDB – a 50% reduction in performance error compared to predictions without using the data augmentation method.

2 Backgrounds

A ribonucleic acid (RNA) is a macro-molecule of nucleotides, each consisting of a phosphate group, a sugar, and an aromatic nucleobase either adenine (A), cytosine (C), guanine (G), or uracil (U). A phosphate group is attached to the 3' position of one ribose and the 5' position of the next, forming the phosphodiester bonds that chain the nucleotides of an RNA molecule into a linear sequence. Nucleotides can interact through chemical bonds, bringing them to the spatial proximity of each other. The most compelling base-base interactions, called *canonical base pairs*, are Watson-Crick pairs A-U, C-G, and wobble pair G-U via hydrogen bonds. Canonical base pairs do not work alone; instead they form a stacked, ladder shape of group, bringing two individual contiguous strands together into a base paired region, called *stem*. They, together with unpaired strands, called *loops*, constitute the *secondary structure* of RNA molecules.

The secondary structure scaffolds the 3D structure, with stems corresponding to double helices in the 3D space. While double helices are the main force to scaffold the 3D structure of an RNA molecule, unpaired loops connect these double helices. Together with other types of nucleotide-nucleotide interactions, they form higher order structures and ultimately the 3D conformation. This process involves not only the canonical base pairs but also non-canonical *tertiary* nucleotide-nucleotide interactions between two bases, a base and a phosphate, and a base and a ribose [5,10,17,30,41]. For example, tertiary interactions in a tRNA, which, together with canonical base pairs, are the force behind the folding of tRNA molecule into the L-shape 3D structure [29].

2.1 RNA-LLMs and RNAErnie

Built with deep learning techniques, RNA-LLMs are RNA language models pre-trained on large RNA data repositories [1,7,13], with sequences, alignments, and long-range interactions of nucleotides being tokenized and annotated. To date, several RNA-LLMs have demonstrated potentials to tackle important tasks related to analysis of RNA sequences. In particular, they have been fine-tuned for specific tasks, such as splice site prediction, therapeutic RNA design, and functional prediction [11,37,40]. Through integration of structural data, e.g., well-defined secondary structure annotations, RNA-LLMs can contribute significantly to success in secondary structure prediction [39]. Due to the scarcity in high-quality annotations of secondary structure and the lack of knowledge in RNA folding dynamic process, existing RNA-LLM based RNA 3D structure prediction methods usually incorporate other informatics, for example, evolutionary history and multiple sequence alignments, to improve performance [28].

RNAErnie [39] is a conceptual extension of ERNIE [33] tailored for RNA-related bioinformatics tasks. RNAErnie focuses on integrating biological knowledge for RNA sequence, structure, and function prediction. Compared to other RNA-LLMs, RNAErnie integrates domain-specific RNA knowledge, typically, RNA sequence motifs, secondary and tertiary structural information, and functional roles of RNA molecules. RNAErnie has the advantage in knowledge-enhanced RNA modeling, extensive external data usage, and functional annotation and interaction modeling. We believe these features are valuable for RNA 3D structure prediction given the situation of data scarcity in validated RNA 3D structures.

2.2 Data Augmentation

Data augmentation is crucial for LLMs to be adapted to applications, where fine tuning is typically needed for specific domains in contrast to the model trained in general domain data. With applications in a specific domain, there is often a lack of sufficient labeled or high-quality training data; issues with balance or diversity of data that can cause model overfitting and performance. The augmentation helps bridge the gap between the general-purpose knowledge of LLMs and the specialized language or patterns required for domain-specific tasks. In particular, for biomolecule (i.e., protein and RNA) 3D structure prediction, more specialized augmentation methods aim at increasing the diversity and quantity of training 3D structures. Due to high dimensionality of biomolecule 3D structures involve complex geometric, spatial, and energetic interactions, augmentation techniques have been more focused on geometric transformation of 3D data, small random perturbation to to atomic positions, generation of synthetic sequences with minor mutations that conserve structure, random recombination of fragments into into unseen yet plausible structures, and random sampling of lower-energy conformations. Among these, generation of new structures from extracted 3D motif fragments is most appealing as it can produce unseen training data, easing issues of data scarcity, structural diversity, and overfitting. This is potentially viable for RNA 3D structure prediction since it may help resolve the data scarcity issue. However, how to make (deep learning) generative model to only produce plausible synthetic 3D models remains a challenging task.

3 Method for Data Augmentation

We now introduce crucial techniques for data augmentation that make it possible to leverage RNA LLMs for 3D structure prediction given the limited availability of experimentally validated RNA 3D structures.

3.1 Dataset

The initial dataset we collected contains 147 experimentally determined RNA 3D models obtained from the PDB [1]. These models are high-resolution (≤ 3.0Å) 3D structures determined by X-ray crystallography as well as models determined by Nuclear Magnetic Resonance (NMR) with a root-mean-square deviation (RMSD) of ≤ 2.0Å to ensure data quality. Although our tests are focused on RNA sequences of 50 nucleotides or less, the collected dataset of 3D models are from various non-coding RNA families, such as transfer RNA (tRNA), ribosomal RNA (rRNA), and ribozymes, of diverse structures for training.

In this paper, we focus on predicting coarse grain models of RNA structure backbone, where every nucleotide is represented by single-atom whose coordinates are those of atom phosphorus (P) in the phosphate group of the nucleotide. Occasionally, atom P may be missing from some RNA 3D model obtained from the PDB (which is most common at the beginning of sequences). In this case, we substitute it with the position of atom O5' or O3' in the group.

3.2 Data Augmentation Techniques

In order to deal with limited availability of RNA 3D structures for training, we propose four complementary data augmentation strategies as follows.

1. **Noise Addition.** Gaussian noise was added globally to the coordinates of atoms P to enhance the model's robustness against minor structural variations. As shown in Fig. 1 (a), this "global noise" (purple) is applied uniformly across the entire RNA molecule. To avoid disrupting the overall conformation, we constrained the root-mean-square deviation (RMSD) between the original and noisy structures to remain below 1.0Å. A 3D comparison between the original and globally noised structures is shown in Fig. 1 (b).
2. **Region-Based Noise Addition.** While the first technique introduces global noise, RNA structures are inherently region-oriented. Hence, we also applied Gaussian noise *only* to continuous subregions rather than the entire structure. In Fig. 1 (a), these targeted "random regions" are shown in orange color, illustrating how small, localized perturbations can affect some parts of the chain. Once again, we ensured that the RMSD remained below 1.0Å to maintain the overall structural integrity.
3. **Attention-Guided Region Selection.** We leverage the self-attention mechanism of our RNA language model to quantify each nucleotide's "importance" based on how strongly it is attended to by the rest of the sequence. Concretely, let $A_h^{(L)}(i \leftarrow j)$ be the attention weight from nucleotide j (as query) to nucleotide i (as key) in head h of the last transformer layer L. We average over all H heads and sum over $j = 1, \ldots, N$ to obtain

$$\alpha_i = \sum_{j=1}^{N} \Big(\frac{1}{H} \sum_{h=1}^{H} A_h^{(L)}(i \leftarrow j) \Big). \tag{1}$$

   This α_i measures how much attention nucleotide i receives from the entire sequence. Let $\bar{\alpha} = \frac{1}{N} \sum_{i=1}^{N} \alpha_i$ denote the global mean of "attentions" over all nucleotides. We label nucleotide i as "important" if $\alpha_i > \bar{\alpha}$ and "unimportant" otherwise.
4. Finally, we add mild Gaussian noise *either* to these important positions (shown in red) *or* to unimportant ones (shown in green), ensuring the RMSD remains below 1.0 Å. By perturbing both high- and low-attention regions, the model learns how local structural changes in different contexts can influence the overall 3D conformation.

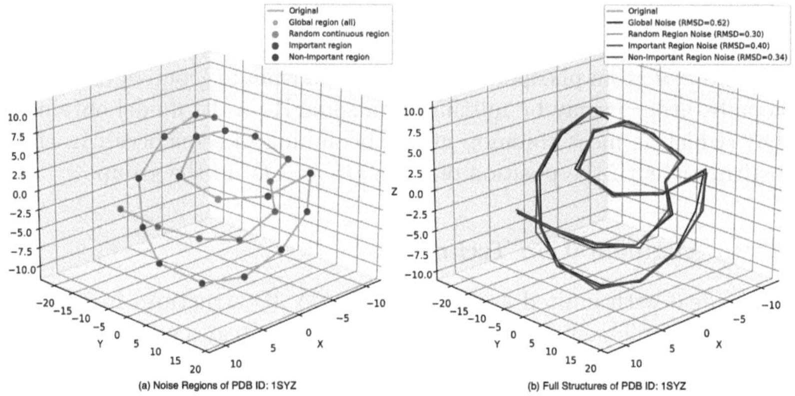

Fig. 1. Illustration of data-augmentation strategies with an RNA structure example (PDB Id: 1SYZ). (a) A zoomed-in view comparing the local atomic positions under each noise method: global noise (purple), random region noise (orange), important region noise (red), and unimportant region noise (green). (b) A broader 3D perspective showing the overall RNA backbone under each of these noise conditions. All noise additions are constrained such that RMSD $\leq$ 1.0Å. (Color figure online)

Data augmentation was applied exclusively to the training set, where for each 3D structure we generated 100 samples using global noise, 100 using region-based noise, 100 with noise added to important regions, and 100 with noise added to unimportant regions, thereby substantially increasing the dataset size and enhancing the model's generalization for unseen RNA sequences (see Sect. 5).

4 Prediction Model

In this study, we leverage the transfer learning capability of RNAErnie [39] to streamline RNA 3D structure prediction. As the pre-trained RNA LLM, RNAErnie can effectively capture bi-directional dependencies within RNA sequences, RNAErnie transforms input RNA sequences into high-dimensional embeddings, encapsulating essential sequential, secondary structural, and functional information. The following summarizes key tasks and components in our prediction model.

4.1 System Settings

Our overall model for RNA 3D structure prediction is formulated as:

$$\text{TertiaryStructurePrediction}(\text{RNA LLM}(\text{Seqs})) \tag{2}$$

where the 3D structure prediction module utilizes self-attention mechanisms and comprises several branches constructed from transformer blocks [36], which consist of multi-head self-attention layers. These capture context from the outputs of RNAErnie, preserve both the overall and relative RNA structure, and take the RNAErnie embeddings as input. These transformer blocks are also designed to model complex interactions and predict the 3-dimensional conformation of RNA molecules based on the enriched feature representations. In addition, our prediction model learns a scale factor based on the input sequence. This factor is particularly important because the average mean square error (MSE) is calculated per structure; by reducing the magnitude of the predicted structures, a smaller scale factor naturally leads to a relatively smaller loss. Here are the details.

1. Before applying our model, we process the output embeddings from RNAErnie, denoted as $\mathbf{X}$. These embeddings are vector representations of each position in the input RNA sequence, capturing both local and global contextual information.
2. To ensure stable training, we first normalize the embeddings using LayerNorm, which standardizes their distribution. We then apply dropout [32], a regularization technique that randomly zeroes out portions of the embeddings to prevent overfitting: $\tilde{\mathbf{X}} = \text{Dropout}\,(\text{LayerNorm}(\mathbf{X}))$, where $\mathbf{X} = \text{RNA LLM(Seqs)}$.
3. Positional embeddings are then added based on the technique as introduced by [36]: $\hat{\mathbf{X}} = \tilde{\mathbf{X}} + \mathbf{E}_{\text{pos}}$, where $\mathbf{E}_{\text{pos}}$ is the learnable positional encoding.
4. Next, we apply a two-layer multilayer perceptron (MLP) to $\hat{\mathbf{X}}$ in a *residual manner*: $\mathbf{X}' = \hat{\mathbf{X}} + \text{MLP}(\hat{\mathbf{X}})$. The MLP consists of fully connected layers with nonlinear activations, enabling the model to further learn complex transformations of the embeddings. By using a residual connection, we allow the original embeddings $\hat{\mathbf{X}}$ to be preserved and combined with the MLP's output. This approach enhances the representational capacity while promoting stable training, as the model can choose to retain the original information if no further transformation is needed.
5. The transformed embeddings are passed through multiple standard transformer encoder blocks with GELU: $\mathbf{H}_i = \text{TransformerEncoder}_i(\mathbf{X}')$, where $\mathbf{H}_i$ denotes the output from the i-th transformer encoder branch. However, instead of a single transformer encoder, we split the processing into different branches in a self-weighted manner: $\mathbf{H} = \sum_{i=1}^{N} w_i \mathbf{H}_i$ where w_i are the learned weight for branch i, normalized via a softmax function such that $\sum_{i=1}^{N} w_i = 1$, and N is the number of branches.
6. After processing through the transformer branches, a two-layer MLP is used as the coordinate projection layer to project the inner embeddings to the 3D coordinates sequence: $\text{Coordinates} = \text{CoordinateProj}(\mathbf{H})$.
7. Finally, since the model learns a normalized representation and to prevent it from predicting overly small structures, we apply a learned scalar factor to scale the coordinates back to their original scale. Specifically, we learn a distinct scalar factor for each spatial dimension (x, y, z) based on the input sequence embeddings: $\text{Scalar(RNA LLM(Seqs))} = (s_x, s_y, s_z)$. These scale factors are predicted by the two-layer multilayer perceptron (MLP), referred

to as the `scale_net` component of the model, which processes the global features of the embeddings. The final coordinates are obtained by multiplying the predicted normalized coordinates with the corresponding scalar factors: Final Coordinates = Coordinates $\times$ (s_x, s_y, s_z).

4.2 Training Procedure

Training Dataset. We first split the dataset of 147 experimentally determined RAN 3D model obtained from PDB (see Sect. 4) into two of the 90/10 ratio. This resulted in a test dataset of 14 RNA 3D structures (unseen data during training) for evaluation. Then, on the training dataset, we performed data augmentation and subsequently split it into 80/20, with 20% used for validation for early stopping.

Cross-Validation for Robustness. Given the relatively small size of our dataset, we additionally performed a 10-fold cross-validation to obtain a more robust estimate of the generalization error. In each fold, we trained on 90% of the data and tested on the remaining 10%. We then computed metrics (e.g., RMSD) on the held-out test subset, which was never used in any training fold.

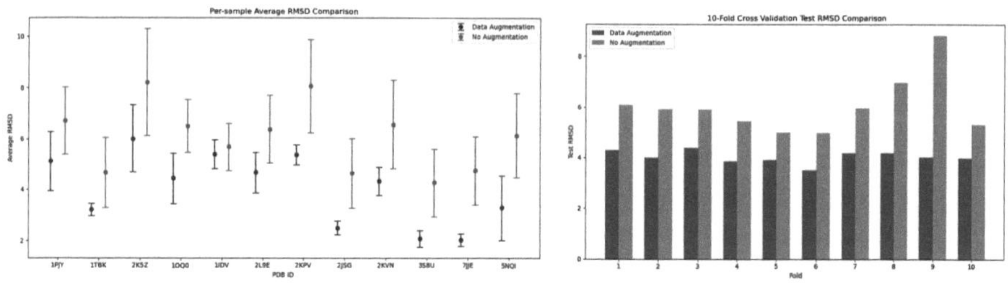

Fig. 2. *Left*: Per-sample average RMSD comparison with and without data augmentation. Each data point represents the mean RMSD (error bars denote standard deviation) over multiple runs. *Right*: 10-fold cross-validation test RMSD comparison. Each bar shows the average test RMSD for a particular fold, with and without augmentation.

These cross-validation experiments have confirmed that our augmentation strategies significantly improve performance. As shown in Fig. 2, the augmented models (blue bars/points) consistently achieve lower RMSD values than their non-augmented counterparts (orange/red), demonstrating the effectiveness of our data augmentation in enhancing the robustness and generalization of RNA 3D structure prediction.

Loss Function. We train our model using a combined loss that includes the mean squared error (MSE) term and the adjacent phosphate distance (P-P) term: $\mathcal{L} = \text{MSE} + \alpha \cdot \text{P-P Loss}$.

A larger α emphasizes preserving local bond distances, while a smaller α favors minimizing the coordinate-wise MSE. Following our tuning experiments (detailed below), we set $\alpha = 6.5$ by default, as it yielded the lowest RMSD across the validation set.

Tuning α To determine an optimal balance between the MSE and P–P terms, we performed a grid search over α from 0.1 up to 20.0, evaluating validation RMSD at each step. As shown in Fig. 3, $\alpha = 6.5$ achieved the best validation performance. Unless otherwise specified, all subsequent experiments have been conducted with $\alpha = 6.5$.

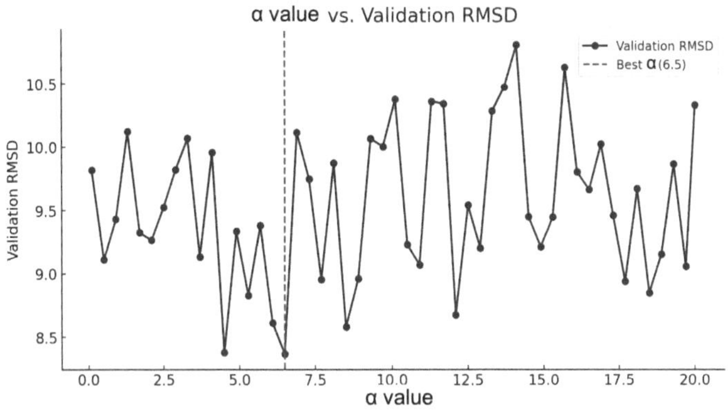

Fig. 3. Validation RMSD versus α. Each point represents the validation RMSD for a particular value of α. The red dashed line indicates the best $\alpha = 6.5$. (Color figure online)

Kabsch Alignment. To ensure the loss focuses on internal structural deviations rather than global rotation/translation, we align each predicted structure $\mathbf{p}_j$ to the native structure $\mathbf{t}_j$ using the Kabsch algorithm [14]. Specifically, the aligned structure is $\mathbf{p}_j^{\text{aligned}} = \mathbf{A}(\mathbf{p}_j)$, where $\mathbf{A}$ is the rotation/translation that best superimposes $\mathbf{p}_j$ onto $\mathbf{t}_j$. This alignment removes extraneous rotational and translational differences from the RMSD computation.

P-P Loss. Since using the average MSE to measure structural accuracy can lead to scaling issues, we also consider the distances between adjacent phosphorus atoms. Specifically, we examine adjacent pairs $(i, i+1)$ for $i = 1, \ldots, N-1$.

After alignment, we have

$$\text{P-P Loss} = \frac{1}{B}\sum_{j=1}^{B}\frac{1}{N-1}\sum_{i=1}^{N-1}\left(\left\|\mathbf{p}_{ji}^{\text{aligned}} - \mathbf{p}_{j,i+1}^{\text{aligned}}\right\| - d_i\right)^2,$$

where $\mathbf{p}_{ji}^{\text{aligned}}$ and $\mathbf{p}_{j,i+1}^{\text{aligned}}$ are the aligned predicted coordinates of the i-th and $(i+1)$-th phosphorus atoms in the j-th structure, respectively and d_i is the ground-truth bond length for the adjacent pair $(i, i+1)$.

Hyperparameter Settings. We employed the AdamW optimizer [19] with an initial learning rate of 5×10^{-5}, a weight decay of 1×10^{-4}, and a batch size of 512. The model was trained for 200 epochs, applying early stopping if the validation loss did not improve for 20 consecutive epochs. A learning rate scheduler (ReduceLROnPlateau) with a factor of 0.5 and a patience of 10 epochs was utilized to reduce the learning rate when the validation loss plateaued. We set the dropout rate to 0.2 throughout the layers to further mitigate overfitting.

5 Performance Evaluation

5.1 Impact of Data Augmentation

To evaluate the effect of data augmentation on the model's performance, we trained the same neural architecture twice: once on a non-augmented training dataset, and once on an augmented dataset. Both datasets contained the same base set of RNA structures, but the augmented dataset included additional transformations (e.g., global noise, regional noise, and random rotations in important vs. non-important regions) to increase diversity. The model architecture consisted of 4 transformer layers, 8 heads, and 4 transformer encoder branches.

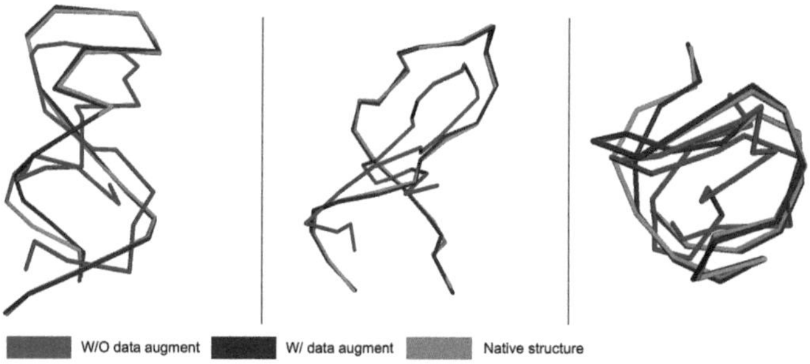

Fig. 4. Three different perspectives of predicted 3D models (trained with and without the proposed data augmentation techniques) against the native structure (green) for PDB ID 7JJE (Color figure online). The prediction without data augmentation (magenta) has an RMSD of 5.833Å, while the prediction with data augmentation (blue) has an RMSD of 0.436Å.

Our model's performance on unseen RNAs significantly improves with data augmentation, from average RMSD dropping from 8.219 ± 1.759 Å(range: 6.264 – 12.702 Å) to 3.851 ± 2.044 Å(range: 0.661 – 6.654 Å). Figure 4 illustrates this marked enhancement for RNA 7JJE.

5.2 Effect of the Scale Factor Component

In this section, we discuss the importance of the scale factor component in the model. Without the scale factor prediction component, the model struggles to train effectively, resulting in poor performance or converging to a suboptimal, under scaled solution. Figure 5 shows an example of the model trained with and without the scale factor component (configured with 4 transformer layers, 16 heads, and 3 transformer encoder branches). As illustrated, removing the scale factor component leads to noticeable scaling issues in the predicted model.

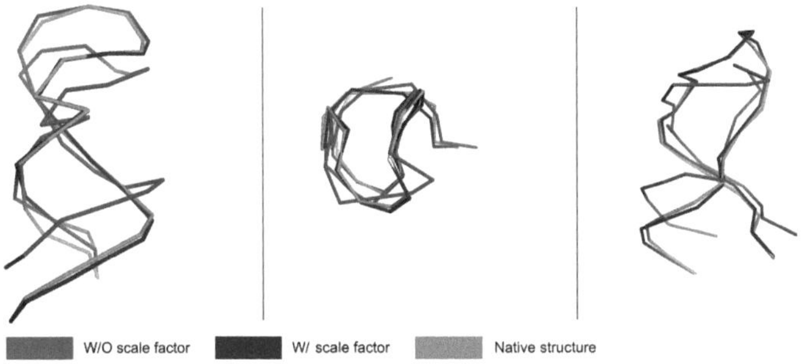

Fig. 5. Three different perspectives of the predicted 3D models in training (with and without the scale factor component) compared to the native structure (green) for PDB ID `5NQI` (Color figure online). The prediction without the scale factor (magenta) has an RMSD of 4.488Å, whereas the prediction with the scale factor (blue) has an RMSD of 0.89Å.

5.3 Performance on the Test Set and Visualization of Predictions

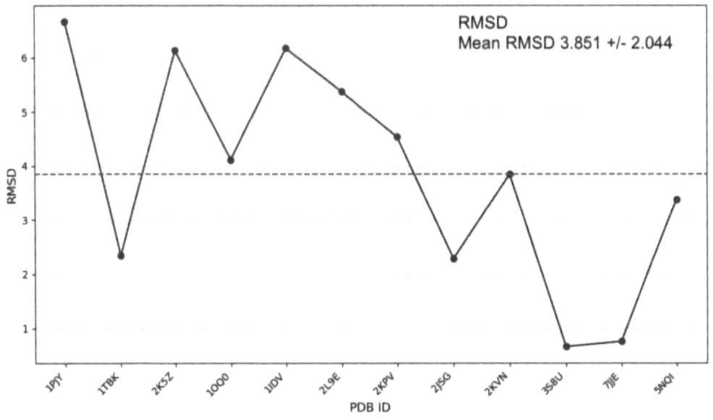

Fig. 6. Performance of 3D structure predictions on the 12 RNA structures in the test dataset, measured with RMSD between the predicted and the native structures, with a mean RMSD of 3.851 ± 2.044Å.

To assess how well the model generalizes to unseen data, we compiled a test set of 12 RNA structures, each with fewer than 50 nucleotides. Figure 6 summarizes the performance across these 12 RNAs, with the red line indicating the mean RMSD (3.851 ± 2.044Å). Overall, these results are comparable to or better than the performance on the training set, suggesting that the model can effectively predict RNAs that the model is not explicitly trained on.

To provide a concrete example of the model's predictive accuracy, Fig. 7 visualizes the predicted structure (blue) against the experimentally determined coordinates (green) for PDB ID 3S8U. In this case, the model achieves a notably low RMSD of 0.661Å. The final model demonstrates excellent accuracy, particularly in predicting the bulge region of the RNA structure.

These visual inspections reinforce the quantitative metrics: most RNAs in the test set are reconstructed with minimal RMSD, although a few challenging examples show slightly higher values. Nonetheless, the consistently low RMSDs indicate that the model captures both global structural features and many local details.

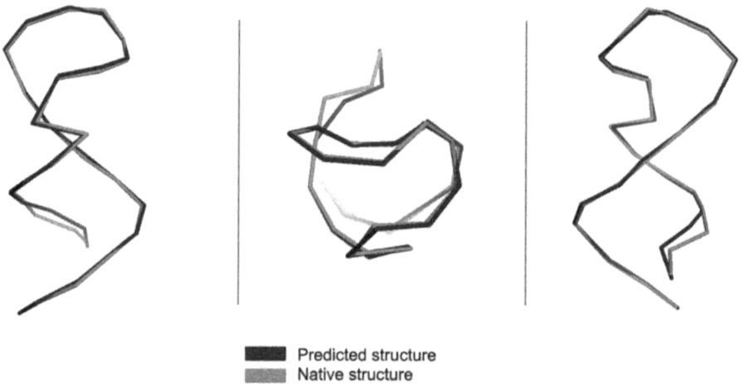

Fig. 7. Three perspectives on the superimposition of predicted (blue) and validated (green) P atom positions of the RNA with PDB Id 3S8U (Color figure online).

References

1. Berman, H.M., et al.: The Protein Data Bank. Nucleic Acids Res. **28**, 235–242 (2000)
2. Bernard, C., Postic, G., Ghannay, S., Tahi, F.: State-of-the-RNArt: benchmarking current methods for RNA 3D structure prediction. NAR Genom. Bioinfo. **6**(2) (2024)
3. Boniecki, M.J., Lach, G., Dawson, W.K., et al.: SimRNA: a coarse-grained method for RNA folding simulations and 3D structure prediction. Nucleic Acids Res. **44**(7), e63 (2016)
4. Case, D.A., Darden, T.A., Cheatham, T.E., et al.: The AMBER biomolecular simulation programs. J. Comput. Chem. **26**(16), 1668–1688 (2005)
5. Cech, T.R., Steitz, J.A.: The noncoding RNA revolution-trashing old rules to forge new ones. Cell **157**(1), 77–94 (2014)
6. Cheng, C.Y., Chou, F.C., Das, R.: Modeling complex RNA tertiary folds with Rosetta. *Meth. Enzymol.* **553**, 35–64 (2015)

7. Clare, A., et al.: RNAcentral: a comprehensive database of non-coding RNA sequences. Nucleic Acids Res. **44**(D1), D114–D118 (2016)
8. Das, R., Baker, D.: Automated de novo prediction of native-like RNA tertiary structures. Proc. Nat. Acad. Sci. USA **104**(37), 14664–14669 (2007)
9. Devlin, J., Chang, M.W., Lee, K., Toutanova, K.: BERT: pre-training of deep bidirectional transformers for language understanding. In: *Proceedings of NAACL-HLT* (2018)
10. Doudna, J.A., Cech, T.R.: The chemical repertoire of natural ribozymes. Nature **418**, 222–228 (2002)
11. Franke, J.H., Runge F., Hutte, F.: Scalable deep learning for RNA secondary structure prediction. In: *Proceedings of ICML Workshop on Computational Biology* (2023)
12. Gingeras, T.R.: Current frontiers in RNA research. Front. RNA Res. **1**, 1152146 (2023)
13. Griffiths-Jones, S., et al.: Rfam: an RNA family database. Nucleic Acids Res. **31**(1), 382–384 (2003)
14. Kabsch, W.: A solution for the best rotation to relate two sets of vectors. Acta Crystallograph. Sect. A Crystal Phys. Diffract. Theor. General Cryst. **32**(5), 922–923 (1976)
15. Keating, K.S., Pyle, A.M.: Ribonucleic acid structure: folding and modeling. Curr. Opin. Struct. Biol. **22**(3), 317–324 (2012)
16. Krahn, N., Fisher, J.T., Söll, D.: Naturally occurring tRNAs with non-canonical structures'. *Front. Microbiol.* Sec. Microb. Phys. Metab. **11** (2020)
17. Leontis, N.B., Lescoute, A., Westhof, E.: The building blocks and motifs of RNA architecture. Current Opin. Struct. Biol. **16**(3), 279–287 (2006)
18. Liu, Y., Hao, Y.: Exploring the Possibility of RNA in Diverse Biological Processes. Int. J. Mol. Sci. **24**(13), 10674 (2023)
19. Loshchilov, I., Hutter, F.: Fixing weight decay regularization in Adam. *CoRR*, vol. abs/1711.05101 (2017)
20. Mukherjee, S., Bujnicki, J, M., et al.: Advances in the field of RNA 3D structure prediction and modeling, with purely theoretical approaches, and with the use of experimental data. *Structure*, Cell Press **32** (11) (2024)
21. Nijkamp, E., Ruffolo, J.A., Weinstein, E.N., Naik, N., Madani, A.: ProGEN2: exploring the boundaries of protein language models. Cell Syst. **14**(11), 968–973 (2023)
22. Nithin, C., Kmiecik, S., Błaszczyk, R., Nowicka, J., Tuszyńska, I.: Comparative analysis of RNA 3D structure prediction methods: towards enhanced modeling of RNA–ligand interactions. *Nucleic Acids Res.* **52**(13), 7465–7486 (2024)
23. Parisien, M., Major, F.: The MC-Fold and MC-Sym pipeline infers RNA structure from sequence data. Nature **452**, 51–55 (2008)
24. Petrov, A.I.: RNA 3D Hub: a database of RNA structural motifs and their interactions. Nucleic Acids Res. **41**, D315–D321 (2013)
25. Popenda, M., et al.: Automated 3D structure composition for large RNAs. Nucleic Acids Res. **40**, e112 (2012)
26. Rother, K., Rother, M.: RNA structure prediction tools and their usage in homologous modeling. Meth. Mol. Biol. **1704**, 315–336 (2018)
27. Sanabria, M., Hirsch, J., Joubert, P.M., Poetsch, A.R.: DNA language model GROVER learns sequence context in the human genome. Nat. Mach. Intell. **6**(8), 911–923 (2024)
28. Shen, T., Hu, Z., Sun, S., et al.: Accurate RNA 3D structure prediction using a language model-based deep learning approach. *Nat. Meth.* (2024)

29. Shi, H., Moore, P.B.: The crystal structure of yeast phenylalanine tRNA at 1.93 Åresolution: a classic structure revisited. RNA **6**(8), 1091–1105 (2000)
30. Simons, R. W., Grunberg-Manago, M.: (Ed) *RNA Structure and Function*, Cold Spring Harbor Monograph Series (1998)
31. Spitale, R.C., Incarnato, D.: Probing the dynamic RNA structurome and its functions. Nat. Rev. Genet. **24**(3), 178–196 (2023)
32. Srivastava, N., Hinton, G., Krizhevsky, A., Sutskever, I., Salakhutdinov, R.: Dropout: a simple way to prevent neural networks from overfitting. J. Mach. Learn. Res. **15**(1), 1929–1958 (2014)
33. Sun, T., et al.: ERNIE: enhanced representation through knowledge integration. In: *Proceedings of the 28th International Joint Conference on Artificial Intelligence* (IJCAI) (2019)
34. Szikszai, M., Mathews, D.H., et al.: Learning models for RNA secondary structure prediction (probably) do not generalise across families. Bioinformatics **38**, 3892–3899 (2022)
35. Townshend, R.J.L., Eismann, S., Watkins, A.M.: Geometric deep learning of RNA structure. Science **373**(6558), 1047–1051 (2021)
36. Vaswani, A., et al.: Attention is all you need. *CoRR*, vol. abs/1706.03762 (2017). [Online]. Available: http://arxiv.org/abs/1706.03762
37. Wang, J., Mao, Y., Zeng, J.: DeepFoldRNA: end-to-end deep learning method for RNA secondary structure prediction and folding. Nucleic Acids Res. **50**(10), e58 (2022)
38. Wang, X., et al.: RNA 3D structure prediction: progress and perspective. Molecules **28**(14), 5532 (2023)
39. Wang, N., et al.: Multi-purpose RNA language modelling with motif-aware pretraining and type-guided fine-tuning. *Nat. Mach. Intell.*, pp. 1–10 (2024)
40. Wayment-Steele, H.K., et al.: RNA secondary structure packages evaluated and improved by high-throughput experiments. Nat. Methods **19**(12), 1234–1242 (2022)
41. Westhof, E., Leontis, N.B.: An RNA-centric historical narrative around the Protein Data Bank. J. Biol. Chem. **296**, 100555 (2023)
42. Yuan, J., Tang, R., Jiang, X., Hu.: Large language models for healthcare data augmentation: an example on patient-trial matching. *AMIA Annual Symposium Proc.* (2024)
43. Zhang, C., Sun, X., Baturka, M., Case, D.A.: Force field optimization for RNA simulations with explicit solvent and metal ions. RNA **27**(6), 687–702 (2021)
44. Zhao, Y., Gong, Z., Zhang, H., Zhou, Y.: 3dRNA: 3D structure prediction from linear to tertiary structures. *Nucleic Acids Res.* **40**(14), e112 (2012)

EfficientNet in Digital Twin-Based Cardiac Arrest Prediction and Analysis

Qasim Zia[1], Avais Jan[1], Zafar Iqbal[1], Muhammad Mumtaz Ali[2], Mukarram Ali[3], and Murray Patterson[1](✉)

[1] Georgia State University, Atlanta, GA 30303, USA
zia.qasim3@gmail.com,
{ajan3,ziqbal5}@student.gsu.edu, mpatterson30@gsu.edu

[2] Zhengzhou University, Zhongyuan District, Zhengzhou 450001, Henan, China
muali@gs.zzu.edu.cn

[3] Dublin Business School, 13/14 Aungier St, Dublin D02 WC04, Ireland
mukarram063@gmail.com

Abstract. Cardiac arrest is one of the biggest global health problems, and early identification and management are key to enhancing the patient's prognosis. In this paper, we propose a novel framework that combines an EfficientNet-based deep learning model with a digital twin system to improve the early detection and analysis of cardiac arrest. We use compound scaling and EfficientNet to learn the features of cardiovascular images. In parallel, the digital twin creates a realistic and individualized cardiovascular system model of the patient based on data received from the Internet of Things (IoT) devices attached to the patient, which can help in the constant assessment of the patient and the impact of possible treatment plans. As shown by our experiments, the proposed system is highly accurate in its prediction abilities and, at the same time, efficient. Combining highly advanced techniques such as deep learning and digital twin (DT) technology presents the possibility of using an active and individual approach to predicting cardiac disease.

Keywords: EfficientNet · Digital Twin · Cardiac Arrest Prediction · Cardiovascular · Real-time Monitoring · Federated Transfer Learning

1 Introduction

Cardiovascular disease (CVD) is the leading cause of death worldwide, and arrest is one of the most deadly manifestations [1]. Arrest is a significant and serious complication of ischemic heart disease, and early and accurate prediction is crucial for effective management and potential improvement in patient outcomes [2]. The application of artificial Intelligence (AI) and Digital Twin (DT) in healthcare has opened new opportunities for patient monitoring and predictive analysis in real time [3]. DT emulates a patient's cardiovascular system based on real-time analysis of patient data, helping to provide an accurate risk assessment

M. Alser et al. (Eds.): ICCABS 2025, LNCS 15599, pp. 317–328, 2026.
https://doi.org/10.1007/978-3-032-02489-3_24

and early identification of arrest events [4]. However, traditional machine learning and deep learning models are limited by their computational complexity and inefficiency in real-time analysis, which hinders their practical application in critical healthcare settings [5].

Due to its optimal architecture and the ability to scale, EfficientNet, a state-of-the-art deep learning model, has shown excellent performance in medical image analysis [6]. CNNs, conventional convolutional neural networks, lack compound scaling, a systematic balance of depth, width, and resolution that EfficientNet employs to improve predictive accuracy while maintaining computational efficiency [7]. This is ideal for cardiac arrest prediction and analysis in the DT environment, where real-time performance and accuracy are crucial [8].

Although EfficientNet has not been applied to digital twins to predict cardiac arrest, there is a gap in the literature regarding their integrated application. Using EfficientNet with high accuracy and efficiency in a digital twin framework can provide a new way of real-time risk assessment and simulation of cardiac events. This integration could improve predictive performance by combining robust image analysis with dynamic, patient-specific modeling, particularly with emerging federated learning and privacy-preserving data-sharing techniques.

The primary objective of this study is to create an EfficientNet framework to predict and analyze cardiac arrest in the DT environment as mentioned below:

- To develop an EfficientNet-based deep learning model for the prediction of cardiac arrest from real-time patient data within a Digital Twin system.
- Apply DT Technology to build a dynamic and individual virtual model of the cardiovascular system of a patient to improve predictive accuracy.
- Assess the effectiveness of the proposed framework by comparing the performance of the proposed framework with conventional deep learning models in accuracy, precision, recall, and F1 score.
- Improve computational efficiency and model interpretability for real-time clinical decision-making. Process data securely and reliably for cardiac health monitoring through private smart sensing through the DT framework.

This paper applies EfficientNet in a DT framework to predict cardiac arrest. Federated Transfer Learning, privacy-preserving analysis, and real-time data processing are used to increase diagnostic accuracy with reduced computational costs. The study indicates that EfficientNet could revolutionize predictive healthcare and improve patient outcomes through an early and precise cardiac risk assessment. The main contributions of this work are:

- Propose DT-based EfficientNet framework for Cardiac Arrest Prediction.
- Optimization model to improve cardiac arrest prediction without wasting computational resources and time.
- Comparisons with other CNN models to highlight the accuracy and performance benefits of EfficientNet.

This paper is organized as follows. Section 2 explains related work. The architecture of a prediction of cardiac arrest based on EfficientNet is suggested in

Sect. 3. The experiments are then carried out comprehensively and assessed in Sect. 4 to determine the effectiveness and efficiency of our suggested architecture. We finally arrive at our conclusions in Sect. 5.

2 Related Works

Cardiac arrest is one of the most important global health problems and survival rates depend on early prediction and management of the disease. Conventional machine learning and deep learning models have been extensively applied for cardiovascular disease detection; however, their integration with DT technology and real-time assessment is novel. The Digital Twin framework offers real-time feedback, emulation, and prognostic evaluation of a patient's cardiac condition from patient to patient, thus assisting the clinician in decision-making. Several studies explored deep learning applications in cardiac health screening.

2.1 Deep Learning for Cardiac Arrest Prediction

Recent studies have found that deep learning can significantly improve the prediction of cardiac events. Handcrafted features-based classical machine learning models such as logistic regression or support vector machines have been replaced by deep neural networks in recent times. Elola et al. [9] used convolutional neural networks (CNN) and recurrent neural networks (RNN) to successfully extract spatial and temporal features from multimodal data such as electrocardiograms ECG and other physiological signals to improve the early detection and prognosis of cardiac arrest using them. Hannun et al. (2019) [10] found that Deep Neural Networks (DNNs) were able to identify arrhythmias with the accuracy of experts which opens up the possibility of applying AI-driven approaches to cardiac event prediction. Similarly, Rajpurkar et al. (2017) [11] proposed the application of convolutional neural networks (CNNs) for the identification of abnormalities in ECG signals, thus supporting the notion of AI-based screening

2.2 EfficientNet in Medical Image Analysis

EfficientNet, a family of convolutional neural networks, was designed to achieve high accuracy with minimal computational cost and can be applied effectively to medical image analysis problems. EfficientNet was developed by Tan and Le [12] and has revolutionized CNN architecture design using a compound scaling method that adjusts the depth, width, and resolution of the network uniformly. This innovative technique allows EfficientNet to achieve top-tier accuracy using fewer parameters and requiring fewer computational resources than previous models such as ResNet and DenseNet. Its success has been confirmed in medical imaging applications, such as tumor identification [13] and organ segmentation [15], making it an attractive option to achieve performance and efficiency. Recent research, including that of Kaba et al. (2023) [14], has also investigated the use of EfficientNet in cardiovascular imaging and has established that it provides better classification results than other conventional models. However, the integration of EfficientNet for the digital twin model of cardiac arrest remains obscure.

2.3 Digital Twin Technology in Healthcare

The application of digital twin technology implies the development of a virtual model of a system, specifically for continuous monitoring, simulation, and predictive analysis of a given physical system; in this case, it is the cardiovascular system of a patient [16]. Zia et al. [17] use a priority algorithm in a network based on Digital Twins (DT) to enhance communication. Awasthi et al. [18] discuss that Digital twins are currently employed in healthcare to model patient-specific dynamics from real-time data collected by wearable sensors, imaging modalities, and electronic health records. It enables personal treatment planning and real-time decision support in treatment, thus improving diagnostic accuracy and potential for early intervention in critical conditions such as cardiac arrest.[19] In this article, Martinez-Velazquez et al. (2019) explained the DT paradigm as a virtual copy of a patient's heart that can be constantly monitored and evaluated. When used in conjunction with rational use of features, this approach can enhance current methods of detection and management of the condition. Based on the studies reviewed above, the present research develops on previous approaches and presents a new concept of combining EfficientNet and Digital Twin technology. This integration is expected to improve the precision and speed of the cardiac arrest prediction model, thus solving a significant missing link in real-time healthcare surveillance.

3 The Proposed EfficientNet in Digital Twin-Based Cardiac Arrest Prediction and Analysis

This section presents an end-to-end framework for integrating EfficientNet in a DT system for real-time cardiac arrest prediction and analysis. The methodology combines data-driven deep learning with patient-specific simulation to enable proactive clinical decision-making.

3.1 Dataset

We used the Cardiac MRI images[1] [20] dataset for our simulations. This is one of the most comprehensive Cardiac Coronary Artery Disease(CAD) image datasets. This dataset contains CAD disease images, which often lead to cardiac arrest. Figure 1 shows the images with CAD disease and normal images. (a-c) cardiac magnetic resonance images in Fig. 1 are from patients with coronary artery disease and (d-f) are healthy person images. The dataset has more than 60,000 images. The images in the dataset have different image quality. It is harder to identify the CAD in some images. So, data preprocessing is necessary. Training the model on heterogeneous image qualities leads to learning invariant features robust to noise and artifacts. This enhances the model's ability to identify subtle signs of Coronary Artery Disease (CAD) even when the imaging is not the best

[1] (https://www.kaggle.com/datasets/danialsharifrazi/cad-cardiac-mri-dataset/data.)

quality. Testing the model on a dataset with different quality is a more challenging task. According to the studies by Litjens et al. (2017) [21] and Esteva et al. (2019) [22] models that are trained on a large variety of image qualities are more likely to perform well. So such a trained model when tested on digital twin images which are often generated under ideal conditions with minimal noise and artifacts is more likely to perform well. Hence, this dataset provides a better platform to evaluate models than digital twin images, which are simple to detect and predict any disease.

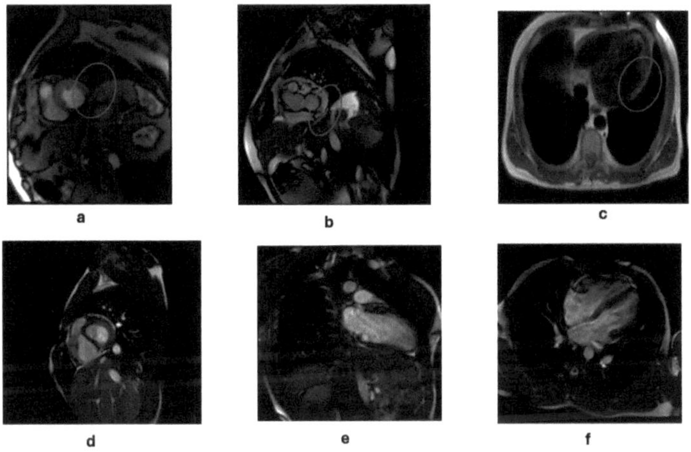

Fig. 1. Sample Cardiac MRI images from dataset [20].

3.2 Architecture of the Proposed EfficientNet Model for Cardiac Arrest

This section provides a high-level architecture for using the EfficientNet model in DT-Cardiac Arrest Prediction and analysis, as seen in Fig. 2. The proposed framework consists of two main components:

A digital twin is a virtual replica of the patient's cardiovascular system, and an EfficientNet-based prediction model takes clinical data as input to predict cardiac arrest risk. Data from wearable sensors, clinical imaging, and electronic health records (EHR) are continuously fed into the digital twin, and the EfficientNet model learns the relevant features to make real-time predictions.

The following is the proposed model workflow:

- **Step 1: Patient data (e.g., IoT wearable devices sensor data) is collected and uploaded to the cloud.**
- **Step 2: After Data Preprocessing**, the cloud runs the digital twin simulation software, which creates a 3D model of the cardiovascular system and simulates its behavior

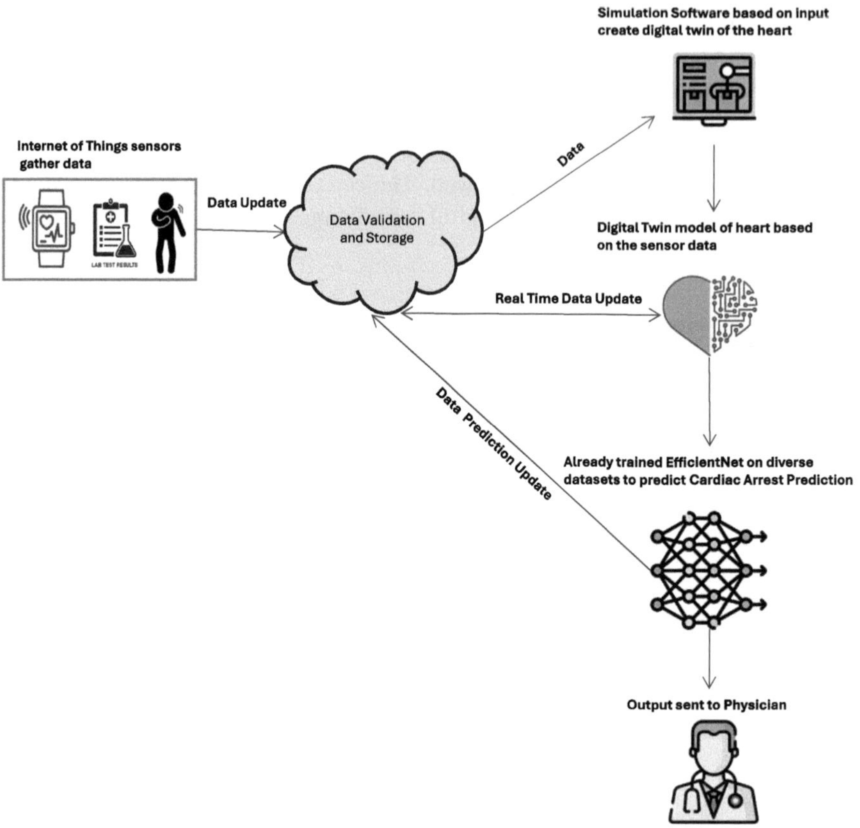

Fig. 2. High-level architecture of using EfficientNet model for digital twin-based Cardiac Arrest Prediction and Analysis.

- **Step 3: The simulation results are passed through EfficientNet-based AI models to predict cardiac arrest risk or other cardiovascular conditions**
- **Step 4: The predictions and insights are sent back to the healthcare providers or hospital staff for further analysis and decision-making.**

3.3 IoT Wearable Device Data Collection and Transmission for Digital Twin-Based Cardiac Arrest Prediction

The IoT Wearable Device Data Collection and Transmission Algorithm is designed to predict cardiac arrest in real time using digital twins. First, it gathers real-time sensor data $S = \{s_1, s_2, \ldots, s_n\}$ from wearable devices like heart rate monitors, blood pressure sensors, and other IoT-based health trackers. This data is combined with other environmental factors like location and activity level.

The data is then sent to the cloud server over an existing communication network (4G, 5G, WiFi, etc.). The transmission process is divided into nested loops to ensure that data from different wearable devices and their neighbors are processed sequentially. For each wearable device W, data collection and transmission take place over a time window t for all the available neighboring devices N. The sensor data $S_W(t)$ that is raw transmitted to the cloud for further analysis. After processing each neighboring device, it guarantees that the next device is updated by $N \leftarrow N+1$. When all the neighboring devices are processed, the next wearable device is considered by $W \leftarrow W+1$.

In this manner, systematic and real-time monitoring and data transmission is enabled to construct digital twin-based cardiac stroke prediction. The data allows cloud-based analysis and thus predicts early cardiac arrest detection and personalized healthcare interventions. The model uses these inputs to update the digital twin and continuously monitor the patient's heart health. Therefore, the algorithm designed for wearable devices and their data collection and transmission is essential for effectively realizing the concept of a digital twin in predicting and managing cardiac arrests.

Algorithm 1. IoT Wearable Device Data Collection and Transmission for Digital Twin-based Cardiac Arrest Prediction

Input: Real-time sensor data $S = \{s_1, s_2, \ldots, s_n\}$ from wearable devices
Output: Preprocessed data sent to the cloud server for further analysis
Step 1: Data Upload Initialization
Each wearable device W collects real-time sensor data S_W (e.g., wearable device, laboratory results, blood pressure)
The wearable device uses the existing communication network (e.g., 4G, 5G, WiFi) for data transfer to the cloud server
Data includes individual readings and environmental context (location, activity level)
Initialize wearable device W, neighbor devices N, and time t
while $W \leq |W|$ **do**
  while $N \leq N$ **do**
    while $t \leq T$ **do**
      Step 2: Transmit Data to Cloud Server
      Send raw data $S_W(t)$ from wearable device W to the cloud for analysis
    end while
    Update for next iteration:
    Update $N \leftarrow N + 1$ for the next neighboring device
  end while
  Update $W \leftarrow W + 1$ for the next wearable device
end while

3.4 EfficientNet in Digital Twin-Based Cardiac Arrest Prediction and Analysis

Cloud Server Side Cardiac Arrest Prediction Using EfficientNet. The Cloud Server Side Cardiac Arrest Prediction Algorithm uses the EfficientNet model and DT technology to predict the risk of cardiac arrest using real-time sensor data from wearable devices.

- Data Preprocessing: This paper uses normalization of the raw sensor data $S_W(t)$ obtained from wearable devices to use the data for analysis. The normalization process guarantees that the data is within a certain range of values, which will help produce more accurate predictions in the subsequent stages.
- Digital Twin Creation: A Digital Twin (DT) is simulated and is created using the processed sensor data. This virtual model uses the incoming sensor data to indicate the real-time condition of the Cardiovascular system, such as blood flow and electrical activities.
- Integration into Digital Twin Model: The sensor data is integrated into the Cardiovascular system Digital Twin model. It enables the simulation of the Cardiovascular system dynamic conditions, which is important for precise cardiac arrest risk prediction.
- Cardiac Arrest Prediction using EfficientNet: The algorithm uses an already trained EfficientNet model for cardiac arrest risk prediction. It processes the cardiovascular system DT model and produces a cardiac arrest risk prediction $D\hat{S}_W(t)$.
- Feed-forward Pass for Prediction: To determine the cardiac arrest risk for each patient, we perform a feed-forward pass through the EfficientNet model, using their respective Digital Twin models as input.
- Cardiac Arrest Prediction Decision: Based on the processed data, a cardiac arrest prediction decision is suggested in real-time for the patient.
- Output Sent to Physician: The cardiac arrest prediction results and the DT model are delivered to the physician for further review. It enables the physician to evaluate the cardiac arrest risk and make appropriate decisions.
- Anomaly Detection: The predicted decision $D\hat{S}_W(t)$ is then compared with the actual cardiac arrest occurrence outcome $DS_W(t)$ to detect anomalies using a 3σ rule, following the receipt of feedback. This compares it over time to improve the prediction model continuously.
- Weighted Decision Making: The decision-making process is further enhanced by incorporating the output of the patient and the physician through a weighted average. The algorithm uses the weights to weigh the feedback from one source or another depending on its relevance. where α is the weight for patient feedback, and w_{W_n} are the weights for doctor feedback.
- Fine-Tuning the EfficientNet Model: The EfficientNet model is fine-tuned based on the feedback received for better accuracy in future predictions.
- Cloud Server Update: The cloud server state is updated depending on the feedback and predictions to optimize the model for the next cardiac arrest risk prediction iteration.

The real-time cardiac arrest prediction algorithm identified above leverages wearable device data, digital twin simulations, and EfficientNet for accurate decision-making and offers significant insights for medical professionals.

Algorithm 2. Cloud Server Side Cardiac Arrest Prediction Using EfficientNet

Input : Raw data from wearable device $S_W(t)$;
Output : Arrest prediction decision $D\hat{S}_W(t)$ for the respective wearable devices W of the patient;

Step 1: Pre-Processing and Managing Data at the DT Layer

$$S_W(t) \leftarrow \frac{S_W(t)}{\sigma}$$

Step 2: Simulation Software to Create a DT of the Cardiovascular system

$$DT_{\mathrm{CVS}}(t) = \mathrm{Simulate}(S_W(t))$$

Step 3: Create DT Model of the Cardiovascular System using Sensor Data

$$DT_{\mathrm{CVS}}(t) \leftarrow \mathrm{Integrate}(S_W(t))$$

Step 4: EfficientNet Model Processing

$$D\hat{S}_W(t) = \text{EFF-NET}(DT_{\mathrm{CVS}}(t))$$

Step 5: Feed-forward Pass for Prediction
Step 6: Suggest Cardiac Arrest Decision
Step 7: Output Sent to Physician

$$\text{Send Output to Physician: } D\hat{S}_W(t), DT_{\mathrm{CVS}}(t)$$

Step 8: Feedback for Anomaly Detection

$$\text{Anomaly Detection: } D\hat{S}_W(t) \text{ vs. } DS_W(t) \text{ based on } 3\sigma \text{ rule}$$

Step 9: Weighted Decision Making

$$D\hat{S}_W(t) = \alpha D\hat{S}_W(t) + (1 - \alpha) \sum_{n=1}^{N} w_{Wn} D\hat{S}_n(t)$$

Step 10: Fine-Tune the EfficientNet Model
Step 11: Update Cloud Server State
Update for next iteration
End of algorithm.

4 Performance Evaluation

This section includes a description of the experimental setup, the metrics used for evaluation, the quantitative results, and the comparison based on EfficientNet for predicting cardiac arrest.

4.1 Baseline Studies

The CNNs VGG16, ResNet50, and EfficientNet will be compared. VGG16 is renowned for its simplicity and efficacy, while ResNet50 has a contemporary architecture and is renowned for its performance and efficiency. TensorFlow is used in this work to implement all models. TensorFlow is used for training all models. The primary goal of our research as mentioned in the Introduction Section is to detect diseases and predict cardiac arrest in real-time. It is not affected by wireless transmission effects (such as network latency or capacity limitations) once we get the images depicting the patient's cardiovascular system. The impact of the CNN model on the decision-making process is challenging to identify if we consider the additional variables introduced by wireless transmission (latency, packet loss, bandwidth restrictions, etc.).

4.2 Experimental and Performance Analysis

Several essential performance criteria will be used to evaluate the effectiveness of the EfficientNet model for predicting cardiac arrest.

Accuracy: The percentage of data our model accurately predicts its accuracy. It is the proportion of accurately predicted values to all expected values.

Recall: The positive cases can be accurately identified by the model.

Specificity: The percentage of accurately identified negatives among all negatives is known as specificity.

F1 score: It determines the model's precision and recall harmonic mean. It calculates the metric by considering both precision and recall. This section will discuss the performance of the proposed EfficientNet model for CAD detection which can lead to cardiac arrest. We also assess the time taken to predict. We will compare the EfficientNet model with VGG16 and ResNet50. EfficientNet performance is compared using various performance metrics that are very important in predicting cardiac arrest, which includes accuracy, recall, F1 score, and Specificity in Table 1.

The EfficientNet model's performance was compared with other models using various performance metrics, including accuracy, precision, recall, F1 score, AUC, and training time. These metrics are essential for decision-making and faster processing.

The results show that EfficientNet outperforms the other model and provides better prediction.

Table 1. Performance Metrics for Different CNN Models

CNN Models	Accuracy (%)	Precision	Recall	F1-Score	Specificity	Auc	Training Time
EfficientNet	93.45	0.925	0.9482	0.9365	0.9203	0.9630	275.24s
ResNet50	91.26	0.9077	0.9192	0.9134	0.906	0.9487	320.40s
VGG16	89.15	0.8903	0.8933	0.8918	0.8896	0.9339	435.24s

The confusion matrix also evaluates EfficientNet, ResNet50, and VGG16 model performance. The details of the confusion matrix can be analyzed in Fig. 3.

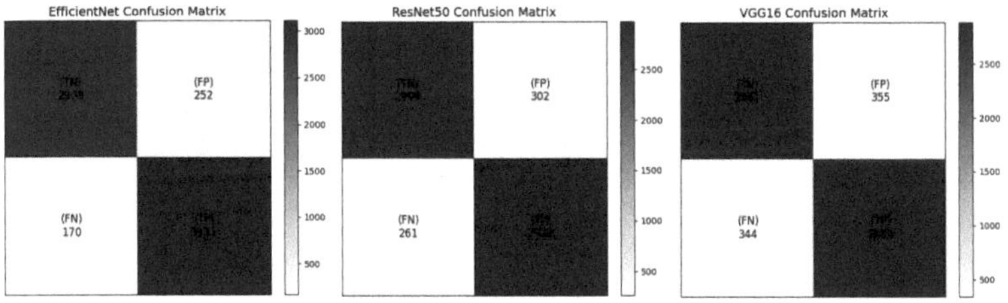

Fig. 3. Comparison of Confusion Matrices for EfficientNet, ResNet50, and VGG16.

5 Conclusion

In this paper, we present a new framework that uses an EfficientNet-based prediction model with a digital twin system to solve the critical problem of detecting early cardiac arrest. As such, we propose using EfficientNet for a dynamic patient-customized digital twin because of its ability to extract advanced features. Our approach creates a digital twin that offers a real-time assessment of the patient's cardiovascular system. The digital twin component, however, establishes a patient's cardiovascular system simulation model through which the patient's current state can be depicted.

The results of our experiments show that EfficientNet can effectively predict the risk of cardiac arrest using cardiac image data. The proposed framework structure enables it to be easily extended to work in distributed healthcare settings. It has the potential to explore other innovations, like federated learning, for improved data privacy and collaboration across different institutions.

Therefore, integrating EfficientNet and digital twin technology can significantly contribute to cardiac arrest prediction. Future work will include creating our digital twin model of the heart to confirm model generalization, optimizing model settings to improve the model's performance, and incorporating more sources of information to improve the system's accuracy.

Acknowledgements. The icons in the figures of this research chapter are taken from the Freepik website.

References

1. Celermajer, D., Chow, C., Marijon, E., Anstey, N., Woo, K.: Cardiovascular disease in the developing world: prevalences, patterns, and the potential of early disease detection. J. Am. Coll. Cardiol. **60**, 1207–1216 (2012)
2. Schultz, A., McCoy, M., Graham, R.: Strategies to improve cardiac arrest survival: a time to act. (National Academies Press) (2015)
3. Jameil, A.: Al-Raweshidy, H. Leveraging AI and Secure Systems for Enhanced Patient Outcomes, A Digital Twin Framework for Real-Time Healthcare Monitoring (2024)

4. JEMAA, M., JEMILI, F., Charfeddine, A., MSOLLI, M., KORBAA, O.: Digital twin for a human heart using deep learning and stream processing platforms (2023)
5. Bian, J., et al.: Machine learning in real-time Internet of Things (IoT) systems: a survey. *IEEE Internet Things J.* **9**, 8364–8386 (2022)
6. Ahmed, W., Massoud, M., El-Bouridy, M.: Optimizing MRI-based medical diagnosis: comparative analysis of efficientnet performance with varying learning rates. J. Egypt. Soc. Tribol. **21**, 105–119 (2024)
7. Lin, C., Yang, P., Wang, Q., Qiu, Z., Lv, W., Wang, Z.: Efficient and accurate compound scaling for convolutional neural networks. Neural Netw. **167**, 787–797 (2023)
8. Shah, A., Ahirrao, S., Pandya, S., Kotecha, K., Rathod, S.: Smart cardiac framework for an early detection of cardiac arrest condition and risk. Front. Public Health **9**, 762303 (2021)
9. Elola, A., et al.: Deep neural networks for ECG-based pulse detection during out-of-hospital cardiac arrest. Entropy **21**, 305 (2019)
10. Hannun, A., et al.: Cardiologist-level arrhythmia detection and classification in ambulatory electrocardiograms using a deep neural network. Nat. Med. **25**, 65–69 (2019)
11. Rajpurkar, P., Hannun, A., Haghpanahi, M., Bourn, C., Ng, A.: Cardiologist-level arrhythmia detection with convolutional neural networks. ArXiv Preprint ArXiv:1707.01836 (2017)
12. Tan, M., Le, Q.: Efficientnet: rethinking model scaling for convolutional neural networks. In: *International Conference On Machine Learning*, pp. 6105–6114 (2019)
13. Tripathy, S., Singh, R., Ray, M.: Automation of brain tumor identification using efficientnet on magnetic resonance images. Proc. Comput. Sci. **218**, 1551–1560 (2023)
14. Kaba, Ş, Haci, H., Isin, A., Ilhan, A., Conkbayir, C.: The application of deep learning for the segmentation and classification of coronary arteries. Diagnostics **13**, 2274 (2023)
15. Chekroun, M., Mourchid, Y., Bessières, I., Lalande, A.: Deep learning based on efficientnet for multiorgan segmentation of thoracic structures on a 0.35 T MR-Linac radiation therapy system. *Algorithms* **16**, 564 (2023)
16. Coorey, G., et al.: The health digital twin to tackle cardiovascular disease—a review of an emerging interdisciplinary field. *NPJ Digit. Med.* **5**, 126 (2022)
17. Zia, Q., Wang, C., Zhu, S., Li, Y.: Priority based inter-twin communication in vehicular digital twin networks, pp. 1–16. International Journal Of Parallel, Emergent And Distributed Systems (2024)
18. Awasthi, M., Raghuvanshi, C., Dudhagara, C., Awasthi, A.: Exploring virtual smart healthcare trends using digital twins. *Digit. Twins Smart Cities Villag.*, pp. 377–406 (2025)
19. Martinez-Velazquez, R., Gamez, R., El Saddik, A.: Cardio Twin: a digital twin of the human heart running on the edge. In: *2019 IEEE International Symposium on Medical Measurements and Applications (MeMeA)*, pp. 1–6 (2019)
20. Sharifrazi, D.: CAD Cardiac MRI Dataset, Version 2. Accessed 23 Jan 2025. from https://www.kaggle.com/datasets/danialsharifrazi/cad-cardiac-mri-dataset/data (2022)
21. Litjens, G., et al.: A survey on deep learning in medical image analysis. Med. Image Anal. **42**, 60–88 (2017)
22. Esteva, A., et al.: A guide to deep learning in healthcare. Nat. Med. **25**, 24–29 (2019)

AmpliconHunter: A Scalable Tool for PCR Amplicon Prediction from Microbiome Samples

Rye Howard-Stone(✉) and Ion I. Măndoiu

School of Computing, University of Connecticut, Storrs, CT 06269, USA
{rye.howard-stone,ion.mandoiu}@uconn.edu

Abstract. Sequencing of PCR amplicons generated using degenerate primers (typically targeting a region of the 16S ribosomal gene) is widely used in metagenomics to profile the taxonomic composition of complex microbial samples. To reduce taxonomic biases in primer selection it is important to conduct *in silico* PCR analyses of the primers against large collections of up to millions of bacterial genomes. However, existing *in silico* PCR tools have impractical running time for analyses of this scale. In this paper we introduce AmpliconHunter, a highly scalable *in silico* PCR package distributed as an open-source command-line tool and publicly available through a user-friendly web interface at https://ah1.engr.uconn.edu/. AmpliconHunter implements an accurate nearest-neighbor model for melting temperature calculations, allowing for primer-template hybridization with mismatches, along with three complementary methods for estimating off-target amplification. By taking advantage of multi-core parallelism and SIMD operations available on modern CPUs, the AmpliconHunter web server can complete *in silico* PCR analyses of commonly used degenerate primer pairs against the 2.4M genomes in the latest AllTheBacteria collection in as few as 6–7 h.

1 Introduction

DNA amplification by Polymerase Chain Reaction (PCR) is a fundamental technique in biomedical research. Commonly, a genomic region of interest is targeted for amplification by identifying a pair of primer sequences that flank the region. Sequencing PCR amplicons (typically variable regions of the 16S ribosomal gene) remains a very popular approach in metagenomic studies, where amplicon sequencing has been shown to enable identification of more taxa at a lower cost compared to shotgun sequencing [1]. To ensure sensitive amplification of the target region even with high-levels of genomic variation in bacterial species, it is common to use degenerate primers and permissive PCR conditions that allow primers to anneal with mismatches [2]. Degenerate primers are represented as sequences over the IUPAC alphabet, in which each letter indicates a subset of one or more nucleotides (e.g., 'N' represents any DNA base, 'R' represents 'A or G', etc.). Experimentally, a degenerate primer is a mixture consisting of all

M. Alser et al. (Eds.): ICCABS 2025, LNCS 15599, pp. 329–344, 2026.
https://doi.org/10.1007/978-3-032-02489-3_25

oligonucleotides that can be formed by selecting compatible DNA bases at each degenerate position of the primer sequence. The degree of degeneracy for commonly used primers varies, with some degenerate primers consisting of hundreds of distinct oligonucleotides. For example, the two degenerate primers targeting the Titan™ region of the ribosomal operon (previously known as StrainID™ [3]) together consist of a total of 960 oligonucleotides.

Primer selection for amplicon metagenomic sequencing involves complex tradeoffs between sensitivity (which increases with primer degeneracy), specificity (which decreases with degeneracy), taxonomic discriminative power (which depends on the length and diversity of the amplified region), and cost (long amplicons require the use of more expensive sequencing technologies), among others. Most existing primer sequences are designed based on highly curated microbial genome sequences available in databases such as the reference genomes in RefSeq [4,5]. This may bias resulting primers towards culturable microbes [6]. The bias could be reduced by designing and validating the primers *in silico* against larger genome collections that are now becoming available, such as the Genome Taxonomy Database (GTDB, currently containing over 580K genomes [7]), the Pathosystems Resource Integration Center database (PATRIC, with nearly 1M genomes) [8], or the AllTheBacteria project (with over 2.4M genomes [9]). Since analysis of genomic collections of this size is impractical with existing tools there is a need for more scalable and accurate *in-silico* PCR software packages.

In this paper, we present AmpliconHunter, a highly scalable open-source *in silico* PCR package available at github.com/rhowardstone/AmpliconHunter. The AmpliconHunter program accepts as input genome sequences in FASTA format, degenerate primer sequences, and PCR parameters such as annealing temperature and allowed number of mismatches and reports detailed information on the predicted PCR amplicons, including their sequences, genomic coordinates, and predicted melting temperatures for primer binding sites. To facilitate the use of AmpliconHunter by users without command-line expertise or access to high-performance computing infrastructure, the tool is also publicly available as a web server at https://ah1.engr.uconn.edu/. The web server can be used to analyze any user specified primer pairs against several pre-loaded bacterial collections.

AmpliconHunter is optimized for the analysis of large genome collections, with the webserver predicting amplicons for the 2.4M AllTheBacteria genomes in 6.5–7 h, depending on primer degeneracy. Amplification predictions are based on accurate primer-target melting temperature estimates using a nearest-neighbor model that allows mismatches [10]. A distinguishing feature of AmpliconHunter is that it includes multiple methods for estimating the amount of off-target amplification using the primer hybridization annotations, scoring for homology using a profile HMM trained on-the-fly using high-confidence target amplicons, as well as analyses based on decoys generated by reversing the genome sequences. Such off-target amplification becomes an increasingly important concern when using highly degenerate primer pairs and permissive PCR annealing conditions,

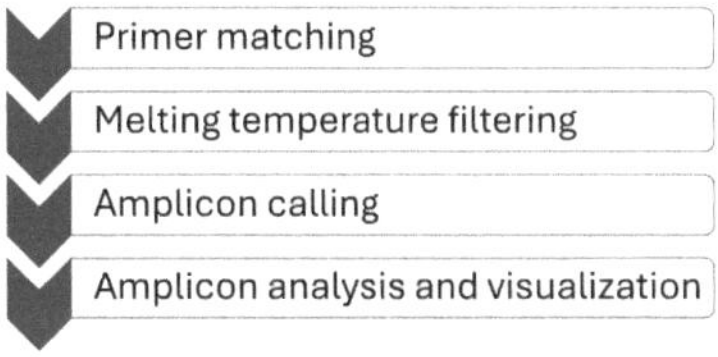

Fig. 1. Main steps of the AmpliconHunter workflow.

particularly for highly complex microbial samples. As shown in Appendix A, off-target amplification predicted by AmpliconHunter is detectable in PacBio HiFi reads generated by sequencing Titan amplicons from a wild mouse gut microbiome sample.

In the rest of the paper we describe the methods used by AmpliconHunter in Sect. 2, present empirical evaluation results in Sect. 3, and conclude with directions for future work in Sect. 4.

2 Methods

2.1 AmpliconHunter Workflow

Figure 1 gives a high-level overview of AmpliconHunter's workflow. The main steps are individually discussed below.

Primer Matching. The core pattern matching step in AmpliconHunter generates a list of candidate primer hybridization sites using Intel's Hyperscan high-performance regular expression matching library, which takes advantage of SIMD capabilities of modern CPUs [11]. In this step, degenerate primers are converted to regular expressions that account for all possible oligonucleotide variants according to the IUPAC nucleotide codes and the maximum allowed number of mismatches. An optional, variable length clamp region may be specified by the user, such that no mismatches are allowed in the 3'-most bases of each primer match. Clamp filtering is conducted as a post-processing step after regular expression matching and prior to melting temperature calculations.

Melting Temperature Filtering. AmpliconHunter uses BioPython's nearest-neighbor model for DNA duplex stability to calculate the melting temperature of each candidate binding site identified in the previous step. Advanced users may specify detailed PCR parameters, including the concentrations of the primer (dnac1), template (dnac2), along with the concentrations of various ions (Na, K, Tris, Mg, and dNTPs), and salt correction method [10]. Users can also specify a minimum melting temperature threshold. This threshold is applied to the *maximum* melting temperature over all oligonucleotides represented by the primer, i.e., a candidate binding site is retained if at least one of the oligonucleotides represented by the primer has a melting temperature above the threshold. Unlike other libraries such as Python's Regex library, Hyperscan does not return the mismatch positions for fuzzy matching, thus all compatible oligonucleotides must be checked individually during melting temperature filtering. The maximum melting temperatures for retained binding sites are saved and included in amplicon header annotations for the sites predicted to yield exponential amplification by the amplicon calling step.

Amplicon Calling. After primer annealing sites have been filtered for 3' clamp and melting temperature considerations, candidate amplicons are generated by pairing primer hybridization sites found on opposite strands and within the user-specified range of amplicon lengths. While shorter amplicons are likely to be amplified with greater efficiency compared to longer amplicons, AmpliconHunter does not seek to predict amplicon yield and instead reports all genome regions that may be amplified and detected after amplicon sequencing.

Amplicon Analysis. AmpliconHunter implements three complementary methods for assessing potential off-target amplification, a feature often neglected by other *in silico* PCR programs:

1. Amplicons are annotated according to their primer pairing - those formed from two of the same type of primer are found and annotated as either 'FF' or 'RR' orientation (versus forward strand amplicons 'FR', or reverse 'RF'). Please see Appendix A for more information.
2. A profile HMM is trained on-the-fly with high-confidence amplicons (exact primer matches only) obtained with the same primers on the RefSeq-complete dataset, then used to score all predicted amplicons called in the previous step.
3. An empirical estimate of the off-target amplification rate is computed by calling amplicons from decoys generated by reversing the given genome sequences (this is an optional feature on the AmpliconHunter web server since performing decoy analysis doubles the required time).

Visualization. The main AmpliconHunter output is a fasta file with predicted amplicon sequences, their genomic coordinates, primer orientations, and maximum annealing temperatures. Additionally, the program generates visualizations of length and orientation statistics. When taxonomy information is provided the program also generates heatmaps of amplicon similarity across species within each genus. The web interface further extends these capabilities with interactive visualizations and the ability to compare results across different parameter settings. All visualizations are available for download.

2.2 Performance Optimization

The tool utilizes Python's multiprocessing library to parallelize searches across available CPU cores. The web server implements a caching system for both profile

Table 1. Summary of databases available on the AmpliconHunter web server.

Database	Genomes	Citation	Description
RefSeq-Complete	5,730	[5]	A curated collection of complete reference genomes for bacteria
RefSeq-All	20,578	[5]	All genomes in RefSeq-Complete, as well as the incomplete bacterial reference genomes
GTDB	581,552	[7]	The Genome Taxonomy Database provides a standardized microbial taxonomy based on genome phylogeny.
PATRIC	989,844	[8]	The Pathosystems Resource Integration Center, a comprehensive genomic database for bacteria.
AllTheBacteria	2,440,377	[9]	A very large dataset of uniformly assembled bacterial genomes

HMM models and amplicon calling results, using content-addressable storage based on input parameter hashes. By caching past runs the web server response time becomes negligible when the same primer pair has been analyzed before with the same search parameters.

To guard against combinatorial explosion of runtime and output size, the server imposes mild limitations on user-specified parameters. No primer may have degeneracy of more than 10,000, the maximum number of mismatches must be at most 10, and each primer must be at least 5bp in length. If the users wish to conduct *in silico* PCR analysis of custom collections of genomes or use parameters outside these ranges, they must download and run the command-line version of AmpliconHunter locally.

2.3 Benchmarking Setup

Five databases [5,7–9] ranging in size from 5.7K to 2.4M genomes are available through the web interface, each with taxonomy to the species level for many entries (see Table 1). These options allow the users to select an appropriate database based on their research needs, balancing between higher quality assemblies (e.g., RefSeq-Complete) and broader taxonomic coverage (e.g., AllTheBacteria).

The RefSeq-complete database was obtained from NCBI using the Datasets command-line tool (version 17.1.0) with the following command:

```
datasets download genome taxon 2 \
  --reference \
  --assembly-level complete \
  --include genome \
  --filename bacteria-complete-refseq.zip
```

Table 2. Feature comparison of compared in-silico PCR tools.

Feature	RibDif2	PrimerEvalPy	AmpliconHunter
Degenerate primers	✓	✓	✓
Melting temperature	×	×	✓
Mismatch parameter	×	✓	✓
Web server available	✓	×	✓
Command line interface	✓	✓	✓
Parallel processing	✓	×	✓
Off-target amplification	✓	×	✓
Programming language	Python/Perl	Python	Python

The --reference flag was included to restrict results to high-quality, curated reference genomes from the RefSeq database. Without this flag, the dataset would include all complete genomes, a much larger set (∼114,000 genomes at the time of writing). The resulting RefSeq-complete database used in this study contained 5,730 genomes.

We conducted benchmarking experiments with four primer pairs targeting different regions of the ribosomal operon:

- V3V4 primers (forward: CCTACGGGNGGCNGCAG, reverse: GACTACNNGGGTATCTAATCC), targeting the V3-V4 subregion of the 16S rRNA gene, with an expected amplicon length of 450bp [12],
- V1V9 primers (forward: AGRGTTYGATYMTGGCTCAG, reverse: RGYTACCTTGTTACGACTT), targeting the 16S rRNA gene with expected amplicon length of 1600bp [13],
- Titan primers (forward: AGRRTTYGATYHTDGYTYAG, reverse: YCNTTCCYTYDYRGTACT), targeting the 16S-23S rRNA region with expected amplicon length of 2455bp [1], and
- mecA primers (forward: GTAGAAATGACTGAACGTCCGATAA, reverse: CCAATTCCACATTGTTTCGGTCTAA) targeting the mecA antibiotic resistance gene with expected amplicon length of 310bp [14, 15]

We assessed AmpliconHunter for both accuracy and computational performance. Tests were conducted on a Ubuntu 22.04 KVM virtual machine configured with 190 virtual cores and 960GB of RAM running on a Dell PowerEdge R7525 server with two AMD EPYC 7552 48-Core Processors and 2TB total RAM. We compared the Hyperscan version of AmpliconHunter (AHv1) to two recently published *in silico* PCR tools, RibDif2 [16] and PrimerEvalPy [17], as well as a preliminary prototype implementation of AmpliconHunter using Python's Regex library, referred to as AHv0. A feature comparison for the compared tools is given in Table 2.

Unless otherwise specified, all methods were run with a time limit of one hour. Accuracy tests were performed using a random subset of 20% of the genomes in the RefSeq-Complete database and timing experiments were performed using random subsets of the AllTheBacteria database.

3 Results

3.1 Accuracy Comparison

Our primary reference dataset consists of all complete bacterial genomes from RefSeq, with corresponding taxonomy and rRNA gene annotations. In order to assess accuracy of all methods, we limited input to a random subset consisting of 20% of these genomes. For all compared methods, predicted amplicons were labeled as true positive amplicons if the primer binding sites (amplicon ends) fall within the expected gene annotation and maintain the correct strand orientation.

Figure 2 gives the F1 scores (defined as the harmonic mean of precision and recall) obtained by running the compared tools using the Titan, V1V9, and V3V4 primer pairs. Whenever possible, the tools were run with varying minimum melting temperatures and maximum number of mismatches (PrimerEvalPy and RibDif2 do not model melting temperatures, and RibDif2 uses a hard-coded threshold of up to 1 mismatch and 1 indel). All three tools achieve comparable F1 maximum scores for all three primer pairs. AmpliconHunter's accuracy is less sensitive to varying the maximum numbers of mismatches than PrimerEvalPy due to its additional filtering based on melting temperatures. This difference is most pronounced when using highly degenerate primers such as the Titan pair.

3.2 Runtime Comparison

We conducted three scaling tests to evaluate computational efficiency. All tests used the V1V9 primer pair with default parameters (2 mismatches allowed, 50°C minimum annealing temperature) unless otherwise specified. A test run was terminated if it did not complete with one hour. In all comparisons, AmpliconHunter dramatically outperforms all other methods, including the preliminary prototype implementation using Python's Regex library.

Input Size Scaling. Figure 3 shows timing results on datasets ranging in size from 100 to 102,400 genomes. As expected, all methods have near-linear scaling with respect to the number of genomes. AmpliconHunter is two orders of magnitude faster than PrimerEvalPy, one order of magnitude faster than RibDif2, and more than twice faster than the Regex prototype. AmpliconHunter achieves a throughput of $\sim$100 genomes per second, with the webserver completing the analysis of the full set of 2.4M genomes in 6h 59 m 27 s.

Number of Mismatches. Figure 4 plots the running time of the compared methods on a fixed dataset of 100 genomes as the maximum number of mismatches is varied from 0 to 6. The runtime of AmpliconHunter rises moderately with the number of mismatches, from 2.25 s to 3.82 s for the Hyperscan version,

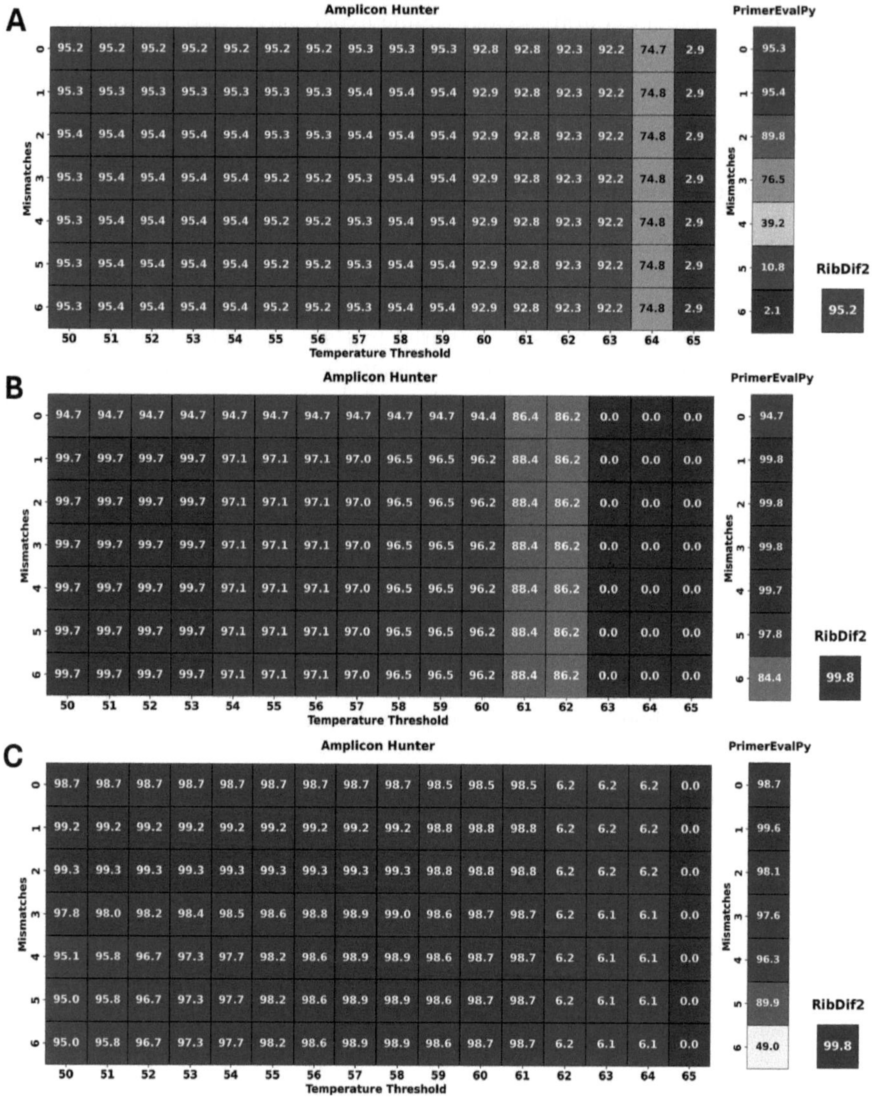

Fig. 2. F1 scores for all methods at various parameter settings on a subset of the RefSeq-Complete database, using three primers pairs A) Titan, B) V1-V9, and C) V3V4 primers.

and from 3.19 s to 7.97 s for the Regex version. In comparison, PrimerEvalPy runtime rises from 16.5 s to 89.6 s over the same range. While RibDif2 takes 25.3 s on the same subset of 100 genomes from the ATB project, it does not expose the number of mismatches as a parameter to be changed. Its primer matching does allow for one indel and one substitution, making it comparable

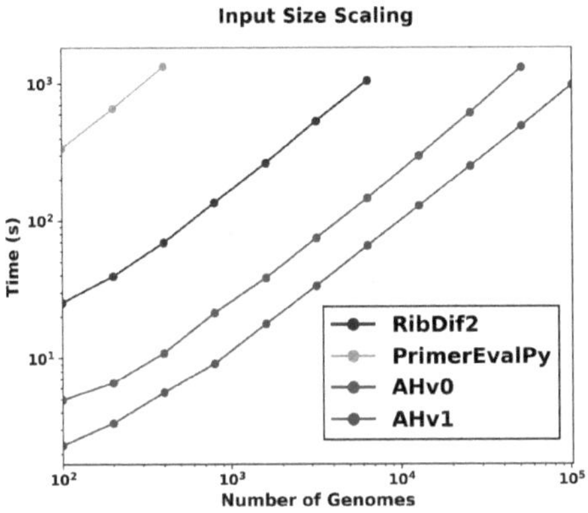

Fig. 3. Runtime on subsets of genomes from the AllTheBacteria database [9] of sizes 100 to 102,400 is shown.

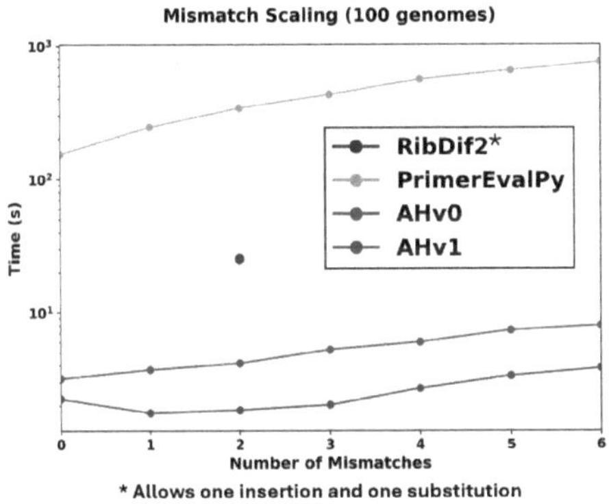

Fig. 4. Impact on runtime as number of permitted mismatches is increased. Ribdif2 does not allow alteration of this parameter and thus is shown as a single measurement at $m = 2$, as it allows one substitution and one insertion.

to two mismatches in runtime. At two mismatches, PrimerEvalPy took 340.96 s on this dataset, while AHv0 and v1 took 4.17 s and 1.86 s respectively.

Primer Degeneracy. To test the effect of increased degeneracy on the running time we progressively replaced bases from the 3′ end of the V1V9 primers with fully degenerate bases ('N'). Figure 5 plots the runtime of the compared methods on 400 random genomes as the number of degenerate bases is increased from 0 to 12 (6 per primer). Processing times were relatively unchanged for all methods, except for AHv0 which doubles the runtime from 6 to 12 added Ns, from 11.9 s to 24.0 s.

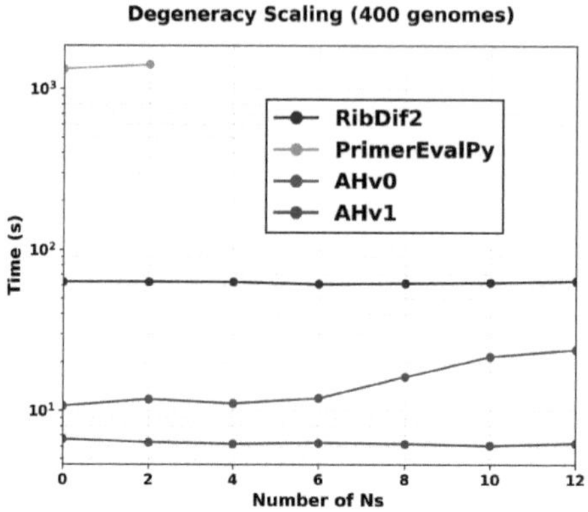

Fig. 5. Adding two Ns at a time (one to the 3' end of each primer) increases degeneracy in a controlled manner, and demonstrates its varying impact on runtime between methods.

3.3 Case Studies

AmpliconHunter's web interface enables exploration of amplicon patterns across taxonomic groups through several visualizations:

Amplicon Pattern Visualization. The tool identifies distinct amplification patterns within genera and displays them as frequency distributions. For a given genus, each pattern (called an 'amplitype') represents a unique combination of copy numbers and sequence variants. Genomes from the genus Escherichia are known to often have 7 copies of the rRNA operon, which contains the 16S gene [18]. In Fig. 6, the amplitypes of the five Escherichia genomes in RefSeq-Complete are shown for three subregions of the rRNA operon. In this example, 4 of 5 genomes contain seven identical copies of the V3-V4 region, with one containing two distinct copies in a 5:2 ratio. The amplitypes of V1-V9 for the same genomes exhibit much less sharing, with each genome forming its own unique amplitype, and many more distinct sequences per genome. The Titan region is yet more distinguishing, with every genome containing at least 3 distinct amplicon sequences.

Species Similarity Analysis. For each genus, AmpliconHunter can generate heatmaps between all member species showing the pairwise overlap in sets of amplicon sequence variants. This analysis serves multiple purposes. It may help users assess the taxonomic resolution power of different primer pairs within their taxa of interest, in the case of rRNA sequencing. It also allows users to directly examine patterns of sharing for genes of interest between closely related taxa, in the case of antibiotic resistance genes for example.

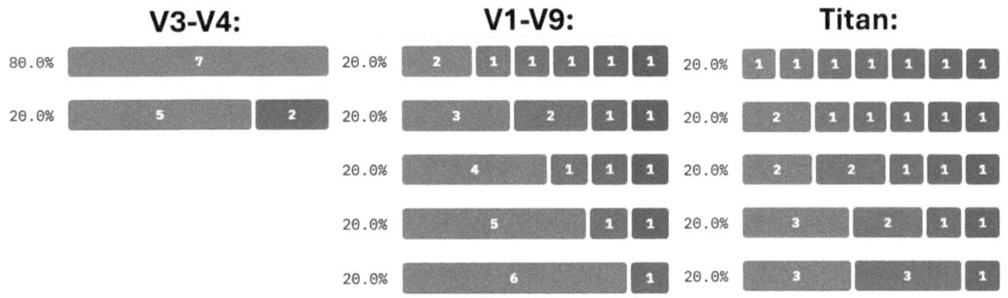

Fig. 6. Example amplitype patterns as generated by the web interface - the five complete Escherichia genomes in Refseq generate different amplitypes for three subregions of the rRNA operon: V3-V4, V1-V9, and the Titan™ region.

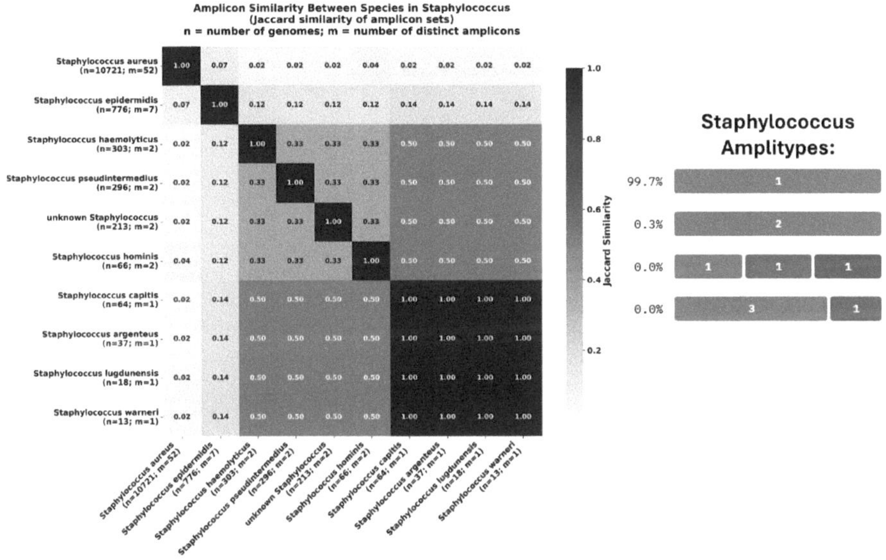

Fig. 7. Species similarity heatmap and amplitypes for the 12,557 Staphylococcus genomes the AmpliconHunter web interface identified from GTDB as containing mecA, a known antibiotic resistance gene (forward: GTAGAAATGACTGAACGTCCGATAA, reverse: CCAATTCCACATTGTTTCGGTCTAA) [14,15]. Jaccard similarity between the ten Staphylococcus species with the greatest number of amplified genomes is depicted. For each species, the number of genomes amplified n, and the number of distinct amplicons identified m is also shown.

Annealing Temperature Optimization. Through the web interface, users can explore the relationship between minimum annealing temperature and amplification specificity. Temperature thresholds can be adjusted while monitoring changes in:

- Total number of amplicons
- Proportion of genomes amplified

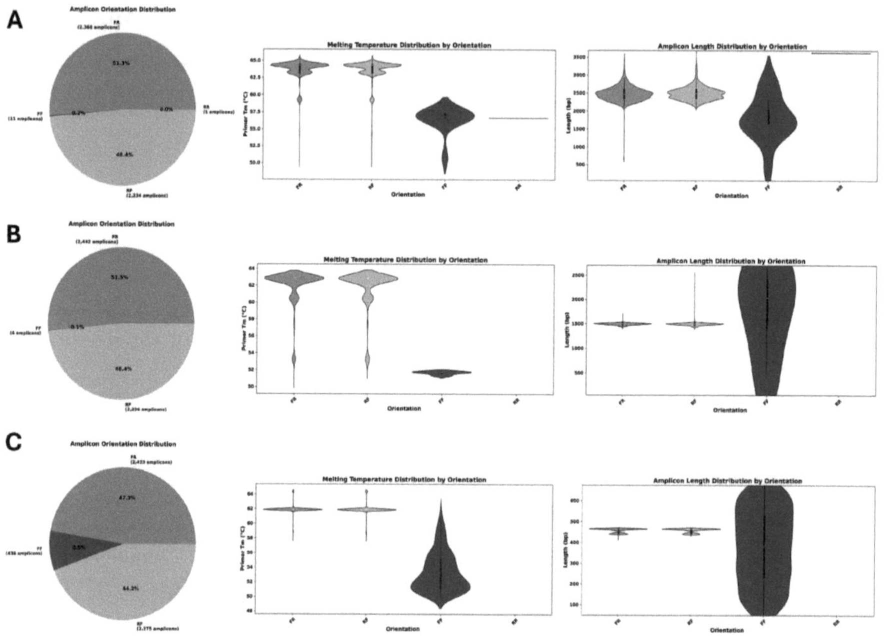

Fig. 8. Automatically generated plots of amplicon orientation, melting temperature, and amplicon length distributions for amplicons extracted from the Refseq-complete database with up to six mismatches and melting temperature of 50 using (A) Titan, (B) V1V9, and (C) V3V4 primers.

- Off-target amplification rates
- Distribution of patterns of amplicon multiplicity (referred to as 'amplitypes')

4 Conclusions and Future Work

In this paper we introduce AmpliconHunter, a highly scalable *in silico* PCR package distributed as open-source command-line tool and publicly available through a user-friendly web interface at https://ah1.engr.uconn.edu/.

In all runtime scaling experiments, AmpliconHunter maintained consistent speed advantages over existing tools while providing additional functionality through its temperature-based filtering and taxonomic analysis features. However, the current implementation searches genomes separately for each pair of primers. As such, genomes must be repeatedly loaded from disk when multiple primer pairs are analyzed. As file input-output represents a significant fraction of the overall runtime, future versions will include functionality to simultaneously analyze multiple primer pairs on a given dataset, dramatically reducing total computational requirements and further improving efficiency.

In ongoing work, we also plan to regularly update and expand the reference databases available through the AmpliconHunter webserver as new genomic

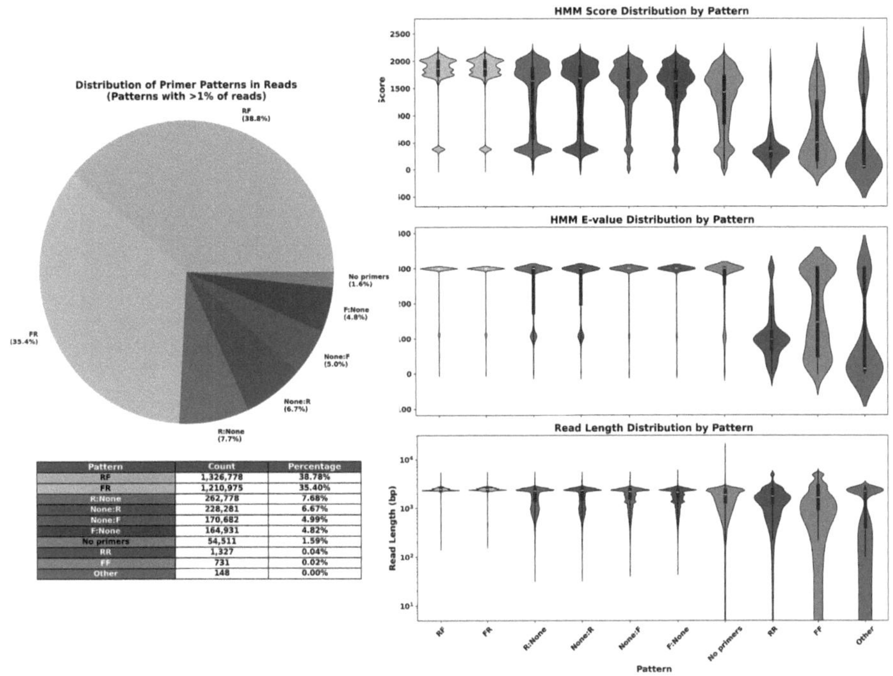

Pattern	Count	Percentage
RF	1,326,778	38.78%
FR	1,210,975	35.40%
R:None	262,778	7.68%
None:R	228,281	6.67%
None:F	170,682	4.99%
F:None	164,931	4.82%
No primers	54,511	1.59%
RR	1,327	0.04%
FF	731	0.02%
Other	148	0.00%

Fig. 9. True primer orientation, as measured using real Titan reads generated via PacBio CCS long-read technology. Patterns are determined through regex matching of primer sequences on the first and last 100BPs of each read, allowing up to two substitutions. No melting temperature or clamp filtering was applied. HMM scoring was conducted using an HMM produced from running AmpliconHunter on the RefSeq-complete database with default parameters.

assemblies become available. We would like to construct an interface that would allow users to easily search subsets of databases, by taxonomy, as well as assembly characteristics in the case of RefSeq. We also plan to add more visualizations and interactivity. Specifically, we would like to leverage the profile HMMs that are already built to automatically align and build trees for amplicon sequences with the same amplitype to analyze sequence similarity within an amplitype. This could be used to identify signatures of horizontal gene transfer in reference databases. We also plan to leverage AmpliconHunter to create databases for machine learning methods to cluster amplicon sequences according to their genome of origin.

A Evidence of Off-Target Amplification in Experimental Data

Off-target amplification is shown to be a small but detectable proportion of output of experimentally obtained PacBio CCS long reads using Titan primers in

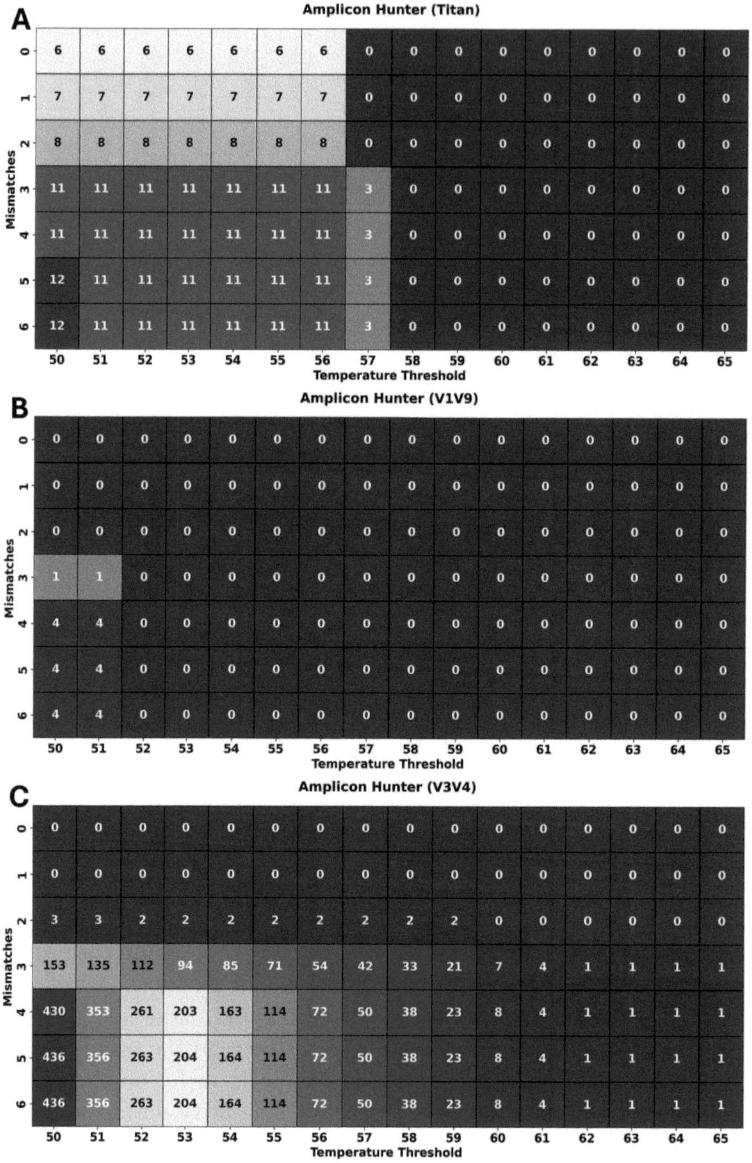

Fig. 10. A heatmap depicting the number of predicted off-target amplicons filtered from accuracy experiments on a subset of the RefSeq-Complete database using primer pairs targeting three different regions: Titan, V1V9, and V3V4.

Fig. 9. However, it is possible that less diligently designed primer pairs may lead to greater incidence of this phenomenon. In our previous accuracy experiments, we filtered out the off-target amplicons generated for the three primer pairs run on the RefSeq-Complete database with various parameter settings, prior to calculating the F1 score. Despite their moderate level of degeneracy, V3V4 primers are predicted to exhibit the highest off-target amplification, whereas the highly degenerate Titan primers show minimal off-target amplification (Fig. 10).

References

1. Gehrig, J.L., et al.: Finding the right fit: evaluation of short-read and long-read sequencing approaches to maximize the utility of clinical microbiome data. Microb. Genomics **8**, 000794 (2022)
2. Abellan-Schneyder, I., et al.: Primer, pipelines, parameters: issues in 16s rRNA gene sequencing. mSphere **6**(1), e01202-20 (2021)
3. Graf, J., et al.: High-resolution differentiation of enteric bacteria in premature infant fecal microbiomes using a novel rRNA amplicon. mBio **12**(1), e03656-20 (2021)
4. Ye, J., Coulouris, G., Zaretskaya, I., Cutcutache, I., Rozen, S., Madden, T.L.: Primer-blast: a tool to design target-specific primers for polymerase chain reaction. BMC Bioinform. **13**(134) (2012)
5. O'Leary, N.A., et al.: Reference sequence (RefSeq) database at NCBI: current status, taxonomic expansion, and functional annotation. Nucleic Acids Res. **44**(D1), D733–D745 (2016)
6. Hug, L.A.: Sizing up the uncultured microbial majority. mSystems **3**(5) (2018)
7. Parks, D.H., Chuvochina, M., Rinke, C., Mussig, A.J., Chaumeil, P.-A., Hugenholtz, P.: GTDB: an ongoing census of bacterial and archaeal diversity through a phylogenetically consistent, rank normalized and complete genome-based taxonomy. Nucleic Acids Res. **50**(D1), D785–D794 (2021)
8. Gillespie, J.J., et al.: PATRIC: the comprehensive bacterial bioinformatics resource with a focus on human pathogenic species. Infect. Immun. **79**(11), 4286–4298 (2011)
9. Hunt, M., Lima, L., Shen, W., Lees, J., Iqbal, Z.: AllTheBacteria: all bacterial genomes assembled, available and searchable. bioRxiv, 2024-03 (2024)
10. Cock, P.J.A., et al.: Biopython: freely available python tools for computational molecular biology and bioinformatics. Bioinformatics **25**(11), 1422–1423 (2009)
11. Wang, X., et al.: Hyperscan: a fast multi-pattern regex matcher for modern CPUs. In: 16th USENIX Symposium on Networked Systems Design and Implementation (NSDI 19), pp. 631–648, Boston, MA, February (2019). USENIX Association
12. Walker, A.W., Martin, J.C., Scott, P., Parkhill, J., Flint, H.J., Scott, K.P.: 16S rRNA gene-based profiling of the human infant gut microbiota is strongly influenced by sample processing and PCR primer choice. Microbiome **3**, 26 (2015)
13. Mason, O.U., Meo-Savoie, C.A., Nostrand, J.D., Zhou, J., Fisk, M.R., Giovannoni, S.J.: Prokaryotic diversity, distribution, and insights into their role in biogeochemical cycling in marine basalts. ISME J. **3**, 231–242 (2009)
14. Hosseini, S.S., et al.: Frequency of genes encoding erythromycin ribosomal methylases among staphylococcus aureus clinical isolates with different d-phenotypes in Tehran, Iran. Iran. J. Microbiol. **8**(3), 161–167 (2016)
15. Gorecki, A., Decewicz, P., Dziurzynski, M., Janeczko, A., Drewniak, L., Dziewit, L.: Literature-based, manually-curated database of PCR primers for the detection of antibiotic resistance genes in various environments. Water Res. **161**, 211–221 (2019)

16. Murphy, R., Strube, M.L.: Ribdif2: expanding amplicon analysis to full genomes. Bioinform. Adv. **3**(1) (2023)
17. Vázquez-González, L., Regueira-Iglesias, A., Balsa-Castro, C., ás Vila-Blanco, N., Tomás, I., Carreira, M.J.: PrimerEvalPy: a tool for in-silico evaluation of primers for targeting the microbiome. BMC Bioinform. 25 (2024)
18. Stoddard, S.F., Smith, B.J., Hein, R., Roller, B.R.K., Schmidt, T.M.: rrnDB: improved tools for interpreting rRNA gene abundance in bacteria and archaea and a new foundation for future development. Nucleic Acids Res. (2014)

Neuromorphic Spiking Neural Network Based Classification of COVID-19 Spike Sequences

Taslim Murad[1], Prakash Chourasia[2], Sarwan Ali[3], Avais Jan[2], and Murray Patterson[2(✉)]

[1] Washington University in St. Louis, St. Louis, MO, USA
murad@wustl.edu
[2] Georgia State University, Atlanta, GA, USA
{pchourasia1,ajan3}@student.gsu.edu, mpatterson30@gsu.edu
[3] Columbia University Irving Medical Center, New York, NY, USA
sa4559@columbia.edu

Abstract. The availability of SARS-CoV-2 (severe acute respiratory syndrome coronavirus 2) virus data post-COVID has reached exponentially to an enormous magnitude, opening research doors to analyze its behavior. Various studies are conducted by researchers to gain a deeper understanding of the virus, like genomic surveillance, etc., so that efficient prevention mechanisms can be developed. However, the unstable nature of the virus (rapid mutations, multiple hosts, etc.) creates challenges in designing analytical systems for it. Therefore, we propose a neural network-based (NN) mechanism to perform an efficient analysis of the SARS-CoV-2 data, as NN portrays generalized behavior upon training. Moreover, rather than using the full-length genome of the virus, we apply our method to its spike region, as this region is known to have predominant mutations and is used to attach to the host cell membrane. In this paper, we introduce a pipeline that first converts the spike protein sequences into a fixed-length numerical representation and then uses Neuromorphic Spiking Neural Network to classify those sequences. We compare the performance of our method with various baselines using real-world SARS-CoV-2 spike sequence data and show that our method is able to achieve higher predictive accuracy compared to the recent baselines.

Keywords: Spiking Neural Network · COVID-19 · Sequence Classification · Spike Sequence

1 Introduction

The COVID-19 disease has affected millions of people across the globe [1]. This disease is caused by the SARS-CoV-2 virus, which possesses the ability to undergo quick mutations and infect various hosts. Figure 1 illustrates the genome

M. Alser et al. (Eds.): ICCABS 2025, LNCS 15599, pp. 345–356, 2026.
https://doi.org/10.1007/978-3-032-02489-3_26

structure of the SARS-CoV-2 virus. The length of this genome is approximately $30kb$ with spike region residing in the $21kb - 25kb$ range. The spike protein region is responsible to attach to a host cell membrane, and also major mutation happens in it [2]. Thus spike region is sufficient to investigate this virus, therefore we have used only spike protein sequences in this paper.

In computational biology, one of the crucial tasks is biological sequence classification which allows researchers to do functional analysis for building an understanding of the sequences like DNA and proteins [3]. Various sequence classification strategies are put forward to study the origin, structure, and behavior of the virus [2,4]. The classification models employed for this purpose include traditional machine learning (ML) approaches, such as support vector machines (SVMs) [5], or artificial neural networks (ANNs) [6].

However, the classification task has a pre-requisite of data being in a fixed-length numerical form. Therefore, several numerical embedding generation methodologies are proposed to convert biological sequences to numerical form. Some of these techniques are feature-engineering-based methods [2,7,8], which contain both alignment-based and alignment-free mechanisms. However, these procedures are domain-specific and less generalized. Moreover, some other methods include signal transformation methods (Spike2Signal [9]), and image transformation methods (RP [10], GAF [10], MTF [10]) but they require additional data transformation steps to get the features, which can be computationally expensive. Furthermore, another embedding generation approach includes the usage of neural networks (NN) to get the embeddings [11]), as NNs portray more generalized behavior upon training so they could be used for heterogeneous sequences. Due to the availability of a large volume of SARS-CoV-2 data post the COVID-19 pandemic, it is more feasible to use an NN-based feature extractor. However, their need for extensive training with a large dataset to achieve good performance is again a computational overhead.

Recently, spiking neural networks (SNNs) [12–14] have emerged as a numerical feature extractor and classifier for biological sequences. SNNs are artificial neural networks that more closely mimic natural neural networks. Unlike ANNs whose neurons exhibit non-linear behavior with being continuous function approximators following a common clock cycle to operate, neurons of SNNs use asynchronous spikes to signal the occurrence of some characteristic event by digital and temporally precise action potentials [15]. In this paper, we propose an SNN-based method to classify the spike protein sequences of the SARS-CoV-2 virus. Given any spike sequence, our method follows an alignment-free pipeline of extracting numerical features from the sequence and using those features to perform classification.

In this paper, our contributions are the following:

1. We propose an alignment-free end-to-end classification pipeline for biological sequences using spiking neural networks (SNN), which first converts the sequences into numerical form and using these numerical features perform Classification.

2. Compared to existing neural network-based baselines, we show that SNN is comparatively more stable in terms of predictive accuracy for biological sequence analysis.
3. We show from the experimental results that we can achieve higher predictive accuracy by using the spike region of the protein sequence only.

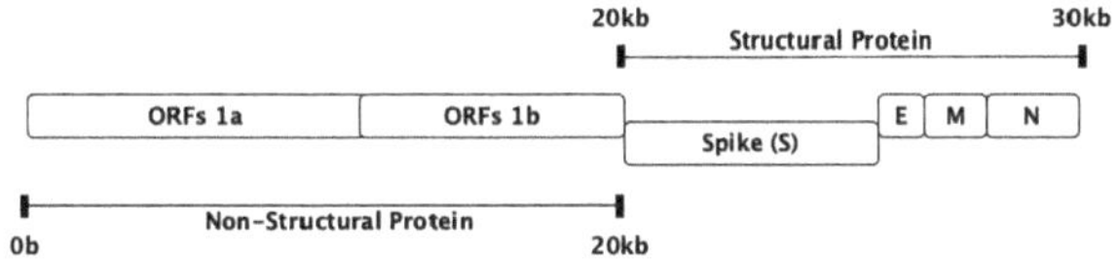

Fig. 1. The genome of SARS-CoV-2 has a length of 30kb, and it consists of non-structural proteins (ORFs 1ab) and structural proteins (E, M, N, S). The S region is important because of its ability to attach to the host cell membrane, and also hold advantageous mutations.

Our work is distributed in the manuscript as follows: Sect. 2 deals with talking about related work, Sect. 3 highlights our proposed method, Sect. 4 discusses the details of experiments, Sect. 5 talks about the results, and Sect. 6 concludes the paper.

2 Related Work

Many works are done in the domain of biological sequence classification. Some researchers have explored traditional ML models, like SVM, for doing this task. The ML-based classification requires mapping the sequences to numerical form first. Henkes et al. [16] proposes a nonlinear regression framework for spiking neural networks on neuromorphic processors. Rafique et al. [17] explores recent technological advancements in biosensors and deep learning methods for diagnosing COVID-19 and its variants. Feature-engineering-based embedding generation methods are popular in this regard, like n-gram-based vector representation [18,19], Sparse [20], Spike2Vec [7], PWM2Vec [21] etc. However, usually, these techniques require sequence alignment, which is a computationally expensive operation. Additionally, feature-engineering-based methods are domain-specific and they may not be able to generalize to heterogeneous (different types) data.

Moreover, ANNs are also gaining popularity for performing biological sequence classification because of their generalizability property and larger sequence data's availability, like protein classification [22], DNA binding site prediction [23] etc. Some of the broader categories of ANN-based classification, depending on the feature extractor method, can be neural networks-based, image transformation-based, and signal transformation based. In neural network-based methods, they employ an ANN model to extract the sequence's features and then perform classification of those features. Like, [11]) uses an auto-encoder to do sequence classification. This network consists of symmetric encoder and decoder parts, and the output for the encoder corresponds to the feature vectors of the sequence. However, they possess a computational overhead of training with a large dataset for achieving good performance.

Furthermore, the image transformation-based approaches e.g. RP, GAF, MTF [10], uses a mechanism to transform the sequence to an image first, and then use this graphical form to perform classification using ANN models. Likewise, a signal transformation-based approach, like Spike2Signal [9], maps a protein sequence to signal-like numerical data by assigning an integer value to each amino acid of the sequence and then uses this signal-like data for classification. It can be noted that both image transformation and signal transformation approaches have an added step of data transformation which is a computationally expensive step.

Spiking neural networks-based sequence classifications are also designed by some of the previous works [13,24] and they illustrate that SNNs can achieve competitive performance compared to traditional machine learning methods while being more efficient and requiring less data [25]. However, these previous studies have mostly focused on simple biological sequences, such as DNA sequences with a fixed length. In this work, we propose the use of SNNs for classifying more complex biological sequences, such as proteins with varying lengths. To the best of our knowledge, this is the first study that investigates the use of SNNs for this task.

3 Proposed Approach

This section highlights our proposed end-to-end alignment-free method to do the classification of SARS-CoV-2 protein spike sequences using a neuromorphic design based on the spiking neural network (SNN). The neuromorphic design field, which has the purpose of building machines that mimic the structure of the brain, has recently taken the realm of machine learning into account. Although the application of neuromorphic architecture to traditional tasks, such as image recognition and logistic regression, is associated with many challenges, the advancements in ML have opened new breakthroughs in this area. Furthermore, the replacement of DL model neurons with spiking neurons can cause neuromorphic computing to improve the efficiency and performance of predictions.

In SNN, based on the exceeded threshold value the neurons emit spikes or electrical impulses in response to the input. In case of input being lower than the threshold, the pre-activation value gradually decreases. This phenomenon is equivalent to a time-dependent version of the ReLU activation function in which at different time steps there is either a spike or no spike. Due to the similarity between the brain's information accumulation and release procedure to SNN, they are considered more biologically realistic than traditional ML and DL models. For input data X, after multiplying it with a weight matrix W the result is passed on to a decayed version of the information inside the neuron. This information exists from the previous time step/time tick (Δt time elapsed). To serve the aim of gradually reducing the inner activation a *decay_multiplier* is used, which prevents the accumulation of stimuli for a long time on a neuron. We use a *decay_multiplier* of 0.9 value in our model. After that, the neuron's activation function is computed based on a threshold. The neuron's inner state is reset on its fire by subtracting the activation from its inner state. This will stop the neuron from firing constantly upon being activated once and also it will isolate each firing event from the other by clipping the gradient through time. Finally, in the classification layer, the spiking neurons values are averaged over the time axis and plugged to the softmax cross-entropy loss function for back-propagation.

Moreover, data processing in SNN contains a temporal dimension and this dimension is incorporated into Artificial Neural Networks (ANNs) by allowing the signal to accumulate over time in a pre-activation phase. We use the SNN for doing classification of SARS-CoV-2 spike sequences.

Our system can be applied to data with varying lengths. We convert the input data to its corresponding one-hot encoding (OHE) vector and pass this OHE-based vector to our model. A zero-padding technique is employed on the OHE vector to handle the sequences of varying lengths. To train our system for experiments, we use cross-entropy loss function, ADAM optimizer, 100 epochs, and 0.001 learning rate.The Selection of 100 epochs was based on preliminary experiments that showed stable convergence within this range. For overfitting, we employed five-fold cross-validation and monitored performance metrics throughout training. Our model consists of two linear layers.

4 Experimental Evaluation

This section discusses the dataset used to perform the experiments, along with all the evaluation metrics. It also highlights the baseline models employed for performance comparison. We follow the $70\% - 30\%$ train-test data split with k-folds ($k = 5$) to conduct the experiments. All experiments are executed using an Intel(R) Xeon(R) CPU E7-4850 v4 @ 2.10GHz having Ubuntu 64-bit OS (16.04.7 LTS Xenial Xerus) with 3023 GB memory.

4.1 Dataset Statistics

We employ a dataset about spike sequences of the SARS-CoV-2 virus taken from GISAID [26] to investigate the performance. This dataset has 7000 sequences and is referred to as the Spike7k dataset. We have selected a subset of $7k$ sequences from the whole GISAID dataset with the aim of mimicking real-world scenarios and preserving the original lineage distribution of the coronavirus to avoid any biases in our results.All 7k sequences from this dataset were used in our experiments. Some of the common lineages are given names by the world health organization, like B.1.1.7 is Alpha, AY.12 is Delta, B.1.429 is Epsilon, AY.4 is

Delta, etc. These 7000 sequences encode details of the 22 Lineage of coronavirus. The detailed distribution is illustrated in Table 1.

Table 1. The distribution of $7k$ sequences across all the Lineages in the Spike7k dataset is shown in this table.

Lineage	Frequency	Lineage	Frequency
B.1.1.7	3369	R.1	32
B.1.617.2	875	AY.4	593
B.1.2	333	B.1	292
B.1.177	243	P.1	194
B.1.1	163	B.1.429	107
B.1.526	104	AY.12	101
B.1.160	92	B.1.351	81
B.1.427	65	B.1.1.214	64
B.1.1.519	56	D.2	55
B.1.221	52	B.1.177.21	47
B.1.258	46	B.1.243	36

4.2 Evaluation Metrics and Classifiers

To measure the performance of our proposed and baseline models we utilize the accuracy, precision, recall, F1 (weighted), F1 (macro), Receiver Operator Characteristic Curve Area Under the Curve (ROC AUC), and training runtime. The reported value for each evaluation metric is an average value over five runs. The ROC AUC is computed using the one-vs-rest method.

The machine learning classifiers used to get the evaluation metrics of the neural network baseline are Support Vector Machine (SVM), Naive Bayes (NB), Multi-Layer Perceptron (MLP), K-Nearest Neighbors (KNN), Random Forest (RF), Logistic Regression (LR), and Decision Tree (DT) classifiers.

The image transformation baselines are evaluated using 3-Layer CNN, 4-Layer CNN, and RESNET34 deep learning models. The 3 Layer CNN and 4 Layer CNN correspond to neural networks with 3 and 4 convolution layers respectively. The RESNET34 refers to a pre-trained RESNET34 [27] model. All these DL models are trained using ADAM optimizer and negative log-likelihood loss function.

The evaluation of signal transformation baseline is done using Fully Convolution Network (FCN) [28], LSTM 3 Layer Bidirect [29], and mWDN networks [30].

4.3 Baseline Methods

We select the baseline methods from 3 different domains to investigate the performance and these domains are:

1. An end-to-end classification pipeline using a neural network, which generates feature embeddings from the raw input and performs classification based on the generated features.
2. Sophisticated image classifiers that take transformed images and classify them.
3. Time sequence classifiers, which first transform sequence into signal-like data and then apply signal NN for classification.

The details of each baseline are as follows:
Autoencoder + Neural Tangent Kernel [11] This approach generates a low-dimensional embedding of protein sequences through an encoder neural network. The architecture of the encoder consists of a stack of dense layers with LeakyReLU activation function, batch normalization, and dropout. It's accompanied by a symmetric decoder for training and a reconstruction loss is used to optimize the whole network. Then the low-dimensional embeddings are utilized to compute Neural Tangent Kernel (NTK). The NTK is a kernel function that measures the similarity between two input sequences based on the geometry of the neural network's decision boundary. This method is efficient for large and high-dimensional data. We have used a 4-layer autoencoder with ADAM optimizer, MSE loss function, and 100 epochs to train this model for our experiments.

Image Transformation. As image-based classifiers are considered state-of-the-art for the classification tasks, therefore we wanted to draw a performance comparison of these classifiers with our proposed model for protein sequence classification which is why we used image transformation methods as baselines.

Image transformation represents a set of methods that transform signal-based sequences into images to perform analysis. The sequences are transformed into signals following the Spike2Signal [9] method. The three models belonging to the category of image transformation reported in the experiments are as follow,

Recurrent Plot (RP) RP [10] is used to get a 4D image of 360x360 size corresponding to a signal-based represented spike sequence. The 4D refers to three color channels and one alpha channel. This image illustrates the distance between the trajectories. Once the trajectories are extracted from the signal, the pairwise distance between them is computed to get a graphical form. For a given signal x_i, it's trajectories with m dimensions and τ time-delay are defined as:

$$\boldsymbol{x}_i = (x_i, x_{i+\tau}, \ldots, x_{i+(m-1)\tau}), \quad \forall i \in \{1, \ldots, n-(m-1)\tau\} \tag{1}$$

Gramian Angular Field (GAF) Given a pair of signal data, GAF [10] extracts their temporal correlation matrix using the formula,

$$\begin{aligned} \tilde{x}_i = a + (b-a) \times \frac{x_i - \min(x)}{\max(x) - \min(x)}, \quad \forall i \in \{1, \ldots, n\} \\ \phi_i = \arccos(\tilde{x}_i), \quad \forall i \in \{1, \ldots, n\} \\ GAF_{i,j} = \cos(\phi_i + \phi_j), \quad \forall i, j \in \{1, \ldots, n\} \end{aligned} \tag{2}$$

where $a < b$ and both fall in the range $[-1, 1]$ and represents the range to rescale the original signal. ϕ_i shows the polar coordinates of the scaled signal.

Markov Transition Field (MTF) The sequence data is discretized into Q quantile bins by MTF [10]. Then a square matrix W with dimensions Q is created from each bin. After that, from the matrix W having transition probability from q_i to q_j amino acids, a matrix M is computed. This M is used for visualization.

Some of the examples of images generated from RP, GAF, and MFT image transformation techniques are illustrated in Fig. 2. It shows a set of images for Lineage AY.12 and B.1.526 from our dataset. For a sequence belonging to a Lineage, we can observe that the image patterns differ corresponding to each transformation technique, which shows that the information is differently captured within the image depending on the underlying transformation technique used, and this can be beneficial for classification.

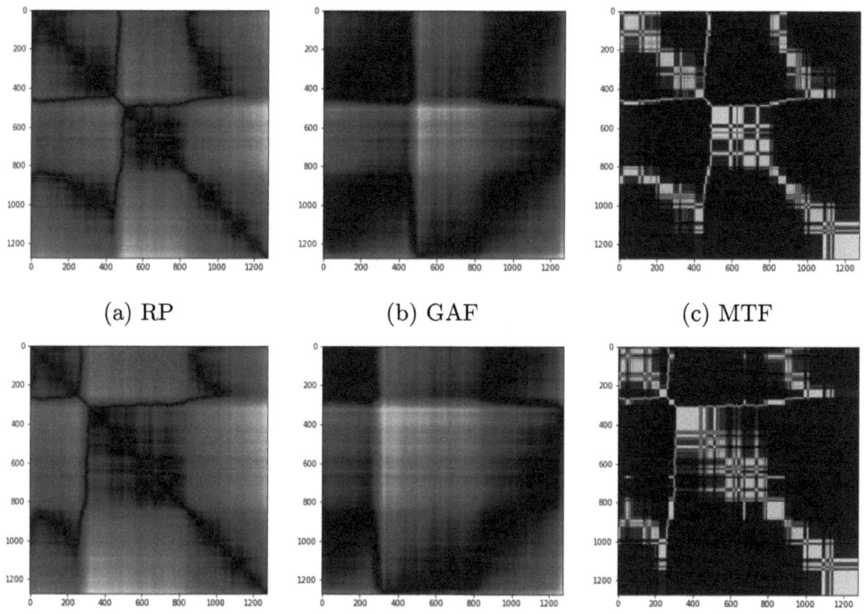

(a) RP (b) GAF (c) MTF

(d) RP (e) GAF (f) MTF

Fig. 2. The images from $(a-c)$ are against a spike sequence corresponding to Lineage AY.12, while $(d-f)$ show images of a sequence from B.1.526 Lineage. They are created using RP, GAF, and MTF approaches.

Signal Transformation. Due to the presence of some NN models specifically designed for signal-based data, we used signal transformation as a baseline because we wanted to explore those NN models' performance on protein sequence classification. The signal transformation includes the Spike2Signal [9] approach which is used to map spike sequences into signal-like numerical representations to make them compatible with deep learning models. This method assigns integer values to amino acids in the spike sequences.

5 Results and Discussion

In this section, we discuss the classification results of our proposed system and compare the results with the baseline methods. Table 2 summarizes the results of the neural network, image transformation, signal transformation, and spiking neural network (our proposed one) based approaches.

The results illustrate that our method has drastically improved the performance for all the evaluation metrics as compared to the neural network-based baseline. This behavior is may be caused by the underlying feature extractor of the baseline being not able to produce optimal features which degrade the classification performance. However, our system uses an advanced SNN model to generate optimal features which can deliver good classification performance.

Similarly, we can view a clear performance improvement for all the metrics using our model over the signal transformation baseline. The low classification performance of the signal transformation baseline can also be associated with using a sub-optimal feature generation mechanism.

Table 2. Classification results for different methods on SARS-CoV-2 data. The best values are shown in bold.

Approach	Embedding	Algo.	Acc.	Prec.	Recall	F1 (Weig.)	F1 (Macro)	ROC AUC	Train Time
Neural Network	Autoencoder + NTK	SVM	0.480	0.503	0.480	0.478	0.146	0.684	0.011 Sec.
		NB	0.507	0.464	0.507	0.458	0.224	0.653	0.002 Sec.
		MLP	0.467	0.460	0.467	0.449	0.194	0.653	0.917 Sec.
		KNN	0.413	0.337	0.413	0.363	0.083	0.581	0.002 Sec.
		RF	0.520	0.490	0.520	0.487	0.199	0.687	0.185 Sec.
		LR	0.507	0.484	0.507	0.482	0.175	0.653	0.009 Sec.
		DT	0.520	0.546	0.520	0.525	0.189	0.694	**0.001** Sec.
Image Transformation	Recurrent Plot	3 Layer CNN	0.780	0.737	**0.777**	0.746	0.398	0.713	1.62 h
	Gramian Angular Field		0.750	0.713	0.754	0.717	0.411	0.706	1.58 h
	Recurrent Plot	4 Layer CNN	0.780	0.750	**0.777**	0.757	**0.468**	**0.752**	1.7 h
	Gramian Angular Field		0.764	0.718	0.764	0.732	0.417	0.716	1.69 h
	Recurrent Plot	RESNET34	0.599	0.587	0.599	0.553	0.172	0.609	12.3 h
	Gramian Angular Field		0.712	0.669	0.712	0.685	0.342	0.674	11.4 h
Signal Transformation	Numerical Representation	mWDN	0.527	0.348	0.527	0.394	0.078	0.523	11 h
		FCN	0.615	0.566	0.615	0.568	0.134	0.563	1.1 h
		LSTM 3 Layer Bidirect.	0.740	0.701	0.740	0.709	0.377	0.682	22 h
Spiking Neural Network (ours)	Numerical Representation	-	**0.810**	**0.790**	0.710	**0.782**	0.426	0.717	3 days

Moreover, we can observe that SNN is performing better in terms of accuracy, precision, and F1 weighted score as compared to the image transformation baseline. However, SNN is not optimized for recall, F1 macro, and ROC AUC score but it yields comparable results for these metrics. Since the image transformation procedures are task-specific, SSNs can portray more generalized behavior after being trained, therefore overall SNNs possess good performance ability among the all mentioned baselines. The good performance of SNN also indicates that it is able to handle class imbalance efficiently.

6 Conclusion

In this paper, we propose using spiking neural networks (SNNs) for biological sequence classification. We describe our proposed method, which uses SNNs to classify proteins with varying lengths, and evaluate its performance on a benchmark dataset. Our results show that our proposed method achieves competitive performance compared to state-of-the-art methods while being more efficient and requiring less data. Overall, our work suggests that SNNs are a promising approach to biological sequence classification, and they can potentially improve our understanding of biological sequences and their functions. Further research is needed to explore the full potential of SNNs in this domain and to develop new methods and techniques for improving their performance. In the future, we can explore the scalability and robustness of the SNNs in terms of protein sequence classification. Applying SNN on nucleotide sequence and evaluating its performance is also an interesting future direction.

References

1. SARS-CoV-2 Cases. https://covid19.who.int/, last accessed 05-May-2022
2. Kuzmin, K., et al.: Machine learning methods accurately predict host specificity of coronaviruses based on spike sequences alone. Biochem. Biophys. Res. Commun. **533**(3), 553–558 (2020)
3. Xia, Z., Cholewa, J., Zhao, Y., Shang, H.Y., Yang, Y.Q., et al.: Targeting inflammation and downstream protein metabolism in sarcopenia. Front. Physiol. **8**, 434 (2017)
4. Ahmed, I., Jeon, G.: Enabling Artificial Intelligence for Genome Sequence Analysis of COVID-19 and Alike Viruses. Interdisciplinary Sciences: Comp. Life Sciences, 1–16. Springer (2021)
5. Cortes, C., Vapnik, V.: Support-vector networks. Mach. Learn. **20**(3), 273–297 (1995)
6. Goodfellow, I., Bengio, Y., Courville, A.: Deep Learning. MIT Press (2016). http://www.deeplearningbook.org
7. Ali, S., Patterson, M.: Spike2vec: an efficient and scalable embedding approach for COVID-19 spike sequences. In: IEEE International Conference on Big Data (Big Data), pp. 1533–1540 (2021)
8. Corso, G., et al.: Neural Distance Embeddings for Biological Sequences. In: Advances in Neural Information Processing Systems 34 (2021)

9. Ali, S., Murad, T., Chourasia, P., Patterson, M.: Spike2Signal: classifying coronavirus spike sequences with deep learning. In: IEEE Eighth International Conference on Big Data Computing Service and Applications (BigDataService), pp. 81–88 (2022)
10. Faouzi, J., Janati, H.: pyts: A Python Package for Time Series Classification. JMLR **21**, 46–1 (2020)
11. Nguyen, T.V., Wong, R.K., Hegde, C.: Benefits of jointly training autoencoders: An improved neural tangent kernel analysis. IEEE Trans. Inf. Theory **67**(7), 4669–4692 (2021)
12. Maass, W.: Networks of spiking neurons: the third generation of neural network models. Neural Netw. **10**(9), 1659–1671 (1997)
13. Ponulak, F., Kasinski, A.: Introduction to spiking neural networks: information processing, learning and applications. Acta Neurobiol. Exp. **71**(4), 409–433 (2011)
14. Guerguiev, J., Lillicrap, T.P., Richards, B.A.: Towards deep learning with segregated dendrites. ELife **6**, e22901 (2017)
15. Pfeiffer, M., Pfeil, T.: Deep learning with spiking neurons: opportunities and challenges. Front. Neurosci. **12**, 774 (2018)
16. Henkes, A., Eshraghian, J., Wessels, H.: Spiking neural networks for nonlinear regression. Royal Society Open Sci. **11**, 231606 (2024)
17. Rafique, Q., et al.: Reviewing methods of deep learning for diagnosing COVID-19, its variants and synergistic medicine combinations. Comput. Biol. Med. **163** 107191 (2023)
18. Ali, S., Sahoo, B., Ullah, N., Zelikovskiy, A., Patterson, M., Khan, I.: A k-mer based approach for SARS-CoV-2 variant identification. In: ISBRA, pp. 153–164 (2021)
19. Ali, S., Ali, T., Khan, M., Khan, I., Patterson, M., et al.: Effective and scalable clustering of SARS-CoV-2 sequences. In: International Conference on Big Data Research (To appear) (2021)
20. Hirst, J.D., Sternberg, M.J.E.: Prediction of structural and functional features of protein and nucleic acid sequences by artificial neural networks. Biochemistry **31**(32), 7211–7218 (1992)
21. Ali, S., Bello, B., Chourasia, P., Punathil, R.T., Zhou, Y., Patterson, M.: Pwm2vec: an efficient embedding approach for viral host specification from coronavirus spike sequences. Biology **11**(3), 418 (2022)
22. Zainuddin, Z., Kumar, M.: Radial basis function neural networks in protein sequence classification. Malaysian J. Math. Sci. **2**(2), 195–204 (2008)
23. Zhou, J., Lu, Q., Xu, R., Gui, L., Wang, H.: Cnnsite: prediction of DNA-binding residues in proteins using convolutional neural network with sequence features. In: IEEE International Conference on Bioinformatics and Biomedicine (BIBM), pp. 78–85. IEEE (2016)
24. Zhou, J., Theesfeld, C.L., Yao, K., Chen, K.M., Wong, A.K., Troyanskaya, O.G.: Deep learning sequence-based ab initio prediction of variant effects on expression and disease risk. Nat. Genet. **50**(8), 1171–1179 (2018)
25. Hunsberger, E., Eliasmith, C.: Training spiking deep networks for neuromorphic hardware. arXiv preprint arXiv:1611.05141 (2016)
26. GISAID Website. https://www.gisaid.org/. Accessed 17 Oct 2022
27. He, K., Zhang, X., Ren, S., Sun, J.: Deep residual learning for image recognition. In: Proceedings of the IEEE Conference on Computer Vision and Pattern Recognition, pp. 770–778 (2016)

28. Wang, Z., Yan, W., Oates, T.: Time series classification from scratch with deep neural networks: a strong baseline. In: International Joint Conference on Neural Networks (IJCNN), pp. 1578–1585 (2017)
29. Hochreiter, S., Schmidhuber, J.: Long short-term memory. Neural Comput. **9**(8), 1735–1780 (1997)
30. Wang, J., Wang, Z., Li, J., Wu, J.: Multilevel wavelet decomposition network for interpretable time series analysis. In: Proceedings of the 24th ACM SIGKDD International Conference on Knowledge Discovery & Data Mining, pp. 2437–2446 (2018)

Author Index

M. Alser et al. (Eds.): ICCABS 2025, LNCS 15599, pp. 357–358, 2026.
https://doi.org/10.1007/978-3-032-02489-3

MIX
Papier aus verantwortungsvollen Quellen
Paper from responsible sources
FSC® C105338

If you have any concerns about our products,
you can contact us on
ProductSafety@springernature.com

In case Publisher is established outside the EU,
the EU authorized representative is:
Springer Nature Customer Service Center GmbH
Europaplatz 3, 69115 Heidelberg, Germany

Printed by Libri Plureos GmbH
in Hamburg, Germany